MARINE TRANSFORMATION

MARINE TRANSFORMATION

Editor
Thomson Dsouza

DPH

DISCOVERY PUBLISHING HOUSE
INDIA

Published by:

DISCOVERY PUBLISHING HOUSE
4383/4B, Ansari Road, Darya Ganj
New Delhi-110 002 (India)
Phone : +91-11-23279245; 23253475; 43596065
Mobile : +91 9811179893 / +91 9871656464
E-mail : discoverybooksindia@gmail.com
orderdphbooks@gmail.com
namitwasan9@gmail.com
web : www.discoverypublishinggroup.com

Marine Transformation
Edited by: **Thomson Dsouza**

International Standard Book Number: **978-81-19205-79-0** (Hardback)

PREFACE

In recent years, companies across all industries have been exploring and implementing new digital technologies to reap their benefits. This process is known as digital transformation (DT), which involves a fundamental change in traditional business practices through the use of digital technology. DT extends beyond simply changing business processes and can lead to the creation of new types of organizations and changes in organizational culture, relationships, value creation, and customer reach. It can also impact market position and enable the redefinition of existing business capabilities, processes, and relationships.

However, integrating and exploiting new digital technologies is a significant challenge that companies face today. The maritime transport sector, which is a critical mode of transportation in international trade, is also moving towards digitalization and DT at different speeds in different domains. Shipping, as part of the logistics chain, is an industry that is in a turbulent condition due to energy price fluctuations, technological immaturity, and upcoming increases in regulations.

Digitalization in the maritime transport sector refers to the implementation of various digital technologies that can enhance productivity, efficiency, sustainability, and transparency in business processes. It may also provide a competitive advantage by connecting all stakeholders involved in the value chain. Ships, seaports, and offshore facilities have become increasingly reliant on information and communication technologies. Despite the potential benefits of digitalization and DT in the maritime transport sector, progress has been slower compared to other transportation sectors.

The book entitled "Marine Transformation" provides a comprehensive response to the challenges that the maritime business faces today. Managing the ever-changing nature and cross-disciplinary challenges of the maritime sector requires a complete understanding of the unique characteristics of the maritime space. The book aims to address the current and future trends, technologies, and strategies in the maritime sector and provides insights into how organizations can successfully navigate the digital transformation process.

The editor has collected reliable and scientifically proven information from various credible sources to provide readers with the latest and most accurate information available. The book is a compilation of contributions from various experts in the field, and the editor expresses gratitude to these contributors for sharing their knowledge and expertise.

Editor

CONTENTS

Chapter 1

On Maritime Digitalization in Emerging Environments

Sanja Bauk

Durban University of Technology, Durban, South Africa

ABSTRACT

The purpose of this chapter is to propose a conceptual framework for implementation of advance info-communication technology and systems (ICT&S) across maritime cluster in emerging environments, with emphasis on some South-East European countries and South Africa. Smart implementation and adoption of the ICT&S in shipping and port management have been considered due to the Holtham\'s & Courtney\'s model (2010). Furthermore, particular attention is given to Blockchain-based Maritime Supply Chain System (BMSCS) conceptual model, which includes distributed relational database, smart contracts, and crypto-currency payment mechanism as main constructs. The document workflow management, financial processes, and device connectivity are considered as key scenarios in the blockchain model. The stakeholders, who play the role of traditional intermediaries in the goods and accompanied documents and data flows, including overall supervision of the processes between end nodes, that is, shippers and customers, are encompassed within the BMSCS scheme. Some strengthens and frailty of BMSCS are highlighted, along with the suggestions for further investigation in this field, with the aim of better understanding prospective benefits and challenges of blockchain more extensive deployment across maritime sector in the future, with emphasize on developing environments.

Keywords: *maritime; digitalization; emerging economies; conceptual framework*

1. INTRODUCTION

Maritime is an important industry. Over 80% of the total transport of all goods takes place by sea because it is the most economical and most massive mode of transport. The world's seas provide free waterways. These are the largest absorbers of carbon dioxide and the largest producers of oxygen. The seas are the main source of food for one-third of the world's population. Oil and diamonds, e.g., are extracted from the seabed. However, the seas are exposed to pollution caused by both natural disasters and human factors. Another paradox related to the seas is that the maritime industry lags significantly behind other industries in terms of digitalization. Some facts that support this statement are as follows: a large number of ships do not comply with the requirements of the Safety of Life at Sea (SOLAS) Convention. Some ships do not have modern electronic navigation aids such as Electronic Chart Display and Information System (ECDIS) nor (Satellite) Automatic Identification System ((*S*)AIS), for instance. Analyzes of accidents at sea have shown that the crew sometimes is not familiar with these devices. Digitization at land is more developed than at sea, and the main reason for this is the lack of profound investigation of Internet connectivity at sea, which can be hampered by sea surface movements, wave occlusions, rough weather, poor coverage, etc. Consequently, inter-organizational information systems (IOS) are used 75% in hinterland and only 25% in maritime. Internet of Things (IoT) is used considerably less at sea than at land. In road and rail transport, it is possible to track cargo at the level of a single unit or a container, while in maritime transport this is still not possible. As an example, we can use transport of dangerous goods. The casks (drums) with radioactive waste (plutonium, e.g.) can be tracked by Radio Frequency Identification (RFID) chips, Internet connection, and security backend web applications at the level of a single freight unit or a drum in the road and in rail transportation [1, 2], but not in maritime [3]. Furthermore, there are a large number of autonomous vehicles on roads (about 1.5 thousand) and in the air, i.e., drones (about 1.5 million), but only one autonomous ship (Yara), and another one is currently under construction [4]. Blockchain technology is not widely accepted yet, since there are various impediments like the lack of trust between stakeholders; government support; legislation; standards; along with the stakeholders' readiness for risky investments in emerging technology. Some extensive desktop studies of academic writings shown that a very small percentage of articles deal with advanced info-communication concepts such as big data, virtual intelligence, robotics, 3D, virtual reality, digital security, etc., in maritime. There is no clear political strategy for further development of info-communication systems in maritime. This complicates maritime digitization in developed countries, and considerably more in developing ones. Concerning the latest, this chapter is organized in the following manner: Section 2 deals with smart adoption of advanced ICT&S in general, and in maritime business. Additionally, a case study has been conducted in several non-EU and EU countries in this respect, based on the Holtham's & Courtney's model. Section 3 considers rational blockchain adoption in maritime, with a focus on developing environments, concerning BMSCS, TradeLens, smart contracts, and Blockshipping. This section also encompasses a case study on blockchain smart adoption in maritime business in emerging economies, with emphasis on South Africa and Montenegro, while Section 4 gives some concluding remarks.

2. SMART ADOPTION OF ICT&S IN MARITIME

We live in a time of massive progress in the field of info-communication technology and systems (ICT&S). The question is do we really need all these innovations and do they always make our lives easier and better. In order to get the best out of ICT&S, we need to know which of these technologies we actually need and how to use them purposely. When it comes to a business environment, it is very important that higher management structures are aware of these needs and discuss them with employees. This is especially important in maritime business, bearing in mind that, stakeholders in maritime are usually conservative and not early new technology adopters. Several studies considered ICT&S rational, intelligent or smart implementation and adoption of advanced ICT&S in maritime [5, 6, 7, 8, 9, 10]. On the Holtham's & Courtney's model, in this chapter, it has been examined how stakeholders in maritime assess key constructs in the model, regarding some developed and developing countries in Europe. The considered countries were treated as the European Union (EU) and non-European Union (non-EU) ones. Before presenting the methodology and the obtained results, the overview of the applied Holtham's & Courtney's model is given.

2.1 The Holtham's & Courtney's model

This model is composed of four constructs: knowledge, ICT&S strategy, system effectiveness, and ICT&S management. These constructs are supported by organizational culture and top manager's mindset (Figure 1).

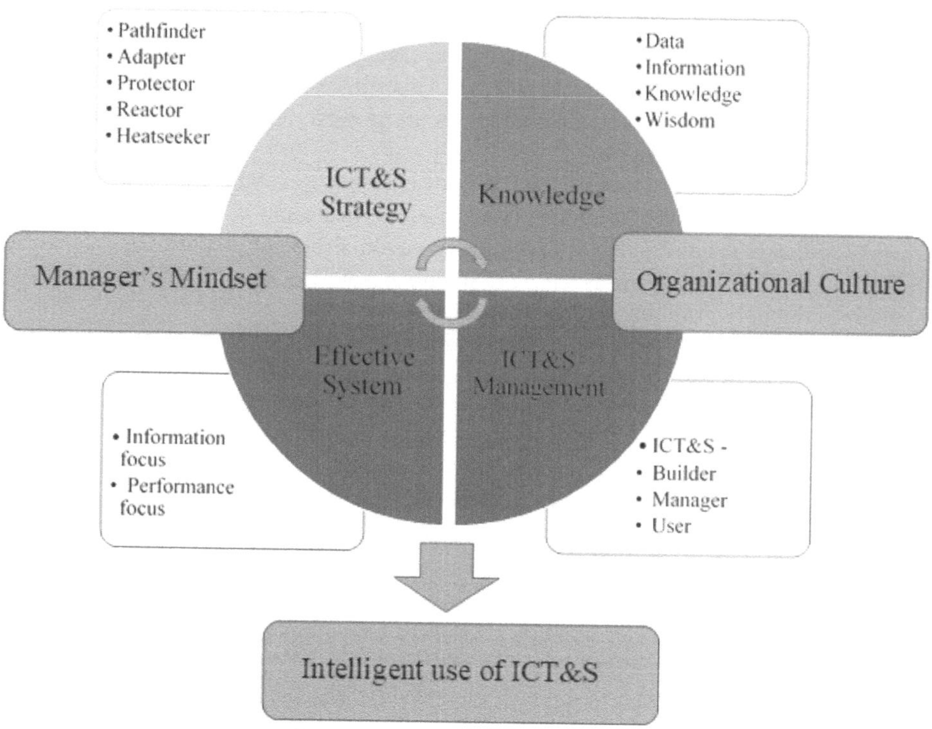

Figure 1. *Adapted Holtham's & Courtney's model (source: Own).*

Knowledge can be described as the understanding of a subject that one gets by experience or study, known either by one person or by people generally; or, as the state of knowing about or being familiar with something [11]. Furthermore, Cambridge dictionary depictures knowledge as awareness, understanding, or information that has been obtained by experience or study, and that either is in a person's mind or possessed by people generally; or, as skill in, understanding of, or information about something, which a person gets by experience or study. Symbolically, knowledge is one of the steps in the so-called ladder of knowledge. This ladder of knowledge encompasses data, information, knowledge itself, and wisdom stairs. The data and information have easier explanations than the concepts of knowledge and wisdom. In this context, the focus will be on knowledge stair, as a key for understanding contemporary ICT&S and their rational implementation in maritime business. Knowledge here means consciousness about advanced ICT&S availability at the market, including the ICT&C purposiveness regarding particular business strategies, processes, and activities.

The ICT&S strategy brings together business and technology. Due to the Holtham's & Courtney's model, there are five different strategic orientations.

Pathfinder: Systematically seeks and selectively exploits relevant ICT&S trends to gain competitive advantage and enable entry into new markets. Pathfinder is willing to experiment with new, advanced ICT&S. Pathfinder constantly seeks a competitive advantage by detecting or sensing emerging ICT&S trends and opportunities.

Adapter: Functions at two types of market: one relatively stable and focused on efficiency, and the other where ICT&S plays an increasingly important role. Adapter applies different rates of technological uptake in each. This split feature is typical for businesses where different rates of technological adoption are present, and adaptability rather than uniform solutions are applicable.

Protector: Meticulously assesses ICT&S investment for its efficiency orientation and applies technology primarily to reduce costs of investments and increase communication processes rather than new market opportunities creation. Protector is control orientated and slow to innovate. These organizations work in domains where core ICT&S-based technologies are commonly available and easily replicable.

Reactor: This is a characteristic of an organization where technology is not perceived as a strategic tool. It responds slowly to change and tends to view ICT&S applications as standalone tools. In this strategic orientation, the ICT&S platforms usually appear to be weak or obsolescent. The risk is that the organization could quickly become non-competitive through a lack of capacity to meet customers' needs.

Heatseeker: Seeks upon ICT&S fashioned instead of strategically analyzing the best ICT&S fit for its business processes. Heatseeker is typical for an organization whose structure is in constant change, experimenting with innovations before obtaining steady business performance. This organization is receptive to ICT&S spending and subsequent partial reversals when intended benefits are not realized quickly.

The system effectiveness can be achieved by setting and communicating critical success factors (CSF) [12] and developing them steadily. The first step is to use technology to create an effective operational platform, primarily with internal information. Then, the CSFs can be widened to foster improved skills to use technology. This will start with employees and then extend to suppliers and customers. Once when these two steps work well, the CSFs can be broadened to encompass external information about markets, customers, and competitors. After these, three steps comes business intelligence, which allows organizations to identify and manage risk while developing new products, services, and markets to ensure a successful future.

The ICT&S management is based on ICT&S builders, ICT&S managers, and ICT&S users. A person or management team that communicates the needs of ICT&S users into ICT&S builders or designers has to be engaged in the organization as a knowledge navigator, or information resource manager. There are business organizations, which recognized this triangle and which are working on filling and improving the personnel skills towards achieving this goal [13, 14].

These four constructs, which form the backbone of the Holtham's & Courtney's model are underpinned by organizational culture and top manager's mindset.

Concerning the organizational culture, there is universal agreement that it exists and plays an important role in shaping behavior in organizations. However, there is little consensus on what organizational culture actually is. Here are quoted several expressions that can be used in absence of universally accepted one [15]: organizational culture is how organizations do things; organizational culture is the sum of values and rituals, which serve as glue to integrate the members of the organization; organizational culture is civilization in the workplace, etc.

Top manager or top management team's role is to weave a fabric of horizontal (information, technology, people, and organization) and vertical (direction, knowledge, process, and climate) threads mutually intertwined. In organizations where knowledge is a core dimension, managers have frequently identified people's skills as the major influence, along with organizational climate. Moving from the information-based to the knowledge-based enterprise is a major challenge for today's companies [16]. Therefore, managers have to combine proper notions from several different domains: organizational behavior, human resource management, big data, analytics, artificial intelligence, etc. Technology is a key enabler, but not usually as significant as skills and climate. Top managers' team mindset covers all considered constructs and it affects intelligent or rational use of ICT&S.

These constructs are used in the following analysis as independent variables. As a control variable is used non-compliance between technology-led potential and its everyday usage, while the dependent variable is intelligent use of ICT&S, which reflexes efficient and smooth communication between tasks, technologies, and employees [4].

2.2 Case study on smart ICT&S adoption

Based on the Holtham's & Courtney's model, a survey was conducted among stakeholders in maritime in terms of how rationally they use the ICT&C in their business. Selected stakeholders from four EU (Croatia, Greece, Italy, and Slovenia) and from four non-EU countries (Albania, Bosnia & Herzegovina (B&H), Montenegro, and Serbia) were included in the survey. Forty experts (five per each considered country) from maritime administrative bodies, agencies, private marine companies, ports, and universities (maritime departments) were asked to evaluate the set of fifteen statements by means of Likert (1–5) scale (Table 1). The last two statements refer to the control and dependent variables in the model, respectively, while the rest of the statements correspond to the independent variables.

Table 1. The survey content.

Construct	Statements
C1: Knowledge	S1.1: Knowledge is important for business success. S1.2: Knowledge and skills of employees are important for efficient and effective use of ICT&S?
C2.1: ICT&S Strategy *Pathfinder*	S2.1: New ICT&S solutions adoption is risky for the organization.
C2.2: ICT&S Strategy *Adapter*	S2.2: Analyzing carefully the existing ICT&S solutions prior to their introduction into the organization is important.
C2.3: ICT&S Strategy *Protector*	S2.3: The ICT&S reduces operational costs of the organization.
C2.4: ICT&S Strategy *Reactor*	S2.4: The available ICT&S solutions can be adapted to the current business needs of your organization.
C2.5: ICT&S Strategy *Heatseeker*	S2.5: The latest ICT&S solutions are the best ones.
C3: System effectiveness	S3.1: Your customers intensively use ICT&S resources of your organization (web site and various online users' apps). S3.2: The ICT&S allows you to become familiar with the current market trends in the area of your business.
C4: ICT&S management	S4.1: The ICT&S functions are important for the successful functioning of the organization and its business success. S4.2: The usage of ICT&S for operational takes within your organization (accounting operations, database of employees, database of business partners, market analysis, etc.) is extensive.
C5: Culture	S5: Positive organizational culture and clime are important for effective use of ICT&S.
C6: Manager's mindset	S6: Manager's mindset is important for effective use of ICT&S.
C7: ICT&S capacities versus exploitation	S7: There is a divergence between ICT&S capacities and their real application on a daily basis in your organization.
C8: ICT&S intelligent exploitation	S8: The ICT&S serves as connective tissue among tasks, technologies, and employees in your organization.

Source: Own.

The level of respondents' agreement with the proposed statements is labeled due to the following scheme: (i) if the average score per group of experts from a certain country is between 1 and 2.5 then the level of experts' agreement is "low"; (ii) if it is between 2.6 and 3.5, then the level of agreement is "moderate", and (iii) if it is between 3.6 and 5, then it is "high". The results of the survey are presented in Table 2.

Table 2. The assessments of the constructs.

Constructs		C1	C2	C3	C4	C5	C6	C7	C8
	Country	**Knowledge**	**ICT&s Strategy**	**System effectiveness**	**ICT&S management**	**Organizational culture**	**Manager's mindset**	**Control variable**	**Dependent variable**
EU	Croatia	High	Adapter	High	High	Moderate	High	Low	High
	Greece	High	Adapter	Moderate	High	Moderate	High	Moderate	High
	Italy	High	Adapter	High	High	High	High	High	Low
	Slovenia	High	Adapter	High	High	High	High	High	Low
Non-EU	Albania	High	Adapter	Moderate	High	High	High	High	Low
	B&H	High	Adapter	High	High	Moderate	High	Moderate	High
	Montenegro	High	Adapter	Moderate	High	Moderate	High	Moderate	High
	Serbia	High	Reactor	High	High	Moderate	High	Moderate	High

Legend: ▲ *High agreement;* ▲ *Moderate agreement;* ▲ *Low agreement*
Source: Own.

According to the results presented in Table 2, it is obvious that all respondents evaluate knowledge, ICT&S management, and manager's mindset as highly important for rational application of ICT&S in maritime organizations. When it comes to the system effectiveness and organizational culture, the experts assessed these constructs as high or moderate important for intelligent implementation of ICT&S. Control variable confirms validity of the assessments assigned to the dependent variable "intelligent use of ICT&S in maritime". Namely, all experts highly agreed with the statement "the ICT&S serve as a connective tissue among tasks, technologies, and employees in your organization", which in fact justify smart exploitation of ICT&S in maritime business. When it comes to ICT&S strategy orientation, the respondents are dominantly *adapters*, while only one of the respondents is *reactor*. The *adaptors* are looking for the third path, while reactors leg behind regarding adopting new ICT&S. Adaptation means that emphasis is on modification rather than fundamental reconfiguring of the existing ICT&S. On the other side, reaction means using weak and obsolescent ICT&S platforms. The reasons behind this orientation should be examined through in-depth interviews with respondents and through further analysis of their business strategies, including the position of the ICT&S in it. This might be the subject of further investigation in the field.

In addition to the analysis of the degree to which ICT&S are used rationally in the maritime business, it was also examined which advanced info-communication platforms are available to the maritime organizations in which the respondents in this study work. The results of this part of the survey are presented in Table 3. It is evident that there are efforts to modernize ICT&S in maritime business, but also that some of the analyzed maritime organizations, i.e., countries are lagging behind, especially those that are not members of the EU. Namely, the non-EU countries have to reconsider their business development strategies and ensure funds for implementing new ICT&S and renewal of the existing ones. These countries should follow actual trajectories and scenarios towards efficient and effective digitalization in maritime [17].

Table 3. Available ICT&S in the examined maritime organizations.

Available ICT&S	**Countries**							
	Non-EU countries				**EU countries**			
	Albania	**B&H**	**Montenegro**	**Serbia**	**Croatia**	**Greece**	**Italy**	**Slovenia**
Electronic Data Interchange (EDI)		×	×		×	×	×	×
Enterprise Resource Planning (ERP)		×	×		×	×	×	×
Customer Relationship Management (CRM) System	×	×	×		×	×	×	×
Electronic Logistics Marketplace (ELM)		×				×	×	×
THETIS (PSC—Port State Control)			×		×	×	×	×
Blockchain-based Maritime Supply Chain System (BMSCS)	▼	▼	▼	▼	▼	▼	▼	▼
Automatic Identification System (AIS)	×	×	×	×	×		×	×
Long-range and tracking (LIRT)	×		×		×	×		×
Vessel Traffic Monitoring Information System (VTMIS)	×		×		×	×		
Sea Traffic Management (STM)	x				×		×	×
e-Navigation					×			×
e-Maritime					×			×
Common Maritime Communication Platform (CMSP)					×			×
Maritime Surveillance Services (MSS)	×				×			×

Available ICT&S	**Countries**							
	Non-EU countries				**EU countries**			
	Albania	**B&H**	**Montenegro**	**Serbia**	**Croatia**	**Greece**	**Italy**	**Slovenia**
SafeSeaNet (SSN)			×		×	×		×
Maritime Single Window (MSW)					×	×	×	×
Automatic Guided Vehicles (AGV)								
Digital twins							×	
Remotely controlled vessels	▼	▼	▼	▼	▼	▼	▼	▼
Unmanned sea or underwater vessels (UxVs)						×	×	
Earth Observation Services —Search and Rescue (SAR) sensors					×	×	×	×
Earth Observation Services —Optical sensors					×	×	×	×
Satellite-based oil spill detection system at sea			×		×			×
Oil spill prediction modeling system			×		×		×	×

Legend: ▼ *There is no ICT&S of such type*
Source: Own.

Through this case study, it is shown that responders, who are employed in maritime administration and business organizations in four EU and four non-EU countries, have similar attitudes towards concerned constructs inherent to intelligent exploitation of contemporary ICT&S. They all identified knowledge, ICT&S management, system efficiency, organizational culture and manager's mindset as key perpetuators of rational and purposeful use of the ICT&S. This speaks in favor of their sound education and awareness about the importance of ICT&S in today's dynamic maritime business environment. Regarding ICT&S strategical orientation, the respondents are cautious, i.e., not prone to take risks of investing in new ICT&S solutions and experimenting in the market.

When it comes to the availability of common and advanced ICT&S in analyzed maritime organizations, it is shown that there are big differences among EU and non-EU countries. For instance, maritime organizations in Slovenia have almost all listed ICT&S except blockchain technology, AGVs, digital twins, are UxVs. The companies in Italy have, e.g., digital twins and UxVs. Croatia and Greece have also quite an extensive list of available ICT&S. On another side, explored non-EU countries are modestly equipped. The examination of the reasons for such difference and how it can be alleviated in order to avoid disruptions in maritime ecosystem and negative economic implications for the non-EU countries should be the subject of further, more profound studies. None of the considered maritime entities does have on disposal Blockchain-based Maritime Supply Chain System (BMSCS), e.g. Since BMSCS is the advanced emerging ICT&S platform in contemporary maritime business, the following text attempts to explain the basic principles of this platform, including its benefits and challenges.

3. THE BMSCS CONCEPTUAL FRAMEWORK

Bockchain in maritime is a far broader system than cryptocurrency-based electronic financial transactions mechanism. In literature, it is named as Blockchain-based Maritime Supply Chain System (BMSCS) [18]. It includes smart contracts and payments, tracking and tracing cargo, empty containers placement, early risk assessment, and services that can create new added values in maritime (Figure 2).

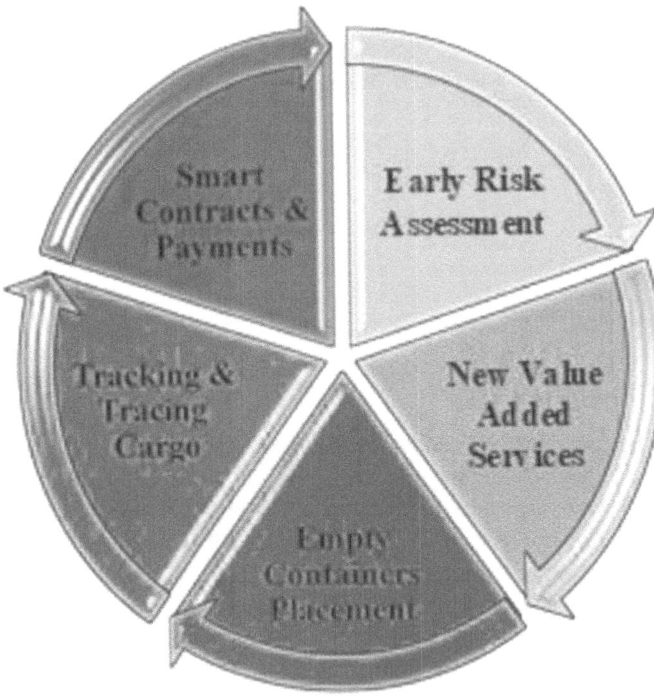

Figure 2. *BMSCS key components (source: Own).*

Maritime is an indispensable link in global supply chains. Hereof, blockchain technology is likely to become unavoidable in shipping and port management, in striving to optimize global supply chains and make these more efficient and effective. The BMSCS should reduce the volume of administrative work, errors that occur due to manual work, delays in the delivery of goods, and consequently overall costs of transportation and delivery of goods. Due to some studies, the costs of global supply chains should be reduced by approximately 15% by implementing blockchain technology [19].

In addition to these advantages, blockchain also has certain disadvantages. Maritime sector is generally risk-averse, tending not to be an early adapter of new potentially risky technology [20]. Some stakeholders in maritime want to keep their data secret, since "competition is fierce" and "a lot of industry actors are basically competing with the same service" [21]. In other words, some partners in the supply chain consider information as a competitive resource and are unwilling to share them. Positional data might be used to track vessels by identifying port locations, fueling locations, and routes [22]. This is particularly the case with tracking dangerous and hazardous goods, pharmaceuticals, or food. The use of blockchain does not guarantee that the information recorded in ledgers is correct and does not prevent tampering data prior to entering it into blockchain ledger, e.g., the contents of a container, fuel production, testing or combustion, and the like [23]. Due to the huge amount of data and traffic generation, including data storage, blockchain requires a wideband like G5 or G6 [24], while the internet speed can be low when the working stage is offshore. Further, blockchain causes high-energy consumption [25]. Blockchain in the maritime sector indicates the potential to reduce transaction costs in a number of areas, including reducing the need for intermediaries such as brokers and courier services and reducing related financial expenses and energy costs. However, one should not forget that this does not include the costs of the overall investment and expenses associated with blockchain implementation and adoption, especially in developing or emerging economies [26, 27].

The present level of awareness, knowledge, and expertise about blockchain is scarce among the stakeholders. Therefore, educational, training, or capacity-building programs are necessary at regulatory, administrative, and operational levels. Higher level of standardization across the global supply chain is necessary as well. The Digital Container Shipping Association (DCSA) conducts efforts in this regard, but further, more extensive, actions are necessary. In general, there is a hesitation by stakeholders in maritime sector to invest in blockchain systems in terms of technological integration, regulatory, organizational, and educational costs, since the maritime sector traditionally relies on its legacy systems. There appears to be a gap between what practitioners in the blockchain area suggest and what has been a range of state-of-the-art approaches in software engineering and information security research and practice [28]. Furthermore, the major liner shipping companies are the most likely parties to benefit from blockchain regarding the complexity of their blockchains and huge requirements on financial resources [29]. This can put other potential actors in the global supply chain at a disadvantage. The last but not the least, the basic attitude should be that technology, in this case, blockchain on the top of the global supply chain should improve the human condition, and not replace humans [30]. Therefore, human and ethical dimensions of blockchain technological development and more extensive deployment, should not be neglected.

In the following text, we shall present blockchain framework in maritime at the example of two applications: TradeLens and Blockshipping. TradeLens is used for tracking and tracing cargo along the global supply chain, early risk assessment, smart contracts, and value-added services created through an open platform. Blockshipping is used for empty containers' optimal placement by autonomous intelligent software agents.

3.1 TradeLens

TradeLens is a new business model in shipping and port management. It enables one-to-many connections for all the actors, all individuals that are involved in a global supply chain instead of bilateral connections. Everybody come together in a maritime industry-neutral, open platform for every participant [31]. Maersk, the world's largest international container shipping and logistics company, and IBM the technology leader in blockchain came together to provide a new, open platform solution underpinned by blockchain to help unlock some of the opportunities for a more efficient global supply chain. Maersk and IBM have a long history of working together, actually decades. In March 2017 these organizations collectively try to improve global trade through digitization. In January 2018, they launched an early adapter program; trials began, and in August 2018, they formally launched the TradeLens limited availability platform, shared among 92 participants. In December 2018, TradeLens is commercially realized, along with 1.5 million events per day published to the platform. Some types of these events are presented in Table 4.

Table 4. TradeLens standardized events.

Actual	Estimated	Planned
Start container tracking	Documentation cutoff: Vessel ETD	Import documents approval
Start shipment tracking	VGM (Verified Gross Mass): Vessel ETD	Discharged from truck
Booking confirmation	Cargo cutoff: Vessel ETD	Loaded on vessel
Stuffing started	Rail ETD	Stuffing started
Vessel ATA	Rail ETA	Stuffing completed
Vessel ATD	Bill of Lading Available	Loading on vessel
Loaded on rail	Vessel ETD	Gate in
Rail ATD	Vessel ETA	Gate out
Rail ATA	Discharged from vessel	Packed container selected for inspection
Loaded on truck	Load on vessel	Packed container passed inspection
+Add more	Custom release	Cargo specific certificate approved
	+Add more	+Add more

Source: [31].

The platform can track 120+ unique consignment shipments, while 60+ network members are onboard or in a process of accessing. TradeLens supports 18+ unique, standardized, trade document types. Some of these documents are listed in Table 5. In February 2019, enhanced document sharing, permissions, and notifications were released. The platform includes half a billion events on annual basis and this number grows with more and more network members.

Table 5. TradeLens standardized documents.

Document	Party
Import documentation approved	Customs House Broker
Customs release	Customs Authority
Cargo geography-specific certificate approved	Customs House Broker
Bill of lading available	Beneficiary Cargo Owner (BCO)
Certificate of origin available	Beneficiary Cargo Owner (BCO)
Packaging list available	Beneficiary Cargo Owner (BCO)
Commercial invoice available	Beneficiary Cargo Owner (BCO)

Source: [31].

More than twenty million containers of cargo information are in the system today, which is roughly 1/5 of global trade and it is growing. The platform involves numerous parties and systems: ocean carriers, ports and terminal operators, inland carriers, shippers, consignees, beneficiary cargo owners, freight forwarders, 3PLs, custom authorities, government agencies, financial and insurance services, transportation management systems, Port Community Systems (PCSs), supply chain validity systems, supply chain, manufacturers, retailers, etc. They all collaborate and share information. TradeLens provides them with comprehensive, real-time visibility and immutability across the end-to-end journey of shipment. In other words, data is available immediately, along with the single simplified view across all shipments. For instance, a terminal operator publishes a piece of information about the fact that a container has been loaded onto a ship that becomes immediately available to everybody else in the supply chain. The idea is to build workflow based on smart contracts using chain code to derive cross-organizational workflow by excluding manual work.

Blockchain on which the platform is based, enables the trust in data that are available on the platform. It is an open and censorship-resistant distributed database model, secured by encryption and decentralization. Blockchain records information in blocks on a shared ledger, storing a synchronized copy of it on all the systems participating in the network, hence assuring its immutability. The trust anchors, which are the blockchain nodes, ensure through consensus algorithms that the information should be written on the platform as approved like valid. All information are auditable, verifiable, and temper proof; so, as soon as a piece of data is published to the blockchain it cannot be edited. The only way to edit a document is to create a new version of the document. Consequently, all the documents are fully auditable. Additionally, cryptographic hash of the data is written to the blockchain, and this is a part of the supply chain. It is important to say that private data remain private. TradeLens as an information-sharing model allows ecosystem partners to have access to the information they should access and vice versa. The platform offers a high level of flexibility through application of RESTful APIs (Application Programming Interfaces), back-end ERP (Enterprise Resource Planning), and secured front/back-end web services.

In the middle or at the very core of TradeLens solution, there is the platform and blockchain behind it. Below the platform is the network. The network is not a physical network.

It is a set of entities that provide the data, including the data itself. The ocean carriers, ports, terminal operators, customs, shippers, inland transporters, etc., provide the data. On the top, above the platform are applications and services, i.e., RESTful APIs, back-end ERP, and secured web that enable people to exchange information. These are based on open published industrial UN/CEFACT standards that are defined at the platform level, so that third parties are allowed to build new value-added services and applications. This is the basic kind of model, through which TradeLens is moving forward as a paradigm shift in information sharing across the whole ecosystem. A conceptual framework of TradeLens as a blockchain-based solution in the global supply chain is presented in Figure 3.

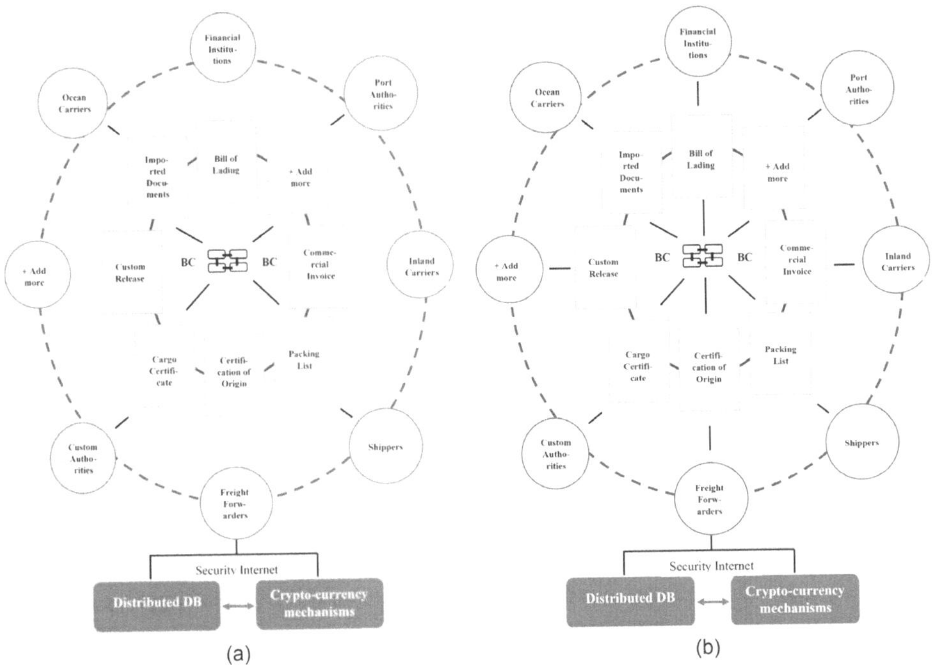

Figure 3. *TradeLens conceptual framework (source: Own).*

What kind of information is shared across the platform, i.e., over the entire supply chain? This information is mostly shipping milestones. The information on: has a container been staffed; has the container be gated; what is the estimated time of arrival (ETA) of the container at the destination, and so on, are in fact shipping milestones. However, it is more than that. It is also the documents in maritime, both structured and unstructured (like PDFs, scans, images, etc.), by making them available to the participants along the supply chain. The documents need to change 'hands'. They need to be approved, updated, and available to build workflow using smart contracts, like the bill of lading, clearance, insurance, etc. This is powerful in terms of driving cross-organizational data flow in maritime.

Within TradeLens, there is sensor data and IoT for referring to the container number, electronic seal, temperature inside it, for instance, etc. All of that is part of the underlined data that is made available to the participants who need that data. There is a whole concept of seamless and permission data-sharing model that is built on the base: what your role is, i.e., are you terminal, ocean carrier, shipper, inland transporter, etc. The default permission model allows people to share information, so that information is made available to those who need it, but it is not available to those who should not see it.

TradeLens is of utmost importance whenever planned actions turn into unplanned. For instance, the ocean carriers' decision has implications not just for them but for all stakeholders further down the supply chain from customs brokers, port authorities, and terminal operators to inland transporters and consignees. With TradeLens, changes to the shipment are reflected immediately allowing supply chain participants to coordinate actions tightly, delivering the consignee's inventory in time. TradeLens allows near-instant logistics adjustments so the disruptions are kept to a minimum. Global trade is an incredibly complex system, but TradeLens and blockchain create an industry-wide and innovative solution to alleviate this complexity and related impediments.

3.2 Smart contracts

Smart contracts are programs stored on a blockchain that run when predetermined conditions met. They are used to automate the execution of an agreement so that all participants can be immediately certain of the outcome, without an intermediary's involvement or time loss. They can also automate a workflow, triggering the next action when conditions are met [32]. Smart contracts work by following simple "if/when (…) then (…)" statements that are written into code on a blockchain. In the other words, a smart contract presents the lines of code that are stored on a blockchain and automatically executes when predetermined terms and conditions are fulfilled [33]. A network of computers executes the actions when predetermined conditions have been met and verified. These actions could include releasing goods, funds, or confirmations in maritime supply chain. The blockchain is updated when the transaction is completed. That means the transaction cannot be changed, and only parties who have been granted permission can see the results. Blockchain network controls access. Within a smart contract, there can be as many stipulations as needed to satisfy the participants, so that the task will be completed correctly. To establish the terms, participants must determine how transactions and their data are represented on the blockchain, agree on the "if/when (…) then (…)" rules that govern those transactions, explore all possible exceptions, and define a framework for resolving disputes. Then, the smart contract can be programmed by a developer, although organizations that use blockchain for business, provide templates, web interfaces, and other online tools to simplify structuring smart contracts. Key benefits of smart contracts are speed, efficiency, accuracy, trust, transparency, and security (blockchain transaction records are encrypted, which makes them very hard to hack; plus, each record is connected to the previous and subsequent records on a distributed ledger, and hackers would have to alter the entire chain to change a single record). In maritime supply chain, sea waybill or bill of lading can be converted into a smart contract, while it requires an agreement between shipper and carrier, and/or any other relevant and permissioned parties to view the consignment, transport equipment, and documents, as permissions allow [34]. The benefits of such smart contract include simplified transmission of shipping instructions; management of document status and versioning; faster submission of shipping instructions for creation of final bill of lading; quick sharing of documents with all permissioned parties; including immutability, traceability, and auditability of the documents involved.

3.3 Blockshipping

Today the container shipping industry accounts for around 60% of all the world seaborne trade. This valuable industry has been troubled for years by challenges like overcapacity, low freight rates, security threats, and increasing environmental regulations. Currently, there are about 27 million containers in the world, which are moved from one destination to another on trucks, container cars, ships, rail, or waiting in the port, container yard, railway station, and the like. About 5 million containers are uncontrolled and nobody knows their precise locations; if they are currently in transit or waiting for collection. Consequently, no one knows if they are empty or loaded, which means that no one knows if a truck or a train is wasting time and energy carrying an empty "metal box" instead of carrying goods [35]. This is a huge waste of energy; it produces additional costs and negatively affects the environment.

Therefore, the global shared container platform (GSCP) is currently under development. As the world's first blockchain-based container registry, it will allow the industry to help real-time track all containers worldwide. The platform will enable the industry players to manage efficiently all kinds of transactions related to container handling. The GSCP has several user groups like shipping lines, leasing companies, banks, financial institutions, blockchain container investment syndicates, transport service providers, beneficiary cargo owners (BCO), container terminals, container depots, repair shops, etc.

Through a secure login, each user group will have a unique set of functionalities that matches their exact needs. For example, if you are a shipping line export user, you can use GSCP platform to find street turn matching opportunities for ensuring that empty containers meet export demands. You will see an inventory list of all export bookings, which require an empty container to the customer location for stuffing, rather than transporting an empty container from the port or the depo. For convenience, the platform enables users to switch between list and map view. The user can apply one or more filters and inventory will update accordingly, for instance, only showing FEUs (40-feet units). Any set of applied filters can be saved in user's filter presets. This way they are quickly accessible whenever the user needs them. Matching export containers with import containers is easy and swift. This enables both importer and exporter to save an empty container haulage trip, plus gate in and gate out fees at the terminals. The system identifies possible matches based on container size, type, boarding date, previous commodities carried, and availability. The platform also enables sending a request to the involved shipping line with the comment. The GSCP provides various ways to import booking and container data. The user can use EDI and API connection with the in-house booking or order platforms [36].

Blockshipping is, in fact, a shared pool of containers, which enables a "just in time containers" situation. Today more than 40% of all containers in transport are empty. Therefore, resources are wasted and costs are increased. With Blockshipping saving potential for the shipping industry might be at least 5.7 billion USD and reduction of CO_2 emission can be 4.6 million tons yearly [37]. Blockshipping platform is a part of the so-called programmable economy. In such an economy, the interactions among different parties will not occur through mediation of a third trusted party, but automatically through autonomous intelligent software agents (AISA). These are also called dApps (distributed Apps) that run on blockchain and are authorized and instructed by the parties involved in the BMSCS to negotiate autonomously on their behalf.

Four key subsystems of Blockshipping are [38]:

- Global Shipping Container Registry (GSCR), that holds real-time information about every container available through Blockshipping;
- Empty Container Repository Engine (ECRE) that continuously calculates the next best-laden transport for each container. The engine also "understands" the position of every truck available to transport the empty container;
- Autonomous Intelligent Software Agents (AISA) that run on blockchain and negotiate all agreements;
- Smart Contracts (SC) that can be treated as rental contracts established through autonomous negotiations, which persist on the blockchain and govern the rental through binding self-enforcing rental agreements.

Blockshipping has developed a unique revenue-sharing model, while it issues two types of tokens:

- Internal utility token, or Container Platform Token (CPT), and
- External revenue share coin, or Container Crypto Coin (CCC).

The CPT will be used for clearing and settlement of transactions between the users of the platform. These transactions will relate to many different services and fees. A percentage of the revenue goes to a revenue share pool and is passed on to the owners of CCC tokens. Blockshipping exchanges the CPTs in the revenue share pool to Ether via USD. Then, Blockshipping uses smart contracts to convert revenue Dutch auction on the Ethereum blockchain in which Blockshipping offers the owners of the CCC tokens price for their tokens. The offered price will increase until all available Ethers are spent. After the auction, Blockshipping distributes the acquired CCC tokens to all the owners of CCC tokens on a pro-rata basis. In this way, token owners are rewarded regardless of their decision to sell or keep their CCC tokens [39].

The processes flow within Blockshipping is based on several simple and fully automated steps. The easiest way to make an explanation is to follow an example. Let us assume that the shipping line needs to rent a container to transport goods from Nairobi (Kenya) to Rotterdam (Netherlands). Blockshipping empty container repository engine identifies the best-positioned empty container in Nairobi and informs the shipping line about the options. The shipping line informs its autonomous intelligent software agents (AISAs) about the containers. The rental negotiations then happen unsupervised between the shipping line and the container owner through their autonomous agents. The agreements established by AISAs are persisted on blockchain in smart contracts that govern the rental in binding self-enforcing rental agreements. Blockshipping container platform tokens CPT are used to pay rental fees, while the fees are transferred from the shipping line wallet, in accordance with the smart contract and reserved payment. Smart contracts can be changed if conditions change. For example, if the rental period is extended when the container reaches its final destination in Rotterdam. Then, the smart Oracle blockchain enforces the smart contract. The rental ends and releases CPTs to the container owner's wallet [40].

In addition to TradeLens and Blockshipping, there are a number of other blockchain applications in maritime. Some of these are given in Table 6. Concerning safety issues, semi-private blockchains are common. The consortium companies' reputation speaks, in fact, about safety. Maritime, as conservative, assesses and recognizes the quality of operation in long run. As we said previously, stakeholders in maritime are not early adopters. However, trust between network participants is a bigger problem than safety. The blockchain is an unorthodox technology and cryptocurrencies are still highly volatile. In such a setting, maritime stakeholders do not like to disclose essential business information about customers, suppliers, and cargo. Many freight forwarders and intermediaries, e.g., earn their profit thanks to information asymmetry. Interoperability will be a smaller problem in terms of technology (since standards have been developing) than in terms of trust and smooth process flows at (inter-)organizational level.

Table 6. Some blockchain applications in maritime.

No.	Blockchain		
	Consortium	**Platform**	**Ledger**
1.	Port of Koper, Slovenia	CargoX	Public
2.	Malaysia's West Port & LPR - Brazilian textile importer	300cubits	Public
3.	Maersk & IBM	TradeLens	Consortium (permissioned)
4.	Abu Dhabi Ports and Port of Antwerp	Silsal	Consortium (permissioned)
5.	EY & Guardtime	Marine Insurance Blockchain	Public
6.	PIL, PSA & IBM	Proof of Concept (POC)	Consortium (permissioned)
7.	Port of Antwerp with Belfruco, Enzafruit, PortApp, 1-Stop and T&G Global	Smart Contracts	Consortium (permissioned)
8.	2021.AI Den Danske Maritime Fond, EUDP, INVICTA	Blockshipping	Public
9.	Port of Malmo & Port of Copenhagen	PortChain	Consortium (permissioned)
10.	AAT, FileVersion Health, CROP	CargoChain	Consortium, (permissioned)

Source: Adapted from [41].

3.4 Case study on blockchain adoption

Within this case study, we explored how maritime stakeholders in two developing countries, South Africa and Montenegro, perceive blockchain technology and its implementation in maritime. Through the methodological framework given in [26, 27] we conducted the survey, which included thirteen closed-ended questions, or statements on blockchain adaption. Concerned statements included the following blockchain dimensions: knowledge, infrastructure, standards, experts, diverse stakeholders, government and regulatory policy, social influence, loss of jobs, computing and storing capacity, complexity, opportunistic behavior, sharing information, and security. The respondents were from maritime companies, agencies, research organizations, governmental bodies, insurance companies, and universities. They are from the executive management level in industry and governmental bodies, and active researchers, professors, and lecturers from universities (10 from South Africa and 10 from Montenegro, all with more than 5 years of research experience). The respondents have had to express their (dis)agreement with the proposed statements via Likert 1–5 scale, where 1 represents the lowest level of (dis) agreement, and 5 the highest level of (dis)agreement. The rest of the offered numerical values are respectively in-between these two extremes. The statements and average values of assessments are given in Table 7. If the average score per group of respondents is between 1 and 2.5, then the level of (dis)agreement is "low"; if it is between 2.6 and 3.5, then the level of (dis)agreement is "moderate", and if it is between 3.6 and 5, then it is "high".

Table 7. The assessments of blockchain adoption.

Statement	Agree	Disagree
S1: The level of awareness and knowledge of BC affects its adoption.	▲ High	▲ Low
S2: The BC adoption is affected by the availability of the infrastructure and functionality to integrate and interoperate within and across the business ecosystem.	▲ High	▲ Low
S3: Standardization and ensuring smooth interoperability is necessary, otherwise, BC can make things difficult instead of making them easier.	▲ Moderate	▲ High
S4: The BC adoption is affected by the availability of skilled and expert resources.	▲ High	▲ Low
S5: The BC adoption is affected by a large number of stakeholders, with different mindsets, organizational culture, and working habits.	▲ Moderate	▲ Moderate
S6: The BC adoption is increased by favorable government and regulatory policies.	▲ High	▲ Low
S7: Social influence positively affects the behavioral intention of using BC.	▲ Low	▲ High
S8: A perception that BC implementation might lead to loss of jobs can be an obstacle in its adoption.	▲ High	▲ Moderate
S9: Development in storage, computing, and cloud infrastructure will affect BC adoption.	▲ High	▲ Moderate
S10: The BC adoption reduces opportunistic behavior (opportunistic behavior means maximization of economic self-interest and occasioned loss of the other partners).	▲ Low	▲ High
S11: The BC adoption is reduced if the information is not shared by the partners, while some stakeholders are hesitant to share information considering it is a competitive advantage.	▲ High	▲ Moderate
S12: Privacy and security of models and data need to be ensured, as BC technology is still immature and vulnerable.	▲ Moderate	▲ Moderate
S13: Blockchain offers a high level of complexity and observability at the same time.	▲ Low	▲ High

Legend: ▲ *High (dis)agreement;* ▲ *Moderate (dis)agreement;* ▲ *Low (dis)agreement; BC – blockchain*
Source: Own.

Five statements with the highest "agree" and "disagree" assessment rates are categorized in different PESTEL (political, economic, social, technological, environmental, and legal) dimensions, along with their rank (Table 8).

Table 8. PESTEL analysis.

P Political	E Economic	S Social	T Technological	E Environmental	L Legal
Respondents "Agree"					
*Favorable government policies (rank 3)	*Hesitancy of sharing information (rank 5)	Awareness and knowledge about BC (rank 1) Skilled and expert resources (rank 4)	Infrastructure (rank 2)	*Hesitancy of sharing information (rank 5)	*Favorable regulatory policies (rank 3)
Respondents "Disagree"					
	Reduction of opportunistic behavior (rank 2)	Social Influence (rank 3)	Complexity and observability (rank 1) *Standardization (rank 4) **Ensuring privacy and security (rank 5)	**Ensuring privacy and security (rank 5)	*Standardization (rank 4) **Ensuring privacy and security (rank 5)

- *Constructs marked with "*" correspond with two different PESTEL dimensions.*
- *Constructs marked with "**"correspond to three different PESTEL dimensions.*
Source: Own.

The respondents consider awareness and knowledge about blockchain as a social dimension of the highest importance for its adoption in maritime. This is understandable, since knowledge is the biggest asset; the only one which grows with exploitation during the time. On the second place is infrastructure, which falls under technological dimension. This is reasonable since, without it, blockchain adoption is practically impossible. In the third places are favorable government and regulatory policies that fall under political and legal dimensions. This is of crucial importance since economic development in South Africa and in Montenegro is controlled by the government. On the fourth place is experts' knowledge, which belongs to the social dimension of PESTEL model, and which is to a certain extent connected with awareness and knowledge, but it can be outsourced in the case of its lack, and under the assumption that awareness and general knowledge about blockchain are in place. The fifth place is the hesitancy of sharing information among the parties, and it falls under both economic and environmental dimensions of PESEL. This is understandable, since once blockchain becomes well established; the impact of this issue will be reduced.

The highest disagreement is observed regarding 'simultaneous' presence of blockchain complexity and observability. Majority of the respondents show suspicion regarding this paradox. Furthermore, the respondents do not agree with the statement that blockchain will reduce opportunistic behavior. South Africa and Montenegro are countries that are for decades in transition and suffer from the permanent reproduction of crisis. Consequently, the responders' rather skeptic attitude towards this statement is completely understandable. The social influence is in the third place. The respondents do not believe that society can affect considerably the implementation of this advanced technology, and this belief is based on their experiences from transitional settings. The statement, which deals with the standardization issue, is "negatively" assessed, but it might be the case due to the experts' belief that standardization must be achieved and that it cannot as such threaten blockchain key advantages. The need for ensuring privacy and security is assessed negatively. This means that some respondents strongly disagree with the statement that blockchain technology is still immature and vulnerable. Due to their response, one can conclude they believe that blockchain technology is at a high level of development and that is less vulnerable than it can appear due to its complexity and deployment at a global scale. This construct can correspond with technological, environmental, and legal PESTEL dimensions at the same time.

Since we collected only twenty survey responses, further research should include in-depth interviews or a survey upon a larger poll of experts and profound discussion on the respondents' assessments, including comments and suggestions. In addition, the following investigation in the field should include experts from other developing and transitional countries (besides South Africa and Montenegro), including a longitudinal approach. Building new knowledge and transfer of existing one on blockchain technological and other crucial dimensions are necessary, particularly in developing countries, which suffer the lack of skilled personnel and expert knowledge, dominantly in the technological domain.

4. CONCLUSION

Maritime is lagging behind other areas of business and industry in terms of digitization. This is especially true in developing countries. Therefore, in this chapter, the results of two case studies on digitalization in maritime, conducted in developing countries, are presented, along with some comparative analyzes with developed countries in certain areas. These studies were conducted on a relatively small sample and in the future, longitudinal studies should be conducted on a larger sample and in a larger number of countries in transition. The results should be presented to ICT&S designers in developed countries in order to find solutions for intelligent design of ICT&S in maritime in perspective. These systems should provide benefits for all. If we strive for sustainable development, then it must ensure development for all. Entire continents cannot be excluded from the plans for further development of ICT&S, artificial intelligence, and virtual reality. This, of course, requires a broader platform, which includes standardization of ports and ships as basic maritime structures onto which ICT&S are built.

General knowledge about new technologies in developing countries exists, but there is a lack of expertise when it comes to hardware, software, network architecture, cyber safety, security, etc. Furthermore, developing countries suffer lack of funds for investments in advanced ICT&S. These countries are excluded from the strategic plans on the development and implementation of new ICT&S solutions. The governments should consequently invest more into attaining sound knowledge in contemporary ICT&S, including disruptive technologies like blockchain and its implementation in global supply chains. Standardization in maritime is the key enabler of faster modernization. In addition, profound research of internet connectivity at sea is of crucial importance for further development of IoT applications in maritime as the key condition for achieving equal presence of ICT&S applications on land and at sea. Maritime industry and high(er) maritime education institutions should cooperate much more closely to alleviate the gap between developing and developed countries in terms of achieving a higher level of graduates' employability, while the sphere of digitalization can provide many opportunities in this respect. Rising awareness about digital transition in maritime and opening discussions among professionals and academics should become common practice. The involvement of legislatures and tight collaboration of key stakeholders in maritime emerging economies are necessary regarding harmonization of ICT&S deployment across global maritime ecosystem and overcoming the existing gap in digitalization.

REFERENCES

1. Tsai HC, Chen K, Liu YY, Shuler JM. Demonstration (DEMO) of radiofrequency identification (RFID) system for tracking and monitoring of nuclear materials. Packaging, Transport, Storage & Security of Radioactive Material. 2010;21(2):91-102. DOI: 10.1179/174650910X12628787557739
2. Tsai HC, Chen K, Liu YY, Norair JP, Bellamy S, Shuler J. Applying RFID technology in nuclear materials management. Packaging, Transport, Storage & Security of Radioactive Material. 2008;19(1):41-46. DOI: 10.1179/174651008X279000
3. Bauk S, Dzankic R. Tracking radioactive materials in sea transportation via RFID technology. In: Proceedings of the 8th Maritime Transport Conference (MT2020); 17–18 September 2020. Barcelona, Spain: UNCTAD; 2020. pp. 91-102
4. Kapidani N, Bauk S, Davidson IE. Digitalization in developing maritime business environments towards ensuring sustainability. Sustainability. 2020;12(21):9235. DOI: 10.3390/su12219235
5. Bauk S, Kapidani N, Schmeink A, Holtham C. Concerning intelligent ICT exploitation in some maritime business organizations: A pilot study. International Journal of Maritime Science & Technology Our Sea. 2017;64(2):63-68. DOI: 10.17818/NM/2017/2.5
6. Bauk S, Kapidani N, Schmeink A. On intelligent use of ICT in some maritime business organizations. Montenegrin Journal of Economics. 2017;13(2):163-173. DOI: 10.14254/1800-5845%2F2017.13-2.11
7. Bauk S, Calvo LJA, Schmeink A, Azam S, Mathar R. On deploying vehicular communication at the developing seaport and related innovation success impediments. Transport. 2019;34(1):126-134. DOI: 10.3846/transport.2019.7987
8. Holtham C, Stace D, Courtney N. Balancing technology, business processes and skills: key determinants of firm-level information systems success? In: ITU and LBS Digital Transformations in the Information Society Conference; 1-2 June 2006. Geneva, Switzerland: WHO; 2006
9. Holtham C, Stace D, Courtney N. Mapping opportunity space: Options for a sustainable E-strategy. Strategic Change. 2004;13(5):237-251. DOI: 10.1002/jsc.691
10. Holtham C, Courtney N. Developing managerial learning styles in the context of the strategic application of information and communication technologies. International Journal of Training and Development. 2001;5(1):23-33. DOI: 10.1111/1468-2419.00119
11. Cambridge Dictionary. Make Your Words Meaningful. n.d. Available from: https://dictionary.cambridge.org/dictionary/english/knowledge [Accessed: 09 Febuary, 2022]

12. Rockart J F. The Changing Role of the Information Systems Executive: A Critical Success Factors Perspective. 1982. Available from: https://dspace.mit.edu/bitstream/handle/1721.1/2010/SWP-1297-08770929-CISR-085.pdf?sequence=1 [Accessed: 13 Febuary, 2022]
13. Earl MJ. Information Management: The Organizational Dimension. New York, NY, USA: Oxford University Press; 1998. p. 514
14. Earl MJ. Management Strategies for Information Technology. London, UK: Prentice Hall; 1989. p. 232
15. Weil P, Ross JW. IT Savvy—What Top Executives Must Know to Go from Pain to Gain. Boston, MA, USA: Harvard Business Press; 2009. p. 182
16. Kadiri SE, Grabot B, Thoben KD, Hribernik K, Emmanouilidis C, Cieminski G. Current trends on ICT technologies for enterprise information systems. Computers in Industry. 2016;79:14-33. DOI: 10.1016/J.COMPIND.2015.06.008
17. Sanchez-Gonzalez PL, Diaz-Gutierrez D, Leo TJ, Nunez-Rivas L. Towards digitalization of maritime transport. Sensors. 2019;19(4):926. DOI: 10.3390/s19040926
18. Liu J, Zhang H, Zhen L. Blockchain technology in maritime supply chains: Applications, architecture and challenges. International Journal of Production Research. 2021:1-15. DOI: 10.1080/00207543.2021.1930239
19. World Economic Forum. Enabling Trade, Valuing Growth, Opportunities. 2013. Available from: https://www3.weforum.org/docs/WEF_SCT_EnablingTrade_Report_2013.pdf [Accessed: 09 Febuary, 2022]
20. Gausdal AH, Czachorowski KV, Solesvik MV. Applying blockchain technology: Evidence from Norwegian companies. Sustainability. 2018;10:1985. DOI: 10.3390/su10061985
21. Ytterstrom J, Lengerg L. What Role Will Blockchain Play Within the Maritime Shipping Industry in Five Years? Gothenberg, Sweden: University of Gothenburg School of Business, Economics and Law; 2019 Available from: https://pdfs.semanticscholar.org/b6d7/cf06c4853e-f1745ab4dd12f65f3cc7f5bcb2.pdf [Accessed: 09 Febuary, 2022]
22. Knirsch F, Unterweger A, Engel D. Privacy-preserving blockchain-based electric vehicle charging with dynamic tariff decisions. Computer Science – Research and Development. 2018;33:71-79. DOI: 10.1007/s00450-017-0348-5
23. Green E H, Carr E W, Winebrake J J, Corbett J J. Blockchain Technology and Maritime Shipping: An Exploration of Use Cases in the U.S. Maritime Transportation Sector. 2020. Available from: https://www.maritime.dot.gov/sites/marad.dot.gov/files/2020-07/Blockchain%20Use%20Cases%20Final%20Report%20%2820200622%29.pdf [Accessed: 09 Febuary, 2022]
24. Bhattacharya P, Saraswat D, Dave A, Acharya M, Tanwar S, Sharma G. Davidson I E, coalition of 6G and blockchain in AR/VR space: Challenges and future directions. IEEE Access. 2021;9:168455-168484. DOI: 10.1109/ACCESS.2021.3136860
25. Vardai Z. Top 5 Blockchain Trends for 2022 and Beyond. 2021. Available from: https://forkast.news/top-5-blockchain-trends-with-lasting-impact/ [Accessed: 11 Febuary, 2022]
26. Kapidani N, Bauk S, Davidson IEA. Developing countries concerns regarding blockchain adoption in maritime. Journal of Marine Science and Engineering. 2021;9(12):1326. DOI: 10.3390/jmse9121326
27. Kapidani N, Bauk S, Davidson IEA. Deploying blockchain in maritime: Developing environments concerns. High Technology Letters. 2021;27(9):942-951
28. Drljevic N, Aranda D, Stantchev A. Perspectives on risks and standards that affect the requirements engineering of blockchain technology. Computer Standards & Interfaces. 2020;69:103409. DOI: 10.1016/j.csi.2019.103409

29. Zhou Y, Soh YS, Loh HS, Yuen KF. The key challenges and critical success factors of blockchain implementation: Policy implications for Singapores maritime industry. Marine Policy. 2020;122:104265. DOI: 10.1016/j.marpol.2020.104265

30. Burgess M. Artificial Intelligence – How Machine Learning Will Shape the Next Decade. London, UK: Penguin Random House; 2021. p. 165

31. TradeLens. Shipping in the Age of Blockchain. 2019. Available from: https://www.youtube.com/watch?v=Xwqo_fwPEJo&t=1925s [Accessed: 11 Febuary, 2022]

32. IBM. What are Smart Contracts on Blockchain? n.d. Available from: https://www.ibm.com/za-en/topics/smart-contracts [Accessed: 01 January, 2022]

33. Jensen T, Henningsson S, Hedman J. Delivering business value with blockchain technology: The long journey of TradeLens. MIS Quarterly Executive. 2019;18(4):221-243. DOI: 10.17705/2msqe.00018

34. IBM and Maersk. TradeLens Documentation: Actionable Doc Flows. 2020. Available from: https://docs.tradelens.com/how_to/actionable_doc_flows/ [Accessed: 10 Febuary, 2022]

35. NewsBTC – Bitcoin & Cryptocurrency News. First view of Blockshipping's Global Shared Container Platform. 2018. Available from: https://www.newsbtc.com/sponsored/first-view-of-blockshippings-global-shared-container-platform/ [Accessed: 11 Febuary, 2022]

36. Blockshipping GSCP. Just in Time Containers. 2018. Available from: https://www.youtube.com/watch?v=BUas6vfWLeo&t=85s [Accessed: 11 Febuary, 2022]

37. Blockshipping GSCP. The GSCP Explained. n.d.. Available from: https://medium.com/@Blockshipping/the-gscp-explained-17804b0b09fc [Accessed: 01 Febuary, 2022]

38. Blockshipping GSCP. The Four Critical Components of the GSCP 2018. Available from: https://www.youtube.com/watch?v=YIz9-e70hmw&t=55s [Accessed: 11 Febuary, 2022]

39. Lennart M. Introducing Blockshipping: The Global Shared Container Platform. 2018. Available from https://medium.com/@lennart_7988/introducing-blockshipping-the-global-shared-container-platform-4da87abf0dba [Accessed: 11 Febuary, 2022]

40. Blockshipping GSCP. The GSCP Flow. 2018. Available from: https://www.youtube.com/watch?v=kMV0A-L-dMw&t=89s [Accessed: 11 Febuary, 2022]

41. Marine Insight. 7 Major Blockchain Technology Developments in Maritime Industry. 2021. Available from: https://www.marineinsight.com/know-more/7-major-blockchain-technology-developments-in-maritime-industry-in-2018/ [Accessed: 09 Febuary, 2022]

Chapter 2

Shipping Digitalization and Automation for the Smart Port

Enna Hirata[1], Daisuke Watanabe[2] and Maria Lambrou[3]

[1] *Kobe University, Japan*

[2] *Tokyo University of Marine Science and Technology, Japan*

[3] *University of the Aegean, Greece*

ABSTRACT

Shipping, like most industries, is undergoing a digital transformation process which influences existing business models and operational practices, in a multifaceted way. Today, the shipping business context has been changing to incorporate further social demands, environmental, innovation and sustainability priorities, into fundamental shipping strategies, while taking advantage of technological advancements. In the era of Industry 4.0, which constitutes a recent evolution of advanced communications and information technologies and further promotes sustainable, human-centric, and resilient business development strategies, shipping and port entities need to embrace a broader perspective and a deeper understanding of various elemental technologies, namely: Artificial Intelligence, Blockchain, Cloud Computing, Big Data, and Physical Internet, in addition to core maritime logistics matters. This chapter proposes a descriptive framework of shipping digitalization and port automation, while providing a review of related technologies and business approaches, also international initiatives, for automation in global ports. Hence the chapter offers insights for business practitioners to steer through the current challenging global environment, also for policy makers to gain a more informed understanding of maritime logistics developments, towards necessary coordination and oversight mechanisms implementation.

Keywords: *shipping; digitalization; automation; smart port; artificial intelligence; blockchain; physical internet*

1. INTRODUCTION

In recent years, major ports around the world have been implementing new technologies to realize "Smart Ports," in order to enhance international competitiveness, reduce environmental impact, and improve the workplace environment. In addition to the automation and labor-saving measures that have been actively pursued so far, smart ports are nowadays expected to benefit from the Internet of Things (IoT), Artificial Intelligence (AI), and Big Data[2], as well as more focused initiatives, in specific the Physical Internet (PI), based on the development of high-speed communication infrastructures, such as the fifth generation mobile communication system (5G). The robust use of the numerous digital technologies, such as the ones above-mentioned, offers substantial business innovation opportunities and requires numerous organizational adjustments [1].

In specific, the digitalization of ports is enabled by 5G networks that provide low latency, high capacity, and increased bandwidth, hence allowing the collection and intelligent processing of vast amounts of data shared over an IoT information network infrastructure. 5G is foremost supporting the PI objectives, namely the widespread adoption of interconnected, sustainable logistics systems and applications.

A smart port is realized by means of wireless devices, smart sensors, actuators, data centers, and other IoT-based systems being connected and exchanging information. The seamless communication of ships, cargoes, and waterway and shore-based facilities, based on 5G and IoT technologies, constitutes the infrastructure of the smart port and shapes the innovation potential of the emergent maritime logistics services and applications. The efficiency, security, and safety of port operations and the achievement of the highly prioritized sustainable development goals are substantially improved by automating port operations, such as vessel management, container terminal operations, and yard automation.

Competitiveness and henceforth the competitive positioning of ports being nodes of multiple, global supply chain networks are of paramount importance and are directly determined by the adopted digital strategy and infrastructures implemented in each port.

This chapter outlines the main technologies' capabilities, as well as the current status of policies and initiatives for the development and deployment of various technologies, primarily for automated terminals and automated ships in ports, as innovations for ships and various automated machinery, mainly for cargo handling, toward the realization of smart ports. Moreover, the chapter introduces a particular technological framework for creating a sustainable next-generation maritime logistics system, by applying PI to digitize maritime logistics networks and BC technology for secure information exchange to improve the overall efficiency of maritime logistics, at ports.

2. INDUSTRY 4.0: TECHNOLOGIES FOR NEXT-GENERATION MARITIME LOGISTICS AND SHIPPING DIGITALIZATION

How does modern maritime logistics look like in the era of Industry 4.0? The concept of Industry 4.0, first formulated by the German government, mainly refers to the automation and digitization of manufacturing processes. Industry 4.0 encompasses cyber-physical systems[3], AI[4], IoT[5], cloud computing[6], cognitive computing[7], smart factories[8], and digital twin[9]. Industry 4.0 has been implemented in many countries besides Germany, including the US, France, the UK, China, Japan, Korea, and Thailand. Particularly in the supply chain area, digitalization, integration of AI and IoT, sharing economy[10], and BC are playing an important role in addition to these core technologies. Logistics has not been associated with high technology for a long time since it was recognized as an industry in the mid of nineteenth century. The situation is changing with the increase of Logistics 4.0 efforts, where innovative technologies such as AI, IoT, and BC are increasingly being implemented in the logistics industry, in parallel with similar efforts to many other industries, presenting even more efficient sustainability and human-centric approaches. However, despite the numerous regional initiatives, there are no adequate frameworks existing for companies in the logistics industry to embrace those technologies to the largest extent. There is a need for guidelines to implement new technologies in the logistics industry for the common good of the entire industry and eventually society as a whole.

Against this background, a closely related development, namely the Physical Internet (PI), has drawn attention from various parties as one of the most effective measures to improve logistics efficiency and reduce greenhouse gas emissions. The features of PI include interoperability, modularity, and standard interfaces and protocols. In order to take advantage of these features, technology to share data while maintaining confidentiality, such as Blockchain (BC) technology, is essential. This technology, also known as a distributed ledger, is difficult to tamper with, requires no administrator, and allows execution of smart contracts. The use of BC technology in PI is expected to dramatically accelerate the construction of a sustainable logistics network. In the following sections, we will present, in more detail, the potential of applying BC technology in the PI network context.

2.1 AI

Artificial Intelligence (AI) is a field of computer science that uses the concept of "computation" and the tool of "computers" to study and implement "intelligence."

AI can be classified into general-purpose AI and specialized AI. General-purpose AI is also referred to as strong AI. It is an AI that is not limited to a specific task but is capable of general-purpose processing; that is, it has the same intelligence as humans. Some examples of general-purpose AI are Doraemon and Astro Boy. None of them exist in the real world yet. Specialized AI is also called weak AI. It is an AI that specializes in performing a specific task. Image recognition, chess, Go, automated driving, human conversation, etc., are all examples of specialized AI.

In a broader sense, AI includes rule-based AI and machine-learning AI. Rule-based AI refers to making decisions according to rules described by humans. It also automates tasks that require hardware and human judgment. It can be described as an office robot. In contrast, machine-learning AI generally does not require a human to write the rules. It has algorithms for self-learning in machine-learning models, and it behaves intelligently based on those algorithms, building the models automatically. Compared with machine-learning AI, rule-based AI has advantages such as faster to automate tasks, a human can train AI, and lower cost. On the other hand, it also has disadvantages such as unable to learn independently, unable to train AI unless it is explicit knowledge, and unable to make decisions on matters it has not been trained on.

Machine learning can be divided into supervised learning, unsupervised learning, and reinforcement learning in general (Figure 1).

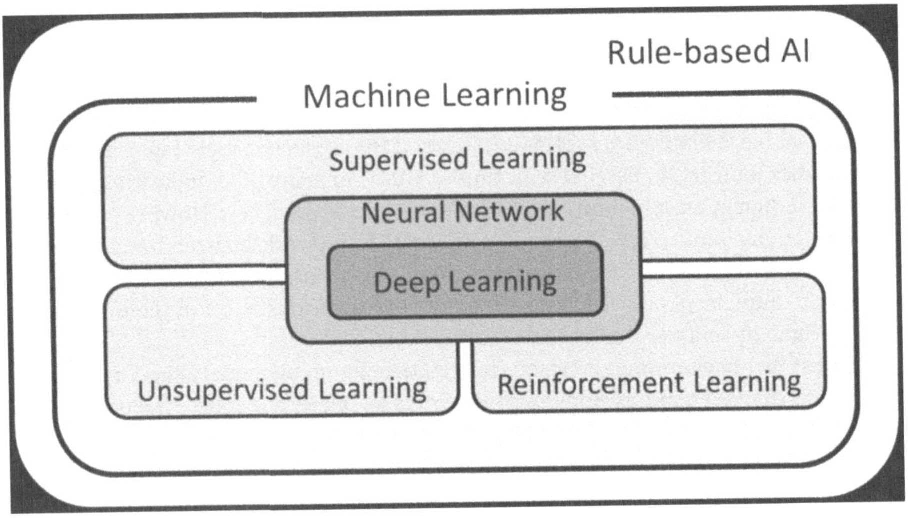

Figure 1. *AI and machine learning relationship diagram.*

2.1.1 Supervised learning

Supervised learning is the process of preparing training data with a set of inputs and correct outputs in advance, and having the computer learn to produce correct outputs when given a certain input. Supervised learning methods are mainly used for prediction and classification (labeling). Examples include image classification. Supervised learning can be applied to forecasting sales and predicting the tendency of customers' churns, etc.

2.1.2 Unsupervised learning

Unsupervised learning is often referred to as self-supervised learning these days. It is used to grasp the inherent structure of data from the input data supplied Unsupervised learning mainly uses clustering techniques to classify data. Examples of applications include product recommendation, customer segmentation, target marketing, etc., based on customer purchase history.

2.1.3 Reinforcement learning

Reinforcement learning is a model that learns to maximize future value instead of giving the correct answer. In other words, it learns to act in a way that maximizes value through trial and error. The problem setting is similar to that of supervised learning, but it is not enough to learn the output of the given correct answer as it is, but it is necessary to learn the behavior that maximizes value in a broader sense.

Tetris game is a good example to understand the scheme of reinforcement learning. When playing a game of Tetris, the problem of getting the highest possible score can be considered in the framework of reinforcement learning. The best way to score at that point is to play in such a way that even a single row can be eliminated immediately, but in the longer term, the score will be higher if you accumulate as much as possible and then eliminate many rows at once.

AlphaGo, which defeated a human player, also incorporates reinforcement learning in some parts of its games. As in the case of Go, it can learn even when humans do not necessarily know the correct answer, so it is expected to acquire the ability to surpass humans.

2.1.4 Artificial neural network

A neural network is a mathematical model inspired by the function of nerve cells (neurons) and their connections, or neural networks in the human brain, called artificial neurons. When the neural network model is properly constructed according to the problem to be solved, it can make a variety of decisions (i.e., outputs), such as the following:

First, image recognition and binary classification. For example, it answers questions such as: Is the object in front of me a ship or a train?

Second, natural language processing, multi-level classification. For example: which is "Emma Maersk" among various images of vessels online?

2.1.5 Deep learning

Deep Learning is a method of machine learning in which neural networks are combined in multiple layers to enhance their representation and learning capabilities. Currently, it is the most commonly used algorithm for AI.

Deep learning, on the other hand, is often used when complex unstructured data is available and is applied in fields such as speech recognition, image recognition, and natural language processing.

There are many cases where conventional machine-learning methods do not work well for classification and regression without complex function approximation, and deep-learning methods are increasingly being used for such problems. In some cases, deep-learning methods have dramatically improved recognition accuracy compared with conventional methods, and deep learning is currently attracting a great deal of attention in the world. Recently, it has been used in a wide range of fields such as recommendation and automated driving.

2.2 Blockchain

BC is an open distributed ledger technology (DLT) based on a peer-to-peer (P2P) approach that allows transactions to be recorded on thousands of servers simultaneously. On the efficient, verifiable, and immutable BC platform, anyone can see the transactions of others in near real time, making it difficult for one user to manipulate the records and control the network [2]. Applying these features, BC facilitates the digitization of traditional economic, legal, and political systems.

In BC, cryptography is used to store records (hash values[11]) of transactions that occur in the network in blocks of records called blocks. In each block, it contains three values, the first is the hash value of the previously generated block, the second is the record of the transaction in the current block, and the third is a new hash value generated by a disposable random value called a nonce. The three values are passed to the next block and the accumulated blocks form a chain of blocks in time series. The name BC comes from this data structure.

BC can be divided into public BC, which allows anonymous participation, and permitted BC, which requires permission to participate. Public BC is mainly applied to cryptocurrencies. Since the permitted type of BC is faster than the public type in handling transactions, it has been applied to various business fields such as supply chain and intellectual property management [3].

There have been five main types of algorithms for consensus building in BC (Table 1). In public BC, PoW and PoS are mainstream; in the PoW consensus, rewards are evaluated by the amount of work done. In PoW consensus, the reward is evaluated by the amount of work done; in other words, the network participant who performs the appropriate computation the fastest receives the reward. In the PoS consensus, rewards are based on both the amount of work and the amount of cryptocurrencies held. In PoS consensus, rewards are based on both the amount of work done and the amount of cryptocurrencies held, easing the fierce competition in PoWs and saving electricity consumption, it suffers the problem that the rich get richer. On the other hand, in permitted BC, PBFT is the mainstream method. PBFT is faster than other consensus methods but has the disadvantage that blocks will not be created if two-thirds or more of the consensus is not obtained.

Table 1. Typical consensus algorithms.

Consensus	PoW (Proof of Work)	PoS (Proof of Stake)	PoI (Proof of Importance)	PoC (Proof of Consensus)	PBFT (Practical Byzantine Fault Tolerance)
Evaluation method	(one's) workload	Workload and coin holdings	Overall evaluation of coin holdings, transaction volume, transaction frequency, etc.	Designated agency	Consensus of two-thirds or more
Benefits	Resistant to transaction tampering	No wasted power consumption	Low power consumption	Fast remittance speed	High speed compared to other methods
Weakness	High-power consumption	How the rich get richer.	Participation is impossible without a certain amount of coins.	centralized	A block will not be created unless a two-thirds majority is agreed.
Representative example	Bitcoin (BTC)	Ethereum (ETH)	Nem (NEM)	Ripple (XRP)	Permitted type consortium type

In the maritime industry, many players from different industries, usually in several countries, are involved in currently operating blockchain platforms. This complexity leads to a lack of transparency in the entire supply chain. In addition, the industry has the disadvantage of high transaction costs for information exchange, the possibility of fraud and theft, and vulnerability to the risk of cyber-attacks. BC offers the possibility to solve these problems [3]. To maximize the capacity and productivity of the digital information space, traditional authentication methods and data structures need to be reformed and modern technologies such as BC need to be actively applied.

2.3 Physical internet (PI)

PI was initially proposed by Montreuil in 2010 [4]. He defined PI as an open, global logistics network that efficiently and sustainably interconnects all elements of the logistics process. The PI includes the complete supply chain including storage, movement, supply, and delivery of goods, and the PI network is composed of various logistic providers. The goal of PI is to create a global logistics system based on the interconnection of existing logistic networks. To achieve this, a standardized set of protocols, modular containers, and smart interfaces are combined modular containers, called PI containers (Figure 2), which come in various sizes and can be combined and loaded to reduce waste. In addition to modular containers, PI-stores, PI-movers, PI-conveyors, and PI-gateways have also been proposed.

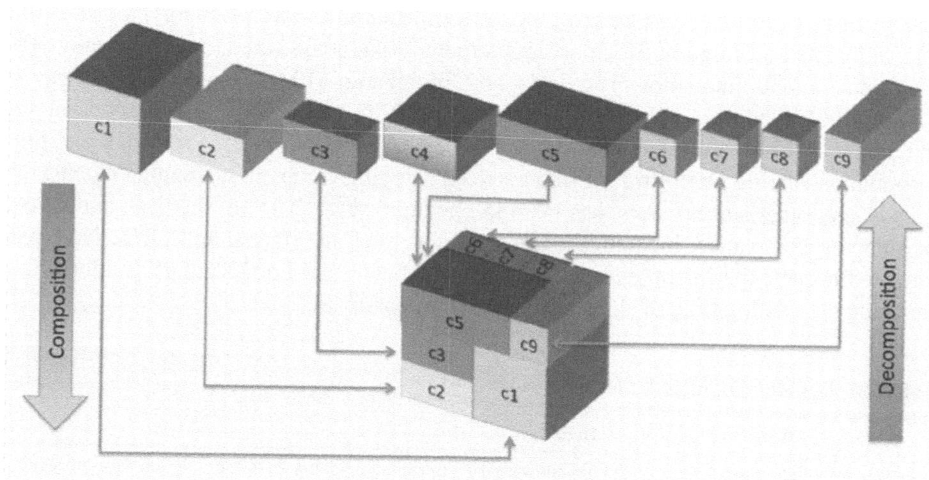

Figure 2. *An example of PI container combination. Source: Montreuil et al., 2010 [4].*

The characteristics of PI include interoperability, modularity, and standard interfaces and protocols. In order to take advantage of these features, technologies to share data while protecting confidentiality, such as BC technology, are essential.

Although it is not difficult to understand the usefulness of PI and BC, specific application measures have not been fully studied. In this chapter, we make a proposal for planning measures to build a BC network in PI, and discuss issues and measures for practical application. Specifically, the next sections also aim to clarify the following two points. First, to clarify the scope of application of BC in PI. Second, we propose a framework for implementing PI and BC technologies.

3. DIGITALIZATION AND AUTOMATION FOR SMART PORTS

3.1 Smart port initiatives in major countries

Various information systems have been introduced in maritime and port-related operations, such as Terminal Operating System (TOS) for internal terminal operations, Port Community System (PCS) for port logistics, Automatic Identification System (AIS), and the Transport Management System (TMS) for land-side operations such as trucking [5]. A smart port is an initiative that aims to improve the efficiency and safety of the port as a whole and to reduce the environmental impact of the port by integrating these systems through innovations in automation and new digital technologies such as IoT, AI, and 5G. Currently, the Port of Rotterdam (Smart Port Initiative) and the Port of Hamburg (smart port) in Europe, Japan (PORT2030), Korea (Smart Maritime Logistics) and Singapore (Sense-making Analytics For maritime Event Recognition: SAFER) and other major ports around the world are working on various initiatives.

The Port of Rotterdam in the Netherlands has a vision of being the "smartest port in the world" and is working on digitalization, energy transformation, and innovation to become carbon neutral. They have identified four levels of "digital maturity" for a port cargo community as shown in Table 2 [8]. The Smart Port Initiative is a roadmap that includes projects in energy and industry like recycling, electrification, renewable energy, logistics like BD, automated driving and BC, port infrastructure like quays, dredging, maritime traffic management, and innovation. The Port Call Optimization (PMO) project is underway as part of Port Collaborative Decision Making (PortCDM), an initiative aimed at optimizing the timing of vessel arrival and departure [9]. PortXChange (formerly Pronto), a real-time information sharing platform for PMO between shipping companies, shipping agents, terminals, and other stakeholders, has been in operation since 2018. In the past, about 75% of shipping companies, including major operators such as Maersk and ONE, have participated in the experiment, and the results have shown that PortXChange is effective in reducing the waiting time of ships, especially for departures.

Table 2. Definition of maturity levels of smart ports.

Level	Name	Overview
1	Digitization of individual activities in the port	The individual organizations operating in the port digitize their processes so that they work more efficiently.
2	Integrated systems in a port community	The digital exchange of information within the port community leads to reliable, efficient, and paperless dataflows, resulting in more efficient port operations. The focus at this level is related to securely sharing data. Cybersecurity and cyber resilience are key.
3	Logistics chain integrated with hinterland	The hinterland four players (importers, exporters, logistics hubs, and domestic transporters) are involved in digital communication with the port community. This integration promotes sharing of real-time information on freight and vessel movements, facilitating better planning.
4	Connected ports in the global logistics chain	The port and its hinterland connections are extended to other ports around the world, forming a global network of interconnected ports. This network will allow further reductions of inefficiencies in the global logistics chains by optimizing the use of port capacities and achieving shorter, more reliable transit times.

Source: Buck et al. 2019 [6] and ADB2020 [7].

In Japan, one of the main measures in the mid-to long-term port policy "PORT2030" announced in July 2018 is to make ports smarter and more resilient by using information and communication technologies. In addition to the complete computerization of ports, which will be called "Cyber Port" through the construction of a port-related data linkage infrastructure, the policy aims to create container terminals with the world's highest level of productivity and a good working environment (AI terminals) by combining AI, IoT, and automation technologies. In the container terminal field, the introduction of remote-controlled cargo handling machinery and automated gate handling is being promoted, terminal operations are being streamlined and optimized using AI and other technologies, automated vessels, and remote-controlled tugboats are being operated, automated guided vehicles are being introduced, and automated trucks are being driven in convoys. In addition, the next generation high standard unit load terminal will be developed. Furthermore, in the next generation of high standard unit load terminals, the use of automatic driving technology for cross-carriage transport and the linkage with automatic navigation and navigation support technology for ships are mentioned.

3.2 Research and development for the realization of the automated ship

Various efforts have been made by shipping companies to develop navigation support technologies to improve the safety and efficiency of ship operations using IoT and big data. In this context, Maritime Autonomous Surface Ships (MASS) have been attracting attention rapidly in recent years. The term MASS generally describes a ship that is highly automated or remotely controlled to perform some or all of the following shipboard tasks: external situational awareness (watchkeeping), monitoring of equipment status, ship operation, engine control, cargo management, and loading/unloading, take-off and landing, and other shipboard tasks by using the latest technologies such as IoT, ICT, and data analysis technologies, various sensors, and land-based monitoring and control centers connected by broadband communications.

A number of projects are underway, mainly in Europe, with the aim of realizing MASS. In December 2018, Rolls-Royce and FinnFerry successfully demonstrated the world's first fully automated ferry. As an example of an international project, One Sea, a consortium launched in Finland in 2016, is developing a roadmap for practical application and discussing the necessary safety standards and international standardization in order to create an environment for MASS operation by 2025.

There is an ongoing international discussion on the legislative framework for safety standards for MASS. With regard to classification societies' certification systems, in February 2017, the British classification society Lloyd's Register published the LR Unmanned Marine Systems Code, which sets out the performance requirements for automated ships. The International Maritime Organization (IMO) has been considering the regulatory aspects of automated ships since May 2018 and has presented a provisional proposal, as shown in Table 3, and is discussing the necessary amendments to IMO rules and new developments. For demonstration tests, the provisional guidelines for safe and efficient demonstration tests of automated ships, jointly proposed by Japan and Norway, have been approved in June 2019. In May 2021, IMO has completed a regulatory scoping exercise on MASS that was designed to assess existing IMO instruments to see how they might apply to ships with varying degrees of automation.

Table 3. Definition of automation levels for MASS.

Level	Name	Overview
1	Ship with automated processes and decision support	Seafarers are on board to operate and control shipboard systems and functions. Some operations may be automated and at times be unsupervised but with seafarers on board ready to take control.
2	Remotely controlled ship with seafarers on board	The ship is controlled and operated from another location. Seafarers are available onboard to take control and to operate the shipboard systems and functions.
3	Remotely controlled ship without seafarers on board	The ship is controlled and operated from another location. There are no seafarers on board.
4	Fully autonomous ship	The operating system of the ship is able to make decisions and determine actions by itself.

Source: IMO [10].

In Japan, industry, government, and academia have been collaborating since FY2017 to develop technologies, develop infrastructure and systems, and study business models for the realization of MASS through demonstration projects of automatic ship operation, remote ship operation, and automatic docking and unloading functions in order to improve the environment, including the formulation of safety requirements. The ClassNK has been working on the development of the technology through demonstration projects. In January 2020, the ClassNK established requirements and procedures for the functional verification of automation and remote-control systems used on ships and remote-control facilities, from the perspective of ensuring safety at each stage of development and design, ship installation, and operation [11]. The Ship Data Centre was established in December 2015 as a platform for the use of ship big data, with the participation of shipping companies, shipbuilders, marine industry operators, and meteorological information companies. The Ship Data Center was established in December 2015 as a foundation for the use of ship big data, with the participation of shipping companies, shipbuilders, marine industry companies, meteorological information companies, etc. Rules for fair and equitable data use have been established to promote the distribution and use of ship big data, and the effective use of accumulated big data is being promoted.

Based on the results of the economic evaluation of the MASS operation system, efforts are being made to commercialize a manned automated ship operation system (corresponding to automation level 1 in Table 3), which is more feasible in the short term. In NYK line, research has been conducted on an action planning system for the decision-making required to execute ship operations, and the world's first demonstration of a manned automated ship based on the provisional guidelines set by the IMO was conducted in September 2019 [12]. Future projections for unmanned automated ships (corresponding to automation level 3 or higher in Table 3) for domestic ships suggest that if 50% of ships are replaced by unmanned automated ships in 2040, the annual economic impact will be approximately 1 trillion yen [13]. In June 2020, the Foundation selected five projects for MEGURI 2040, which aims to realize unmanned automated ships by 2025, by conducting the world's first demonstrations in waters with high vessel traffic, long-distance navigation, and using large vessels.

3.3 Deployment of automated terminals at ports

The unmanned and automated handling of cargo at container terminals is expected to increase productivity per worker, improve the working environment and safety, and reduce the effects of weather conditions such as fog and wind. On the other hand, the introduction of the system is not without its challenges, such as high initial investment costs, maintenance costs (e.g., power consumption), and coordination with trade unions. At the ECT Delta Terminal in the Port of Rotterdam, the world's first automated terminal, an Automated Guided Vehicle (AGV) and an Automated Stacking Crane (ASC) were introduced in 1993. However, the introduction of these systems did not proceed due to technical and economic problems and difficulties in coordinating with labor unions, and there were only a few cases until the mid-2000s. However, since the mid-2000s, and especially since the 2010s, the number of automated terminals has been increasing rapidly, with nearly 60 terminals worldwide having installed the system so far. The number of automated terminals has increased rapidly since the mid-2000s, especially in the last decade. Table 4 summarizes the status of the introduction of automated terminals, focusing on the level of automation. The symbols in the table indicate: ◎: mainstream status with many cases of introduction, ○: diffusion stage with several cases of introduction, and △: early stage with limited cases of introduction. In the case of marshaling yards and land-side container handling, remote control and automation are the basic systems. On the other hand, the manned operation is the mainstream for quay cranes, and full automation has been introduced only recently in a limited number of cases. In the case of horizontal transport within the premises, although there is a high degree of automation, various types of cargo handling machines have been introduced, and the level of automation differs greatly between ports.

Table 4. The introduction of automated terminals with the level of automation.

Type	Activities	Typical cargo handling machine	Manned	Unmanned	
				Remote control	Automation
Waterside transport	Moving containers from ship to shore	STS (Ship To Shore Container Crane)/ QC (Quayside Conatiner Crane)	◎	○	△
Horizontal transport	Moving containers between the quayside and yard storage blocks	AGV (Automated Guided Vehicle), SC (Straddle Carrier), ShC (Shuttle Carrier), Chassis	◎	—	○
Marshaling yard	Moving containers at yard storage blocks	ASC (Automated Stacking Crane), ShC	—	○	◎
Landside transport	Moving containers from the terminal truck gate or intermodal railhead to the marshaling yard	ASC, ShC	—	◎	○

Source: Takahashi 2018 [7] and PEMA 2016 [14, 15].

In Japan, AGVs and remote-controlled ASCs were introduced at the south side container terminal of Tobishima Pier in the port of Nagoya in December 2005, which is a relatively early stage in international perspective, but it is the only automated terminal in Japan at present. Currently, efforts are being made to promote the introduction of remote-controlled ASCs, mainly at strategic international container ports, and to improve the efficiency and optimization of terminal operations by using AI-based on container cargo information.

4. A CASE STUDY OF PI AND BLOCKCHAIN TECHNOLOGY FOR SMART PORTS

In the following, a specific case study of a blockchain technology application, in a Logistics 4.0 Physical Internet environment, is explicated, as a representative system implementation for innovative, digital maritime logistics environments, with automated ships and terminals constituting flagship applications of Industry 4.0.

Three recent representative studies on the application of PI and BC in the supply chain are available [16, 17, 18]. Meyer et al. [16] proposed a conceptual framework for the exchange of value and physical assets in logistics networks that proposed a BC-based conceptual framework and provides a solution to the fundamental barrier of PI. As the main contribution, they identified barriers to transforming current logistics systems into PI networks through case studies. The key barriers included the creation of a network with equal participation, robustness of the framework, assurance of integrity and resilience, rewards in the operational process, and reliable data exchange. By further describing the key features of the technology, they discussed how the BC would address the barriers to PI adoption. They proposed Ethereum BC, implemented smart contracts based on the ERC721 standards[12], and evaluated the transport process in PIs. The authors conclude that BC technology can solve the barriers in PI because it enables a reliable and secure exchange of value in an untrustworthy environment. The authors propose a PoS based BC environment in order to save computational resources. In the case of small-scale PIs, the proposed solution already works, but the scalability[13] problem as a whole needs to be solved before PIs can be widely adopted.

Hassan et al. [17] presented a permitted BC architecture suitable for the integration of BC technology with PI. They discussed how to take advantage of the interoperability[14] between two permitted BCs. They demonstrated the applicability and practicality of the PI architecture to be built on top of a permitted BC and presented a case study of its application. The authors pointed out the scalability of both BC and PI networks as an issue to be solved.

Tan et al. [18] presented a framework of green logistics based on BC to realize sustainable logistics by integrating IoT and big data. The authors propose a framework with seven layers: physical layer, perception layer, network layer, blockchain layer, management layer, application layer, and user layer. The authors pointed out three issues: data storage and transmission, implementation cost, and risk. Then, for future research, the authors suggested to focus on the following: (1) developing a way to effectively connect the physical and perceptual layers to collect logistics data, and (2) designing an incentive mechanism to encourage logistics companies to participate in the BC platform.

In contrast to the various advantages of BC, this technology requires a transformation of digital systems. First, existing processes need to be digitized. Currently, there are many tasks in logistic operations that are done by hand on paper or on computers that are not connected to a network. In order to effectively accumulate and utilize data in these tasks, it is necessary to digitize the tasks themselves or use AI (e.g., Optical Character Reader) services to digitize them.

Next, in order to work with platforms such as BC, existing systems need to have a mechanism to use APIs[15]. In the logistics field, many existing systems are still based on EDI[16], which supports only batch sending and receiving, while APIs support real-time sending and receiving. In the logistics sector, there are still many existing EDI-based systems; EDI supports only batch sending and receiving, while APIs support real-time sending and receiving, and the development cost is higher than APIs [3]. Making EDI-based core systems API compatible is an important task. The issue of standardization is important in the diffusion of APIs. At present, there is a bunch of standards at the level of international organizations, governments, and industries. Some of the standards conflict with each other. These standards need to be unified. Organizations like Digital Container Shipping Association (DCSA) help to expedite the process. DCSA aims to develop digital standards for the containership industry and has compiled and published electronic standards such as vessel schedules, port operations, and electronic B/Ls. The PoV should follow the standards and protocols published by DCSA. In addition, the API is an architectural style that can be easily manipulated and can flexibly respond to the unique standards of countries and industries, for example, 10-ft container, low floor chassis.

Munim et al. [19] identified the main challenges in the practical application of BC in the maritime sector as lack of standardization of data elements, lack of interoperability and scalability between systems, delay in legislation, lack of understanding of the technology, and lack of training facilities and materials. PiChain is facing the same challenges as it uses BC technology. In addition, it is necessary to solve the issues of attracting participating logistics companies and infrastructure development in the implementation of PI. The search for solutions to these issues remains a future task.

These previous studies pointed out the issues of scalability after conducting small-scale demonstrations. In this study, we propose a framework for building innovative Logistics 4.0 systems and applications to solve these issues.

4.1 Scope of application

We propose the following scope of application of BC technology in PI. PI contains three flows, namely physical (logistics) flow, information flow, and financial flow. BC technology is indispensable for two of the three flows: information flow and financial flow (Figure 3).

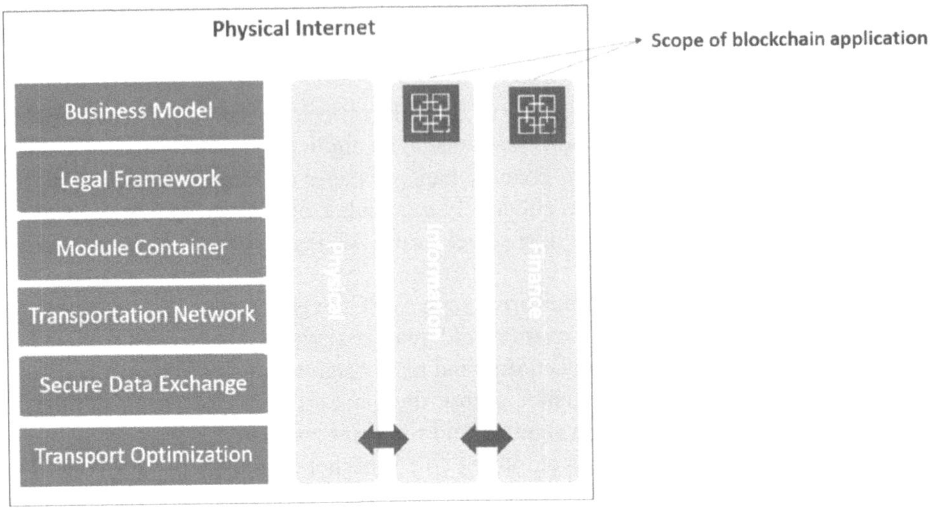

Figure 3. *The scope of BC technology application in PI.*

4.2 Implementation framework

For implementation, this chapter proposes a new framework called PiChain, which consists of five components (Figure 4).

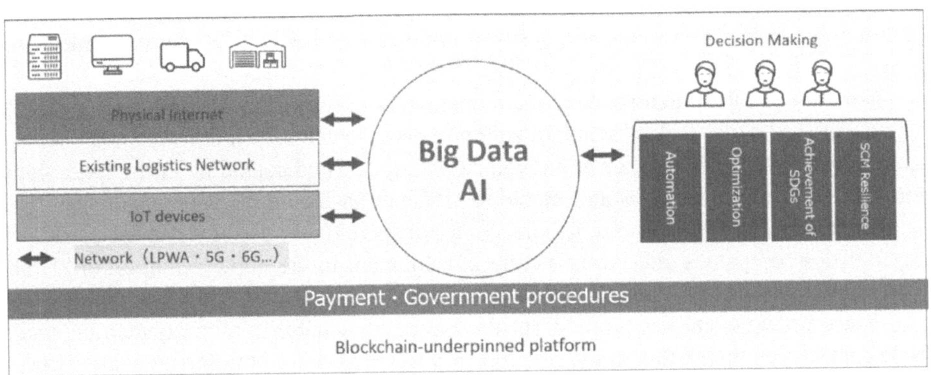

Figure 4. *Overview of the proposed framework "PiChain."*

The first component is the PI that interconnects the maritime logistics networks. The second component includes networks such as LPWA that supports various types of IoT devices, and 5G and 6G that support cloud computing and edge computing. The third component consists of sensors, IoT devices, Internet of Robotic Things (IoRT), and drones. The fourth component includes big data for optimization with AI and visualization with business intelligence (BI). The top layer is the component that refers to the space optimization of business and operational processes implemented goals of logistics and sustainability goals (SDGs). All these components function on a decentralized platform that uses BCs to prevent data modification, which also support traceability and various types of payments.

To give a better idea of how PiChain works, let us look at a few scenarios. In the future cargo transportation, upon booking information received, AI will automatically calculate the optimal transportation route and method constraints such as cargo destination, size, and weight using PI containers that match the size of the cargo. As soon as this information is finalized or updated, it is automatically sent to the shipper. In the shipper's internal system, AI uses the latest transportation information to optimize ordering, warehouse storage, and production planning. As soon as the consignee receives the shipment, the payment managed by BC's smart contract will be automatically executed.

4.3 Choosing a blockchain

To overcome the lack of scalability of BCs, as pointed out in previous studies, this study recommends the adoption of the Avalanche protocol (Table 5) proposed by Rocket 2019 [20].

Table 5. Comparison of BC protocols.

Protocol	**Classic (1980)**	**Nakamoto (2008)**	**Avalanche (2019)**
Typical applications	Byzantine General Problems	Bitcoin	AVAX
Robustness	✕	○	○
Low latency and quick finality[1]	○	✕	○
High processing capacity	○	✕	○
Lightweight	○	✕	○
Low power consumption	○	✕	○
51% attack resistance[2]	✕	✕	○
High scalability	✕	✕	○

[1]*Finality means that the amount of money is certain to be obtained as expected. The Bank of Japan lists the following specific conditions for a finalized settlement: (1) The money received will not later be turned into scrap paper or disappear, and (2) the settlement made will never be reversed later.*
[2]*A 51% attack is the control of 51% (more than 50%) of the hash rate of the entire network by a malicious group or individual to perform fraudulent transactions.*
Source: Prepared by the author based on Sirer 2020 [13].

The Avalanche protocol is one of the most promising platforms for BCs because of its scalability, ease of use, flexibility, and proper governance. Avalanche was initially built to serve the financial markets (cryptocurrency AVAX), but it is interoperable with both public and permitted BC. Therefore, it can be adopted by other industries in addition to finance. Avalanche is also capable of combining public and private BCs. Compared with the existing Classical and Nakamoto protocols, the BFT-based Avalanche protocol is very efficient and robust, and can also achieve high productivity and fast finality. While traditional protocols require any given node to communicate with all other nodes, the Avalanche protocol communicates with a small subset (validators), which dramatically reduces the latency to 1/3600 compared with Bitcoin (Table 6). As a result, Avalanche is able to achieve a performance of 7100 transactions per second, compared with 5 transactions per second for Bitcoin, the Nakamoto protocol equivalent.

Table 6. Comparison of performance.

Performance comparison	Bitcoin	AVAX
Latency	1	1/3600
Speed (transactions per second)	5	7100
Finality	Accomplished in 1 hour.	Accomplished in 1 second.
Number of simultaneous participants	About 20 people	millions of people

Source: Prepared by the authors based on Sirer 2020 [13].

We recommend a small-scale Proof of Value (PoV) to be executed. In the PoV, a small-scale PI network will be constructed by implementing the Avalanche protocol, which is small scale and scalable, in order to connect with the existing logistics network and using BC technology, the collected big data will be analyzed for optimization, etc., using AI while ensuring its confidentiality. The results will be used for management decisions and environmental protection measures. This positive cycle will dramatically improve the efficiency of logistics and promote sustainable development. It will also be possible to measure and optimize the contribution of innovative technologies to the sustainable development of logistics and the SDGs. For example, it will be possible to quantify how much waste in loading and waiting for pickup is eliminated by the use of PI-containers, and how much congestion is eliminated by optimization at ports. Or how much truck driver time was saved by reducing waiting time, or how much CO_2 emissions were reduced, such measurements could be automatically recorded and measured on a reliable BC. The combination of Avalanche and PI in the logistic industry will drive logistics digitalization to a global scale.

4.4 Summary and future prospects

In this chapter, we have overviewed the international efforts for innovation in automation and digitalization for the realization of the Smart Port, including the development of technologies and demonstration experiments for their diffusion. It is necessary not only to automate cargo handling at the terminal, but also to share information in real time regarding pre-and post-processes, such as the arrival and departure of ships and the waiting status of vehicles, in cooperation with various IoT devices, and to prepare in advance based on predictive information using AI. It is expected to establish a de facto standard on a global scale by promoting the packaging of port handling machinery combining hardware and software. On the other hand, the international standardization for automated ships is being discussed internationally mainly by initiated by IMO as de jure standard. In addition, it will be necessary to share real-time information on the movements of ships using AIS and IoT and to make comprehensive efforts in cooperation with port infrastructures, such as coordinated operation with work vessels such as tugboats and refueling vessels, and automatic mooring at the wharf. In August 2020, MASSPorts, a framework for international collaboration on the operation of Maritime Autonomous Surface Ships (MASS) in ports, will be launched to establish guidelines for demonstration operations and to study interoperability in multiple ports around the world.

From the research and development stage to the full-scale deployment of new technologies, it is necessary to consider not only the individual optimization of ports and ships but also the overall optimization involving the ocean, ports, and the inland hinterland. As the need for mutual cooperation between ports and ships increases, new infrastructures, systems, laws and regulations, and international standardization need to be discussed. In addition, it is necessary to address not only the safety and cost aspects of installation and operation, but also cyber security, environmental aspects including renewable energy and electrification, and social acceptability. In order to enhance social acceptability, it is important to build a relationship of trust so that not only the port and maritime industries but also the general public can feel secure.

Furthermore, in this chapter, we have proposed PiChain, a framework for creating a sustainable, next-generation maritime logistics system, by applying PI to digitize maritime logistics networks and BC technology for secure information exchange to improve the overall efficiency of maritime logistics, at ports. The following three effects can be expected from the adoption of PiChain.

4.4.1 Strengthening supply chain resilience

Recently, there have been many disruptions in the supply chain due to unexpected circumstances such as bad weather, cyber-attacks, and port congestion. In particular, the covid-19 pandemic that began in early 2020 led to travel restrictions and border closures in some countries, which in turn caused disruptions in sea, land, and air logistics. The disruption of logistics also resulted in stagnant production in many countries. The framework proposed in this chapter, PiChain, is useful for supply chain management during emergencies and early recovery from disasters, as the BC-based platform provides real-time information and visualization of the entire delivery process, which in return enables quick response to unexpected delays. In addition, since PI containers can be freely combined or split, switching parts of cargo to alternative modes of transportation can be done flexibly when needed. Furthermore, the application of automated delivery methods, such as drones, will make it possible to deliver even in the middle of the night or in hazardous environments. In conclusion, PiChain can strengthen the resilience of the supply chain while managing a wide range of unknown risks.

In addition, cyber-attacks in the supply chain have been dramatically increasing in recent years. With risks increasing day by day, a high level of cybersecurity is vital for supply chain resilience. The PiChain framework uses BC technology to make the entire network highly resistant to cyber-attacks. Even in the event of being attacked, the stolen data are a hash value generated by BC, so it is difficult to recover the original data from the hash value, which significantly reduces the risk of information leakage. In addition, since the data are stored in a distributed manner in the cloud, it can be recovered quickly after an attack.

4.4.2 Port sustainability

Regarding the establishment of a sustainable supply chain within the framework of PiChain, the following three points can be summarized.

First, realizing the SDG Goals. Visualizing, measuring, and optimizing GHG emissions will help to minimize their negative impact on the environment. This will contribute to the realization of SDG Goal 13: Take urgent action to combat climate change and its impacts.

Second, improvement of working conditions. In the framework of PiChain, it is expected to improve the working conditions of workers. For example, AI-guided optimal delivery routes will shorten waiting time hence reducing stress for drivers. In addition, logistics workers will be able to enjoy the benefits of advanced visibility and perform their daily work in a more independent way. This will contribute to SDG Goal 8: Promote sustained, inclusive, and sustainable economic growth, full and productive employment, and decent work for all.

Third, improvement of service quality. Visible logistics management with PiChain will make just-in-time delivery easier. The deployment of drones will also make nighttime delivery possible. This will help improve the quality of logistic operations by enabling a quick response to any unforeseen circumstances that may arise.

4.4.3 Creating new business models

With the spread of innovative technologies, new business models will emerge. For example, the inspection at the yard when returning empty containers used to be done by human eyes. Now, using AI image recognition technology, AI can quickly analyze photos of containers automatically taken by surveillance cameras at the gate, and determine if the container is dirty or damaged with an accuracy of over 98%. This kind of service is already being offered in countries like China.

AI will be able to analyze the big data collected and accumulated from the IoT and other sources to realize end-to-end optimization and real-time visibility of the supply chain. New business models that provide niche services in the supply chain emerge continuously, benefiting from innovative technologies.

In addition, there will be businesses that provide training on technologies such as BC and AI. Not only corporate management but also frontline workers will acquire basic knowledge of the new technologies to work more comfortably and efficiently. Since such services are still few, demand will increase further and new businesses in this field are expected to grow considerably. With the digitization of operations and the creation of new business models, maritime logistics, and its various systems, including smart ports will evolve into the next generation, Industry 4.0 modus operandi.

5. CONCLUSION

As data-centric processes gain more momentum, shipping and port operations are transformed by meticulously embracing innovative business models and Industry 4.0 technologies. Reliable, secure, and efficient cross-collaboration maritime logistic processes are of paramount importance for the promotion of automated ports, as critical nodes in sustainable, digital supply chains.

A decentralized maritime logistics management system, such as a blockchain-based and IoT-aware system that is deployed over a Physical Internet reference architecture, can support and further advance the development of shipping digitalization and ports' automation. AI techniques and respective platforms can exploit real-time information on the movements of ships using AIS and IoT, also port infrastructures and machinery information, and data from vehicles and objects supporting the overall logistics flow, in road or air transport segments, respectively. In this chapter, we outline the technologies and provide a blueprint for building efficient, decentralized, scalable Logistics 4.0 systems, offering a prototype infrastructure model and immediate practice guidance to the next-generation shipping and port community.

Acknowledgments

This work was supported by JSPS KAKENHI Grant Number 21H01564.

REFERENCES

1. Lambrou M, Watanabe D, Iida J. Shipping digitalization management: Conceptualization, typology and antecedents. Journal of Shipping and Trade. 2019;4(1):1-17
2. Morkunas VJ, Paschen J, Boon E. How blockchain technologies impact your business model. Business Horizons. 2019;62(3):295-306
3. Hirata E. The potential of using blockchain technology in shipping and logistics industry. Journal of Logistics Shipping Economics. 2019;53:61-70
4. Montreuil B. Toward a physical internet: Meeting the global logistics sustainability grand challenge. Logistics Research. 2011;3(2):71-87
5. Heilig L, Voß S. Information systems in seaports: A categorization and overview. Information Technology and Management. 2017;18(3):179-201
6. Asian Development Bank. Smart Ports in the Pacific; Asian Development Bank. Manila: Asian Development Bank; 2020
7. Takahashi K. Trend analysis of automated container terminals in the world. Rep. Port Airpt. Res. Institute Sustainability Natl. Inst. Mar. Port Aviat. Technol. Yokosuka Jpn. 2018;56(4)
8. Buck W, Gardeitchik J, van der Dejil A. Move Forward: Step By Step Towards A Digital Port. White Paper. Port of Rotterdam: Rotterdam, Netherlands; 2019. p. 2019
9. United Nations. Digitalizing the Port Call Process, UNCTAD Transport and Trade Facilitation Series. Geneva: United Nations; No. 13. 2020
10. International Maritime Organization. Autonomous shipping.Available from: https://www.imo.org/en/MediaCentre/HotTopics/Pages/Autonomous-shipping.aspx [Accessed 2021-07-17]
11. Mitsubishi Research Institute: Class NK. FUTURE 2040-Unmanned Ships: The Future of Japan in 2040. Tokyo: Nippon Foundation; 2019
12. Kutsuna K, Ando H, Kuwahara S, Nakamura S. Conceptual proposal of action planning system (APS) for manned autonomous ship maneuvering. The Japan Institute of Metals and Materials in English. 2019;54(2):88-92
13. Sirer EG. Open and programmable finance with avalanche. In: Block Chain Jam 2020. 2020. Available from: https://cryptonews.com/videos/open-and-programmable-finance-with-avalanche.htm

14. PEMA. Container Terminal Automation. Port Equipment Manufacturers Association. PEMA Information Paper. London: PEMA; 2016

15. Class NK. Guidelines for Automated/Autonomous Operation on Ships (Version 1). Tokyo: Nippon Kaiji Kyokai; 2020

16. Meyer T, Kuhn M, Hartmann E. Blockchain technology enabling the physical internet: A synergetic application framework. Computers and Industrial Engineering. 2019;136:5-17

17. Hassan LAH, Mahmassani HS, Chen Y. Reinforcement learning framework for freight demand forecasting to support operational planning decisions. Transportation Research Part E: Logistics and Transportation Review. 2020;137:101926

18. Tan BQ, Wang F, Liu J, Kang K, Costa F. A blockchain-based framework for green logistics in supply chains. Sustainability. 2020;12(11):4656

19. Munim ZH, Duru O, Hirata E. Rise, fall, and recovery of blockchains in the maritime technology space. Journal of Marine Science and Engineering. 2021;9(3):266

20. Rocket T, Yin M, Sekniqi K, van Renesse R, Sirer EG. Scalable and Probabilistic Leaderless BFT Consensus through Metastability. ArXiv Prepr. ArXiv190608936 2019

Chapter 3

Mapping the Applications of Vehicular Communications in the Context of Smart Maritime Ports

El Idrissi Ayoub[1], Haidine Abdelfatteh[1], Ait-Allal Abdelmoula[2], Dahbi Aziz[1] and Aqqal Abdelhak[1]

[1] *National School of Applied Sciences, University Chouaib Doukkali, Morocco*

[2] *High Institute of Maritime Studies (ISEM), Morocco*

ABSTRACT

The maritime transport networks play a critical and major role in an increasingly globalized world economy. Within these networks, the maritime ports play the role of hubs. Any disturbances in these hubs will negatively affect the worldwide economy. Therefore, economy players are transforming the ports through an evolutionary process to become smart maritime ports. These smart ports are built through an ensemble of smart domains that adopt sensing, data transmission, and data intelligence to support intelligent decision-making processes. Examples of such smart domains include smart grid/microgrids, smart container management, and smart/automatized terminal operations. In each of these domains, optimal decisions must be met to optimize the use of resources, increase the economy efficiency of the ports, and increase the safety and security for assets, goods, and people. In smart maritime port environment, vehicular applications are adopted everywhere, such as automated guided vehicles to transport containers, unmanned aerial vehicles for different port operations, etc. In this work, we discuss some concrete examples of these vehicular applications in the smart port environment and suggest the adequate and optimal vehicular communication technologies to be deployed to support a reliable data transmission for these applications.

Keywords: *smart maritime ports; vehicular communications; unmanned aerial vehicles; intelligent transport system; mobile communications*

1. INTRODUCTION

In today's maritime world, numerous pressures are exercising on ports; so many things have been changed deeply. Therefore, ports worldwide face challenges every day. Nowadays, we assist to more goods and services that are being shipped, resulting in ship's traffic that is increasing permanently [1]. Ports are crowding because of the growth of population in the world, accentuated by the globalization and the growing maritime industry. So far, the maritime world has been impacted by the global trade, sustainability, and digitalization.

A simplified configuration of maritime transport networks is depicted in Figure 1, showing that such a network can be modeled as graph network G = (V, E), where the set E constitutes the maritime routes, which must be secure, cost-efficient, short-delay, and reliable. The set of nodes (or vertices) is constituted by the world maritime ports/seaports. These nodes must also be secure/safe, cost-efficient, short-delay, reliable, environment friendly, and energy-efficient. Any disturbances in these nodes, that is, ports, will result in dramatic scenes and negative impact on the economy, nationally and as well as internationally in more globalized economy/industries. Such a concrete example is depicted in Figure 1 (bottom) showing an increasing queuing of jam in front of maritime ports. More concretely, port authority recorded shattered 73 container ships stuck waiting off California port [2].

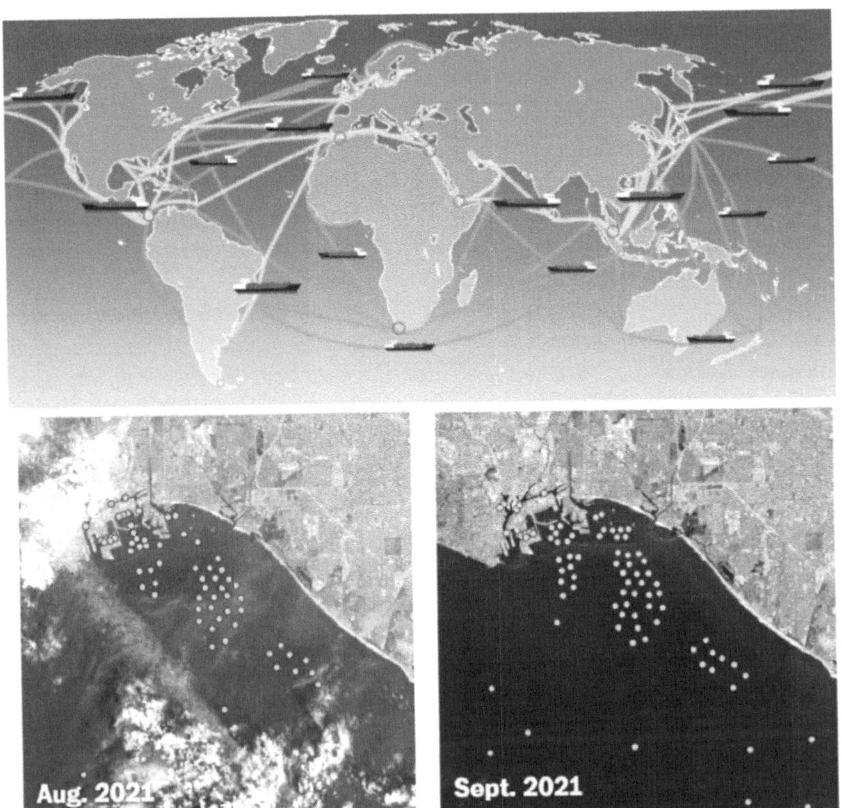

Figure 1. *Simplified presentation of maritime transport network (top, Credits: Dreamstime) and Jam of containerships in front of Los Angeles port resulting in a long delay of containers and goods delivery (bottom, Credits: Washington Post).*

All these maritime challenges and so more could be reduced or even deleted if we focus on the solutions proposed by the new technologies that convert all the drawbacks into opportunities. Nowadays, and in a competitive world, being attractive or even exist needs to innovate with more intelligence and by using the digital way to care more for the future of the industry. Thus, the maritime ports should follow this evolution.

According to this context, and including the intelligence to ports, a new concept has emerged, which is called smart port intended to care for the maritime environment in order to be more effective and performing. Smart ports lead to provide more with less. This means saving time, resources, and money. Furthermore, smart ports is aiming to achieve a better connection for stakeholders using the automated ships, to massively reduce pollution and congestions, while increasing time/economic effectiveness and safety.

Smart ports' solutions and strategies can be integrated to solve most of the maritime challenges. The use of new technologies and intelligence is a way to transform the traditional practices of ports and their weaknesses to smart ports with more benefits. Smart technologies, such as big data, artificial intelligence (AI), Internet of Things (IoT), blockchain technology, and fifth-generation (5G) connection, are means to change the maritime industry. They are used as a decision-making system and a way to monitor processes with developing more security and safety in the working environment.

In this chapter, we discuss the main components for the realization of smart maritime ports; then, we focus more on vehicular applications. These applications should lay on a strong performing and reliable communication layer. Therefore, we analyze different possible transmission technologies for vehicular communications through a qualitative evaluation of different performance criteria, which are more relevant for smart maritime port environment.

2. SMART MARITIME PORTS AND IOT MODEL

2.1 Smart ports: definitions and objectives

Smart is a new concept adapted to the technology setting to follow the modern world. It is a word that contains digital information, computer development, and a way to innovate. According to [3], smartness is defined as the application of automatic computing principles such as self-configuration, self-protection, self-healing, and self-optimization. Smartness has found its way into many industries, especially in urban areas, as a solution to a number of problems such as pollution, increasing cost, problem of congestion, and preservation of historical sites.

Smart ports provide digital port solution consultancy for new and existing ports. Smart ports combine new technologies, automation, and digitalization to optimize the port operations. The use of smart ports is a way to boost digital transformation and make it effective in practice; also, it gathers technology, innovation, and digital tools. Smartness relies on information and digital technologies to become more efficient, sustainable, and collaborative. It encompasses different types of technologies such as big data, artificial intelligence (AI), Internet of Things (IoT), blockchain, and 5G.

Finally, a smart port is a port facility using information and communication technologies (ICT) to increase efficiency and ensure the safety of transshipment operations.

The objective of smart ports is to provide solutions to four key issues: economic issues, ecological issues, energy issues, and social issues. Today, many ports seek to use technology to achieve these objectives and to become a sustainable port responding to economic, ecological, and citizen issues:

- Concerning the economic aspect, the objective of the smart port will be to reduce the costs linked to the increase in flows and goods; many smart ports use artificial intelligence. For instance, AI can be used to predict the arrival of ships or to know the activities of each in real time.
- As for the ecological aspect, and to meet this challenge, smart ports seek above all to reduce their greenhouse gas emission, which also increases the air quality of their city.
- Smart ports aim to address the challenge of energy in two main ways. The first is through the development and use of renewable energy sources, such as solar, wind, and wave power. The second is by reducing overall energy consumption in the ports through the use of energy-efficient technologies and processes.
- From the perspective of the social aspect, ports generally adopt two solutions: education and participatory democracy.

2.2 Smart domains in smart maritime port environment

The evolution of the traditional port into a smart maritime port is complex and involves various research activities. Therefore, the smart maritime port can be seen as a set of (also known as "vertical" domains) smart domains, as depicted in Figure 2. In the following, we describe some of these smart domains.

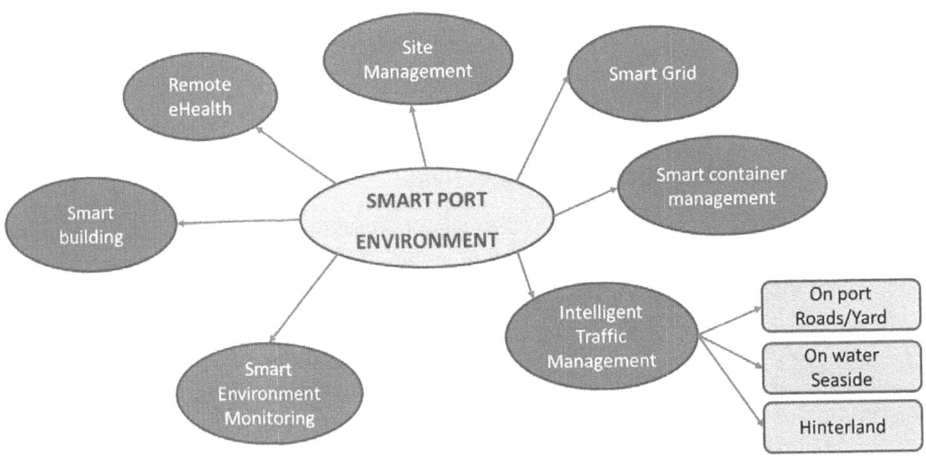

Figure 2. *The smart port environment is constituted of a set of smart domains.*

Smart grid: Smart grid technology represents the next major advance in power grid management. With the help of various technological means, it allows an intelligent control of energy flows, enabling an adequate balance between energy supply and demand, resulting in savings for both consumers and producers. Smart grids also play an important role in the development of renewable energies. However, the high implementation costs and certain data protection issues remain major challenges against this technology.

The major applications and roles of the smart grid can be summarized as follows:

- Balance: It allows for better management of the energy produced and the energy delivered.
- Security: The smart grid improves the stability of the electrical network, reducing the risk of overloads or outages, which can have negative impacts on both individuals and industries.
- Ecology: The smart grid allows a better use of renewable energies in the global energy consumption and, thus, avoids very important losses.
- Economy: A more stable network management allows a drastic reduction of production and consumption costs and lower losses.

Intelligent transport system (ITS): The state-of-the-art solution for the management of truck flows is based on an innovative concept of an intelligent parking lot, which makes it possible to put an end to port congestion and to fluidify the adjacent road traffic.

In this context, having an access permit is no longer sufficient for gaining access to the port. Access to the port is subject to an appointment with the operator through its online software system connected to automated gates and cameras; trucks accessing the port have the option of using the physical parking lot or the virtual online parking schedule. They will only be allowed into the parking lot after a mission has been created and the port operator has confirmed the scheduling of an imminent operation.

The Internet of Things (IoT) and artificial intelligence (AI) are enabling a new class of intelligent transportation systems (ITS) for road, air, rail, and sea. These solutions connect vehicles, traffic signals, toll booths, and other infrastructure to help reduce congestion, prevent accidents, reduce emissions, and make transportation more efficient. Examples include fleet management, intelligent traffic management, vehicle-to-everything (V2X) communication, electric vehicle charging, electronic tolling, and a wide range of other mobility solutions.

Terminal operations: In a nonmechanized seaport, operator travel can take up to 40% of their working time, resulting in a significant drop in productivity, soaring costs, and deterioration in customer service. However, with the multiplication of intelligent solutions, the connection of machines to each other, and the development of information systems as well as the network of connected objects, it is very easy and often less expensive than one might think to automate all or part of its flows.

Today's port professionals need to have a thorough understanding of container terminal management, the developments that impact their development and the challenges and opportunities for container ports: growth in container traffic requiring additional container capacity, ever larger ships calling at your port and terminals, alliances between container lines, automation of port terminal operations, growing importance of intermodality, and increased competition focusing on profitability and service.

Smart container management: The automated container terminal is not a futuristic view. For over a decade, automation has been implemented in terminals, and this trend is continuing. Today, in some terminals, the management of container fleets (including storage and movement) is done with fully automated equipment. Automated vehicles such as automated guided vehicles (AGVs) and straddle carriers are used in some terminals for transporting containers between storage areas and dock gantries.

For terminals, IT management is an element of activity management (storage and movements) in relation to ship planning and inland service tools. The tools for inland services. With the total automation of the terminal, we can imagine totally autonomous terminal managing flows: positioning and provisioning. The “smart terminal” will become, over several hectares, what is done in some logistics warehouses. The future would then be a terminal with artificial intelligence for total management.

2.3 IoT model for smart maritime ports

The deployment of smart environment, such as smart city, smart home, etc., uses the approach of Internet-of-Things (IoT) models. In such models, a set of sensing nodes (or IoT end-devices) is dispatched in this environment to collect different types of measurements/data. These data are transmitted over different communication links and networks and then stored somewhere for further real time or offline treatment, exploration, and exploitation. The large success of the IoT leads engineers to adopt different IoT models, mostly a three-layer model, a four-layer model, or a five-layer model. Because this chapter focuses on vehicular applications in the port environment, we adopt the model presented in Figure 3, which is sometimes called “IoT value chain” and used in the literature to point out the services related to vehicular applications. This includes six levels:

- Sensing layer: This layer is constituted by different end-devices known as sensors/actuators, which sense the environment and collect data. These data can be either sent directly to the upper layer or be submitted to some local but limited processing, such as compression and/or filtering.
- Transport/communications layer: The data collected locally by the sensing layer are given to this layer to be transmitted, generally to far locations where we have some advanced data processing equipment. This layer can be built by different alternative technologies. This can be wired technology like optical fiber and cables, or wireless using IEEE technologies (WiFi, WiMAX, long-range Bluetooth, or ZigBee) or 3GPP transmission technologies, which are categorized in second generation (2G), third generation (3G), fourth generation (4G), or 5G.
- Data layer: The data collected by the sensing layer and transmitted arrive to this destination where they are stored for advanced processing. This layer is comprised of data centers or clouds where advanced big data operations are executed.
- Analytics layer: This layer is characterized by a higher level of intelligence, where algorithms or processes related to artificial intelligence and/or machine learning treat the stored data.
- Control layer: Based on the insights and knowledge gained from the analytics layer, this layer should meet optimal decisions to control the processes or the infrastructure monitored through the considered IoT system.
- Business layer: The IoT system is foreseen to be deployed in industrial and/or professional environments. Therefore, the objective remains generally the achievement of some economic benefits. The responsible processes to guide the entire processes to achieve these objectives and benefits must be englobed in this layer.

Figure 3. *Simple horizontal model for the IoT value chain in vehicular applications.*

3. VEHICULAR APPLICATIONS/SYSTEMS IN SMART MARITIME PORT ENVIRONMENT

3.1 Vehicles in maritime port yard

The maritime ports is a very active area, where heavy goods mostly encapsulated in containers must be handled in different processes, namely, loading, unloading, storage, delivery toward hinterland, received from hinterland, controlling through customs, etc. These operations require the use of different vehicular machines such as cranes, mobile cranes on wheels or rails, cranes operated by drivers, automated cranes, trucks on port yard, hinterland trucks, trains from hinterland, etc. Figure 4 shows some examples of vehicles on port yard, including unmanned aerial vehicles (UAV) or drones. Some ports also use underwater unmanned vehicles (underwater drones) for different control and environment supervising tasks or control the part ship container underwater.

Automated guided vehicles (AGV): An automated guided vehicle (AGV) or automatic guided vehicle is a vehicle used to transports containers (and other items) in the port's area. These vehicles are remotely controllable, but also have an advanced level of intelligence to execute self-driving and other optional decisions, as depicted in Figure 5. Several advantages are guaranteed through the use of AGVs, such as economic/cost effectiveness, time efficiency, increased efficiency of the overall processes in the ports, considerable reduction in accidents, and reduced injuries due to the absence of human traffic, etc.

- **UAV or drone-assisted operation:** Drone or unmanned aerial vehicles (also unmanned aircraft systems—UAS) are gaining a large application fields. Maritime ports are already testing drones for their operation, such as [4]:
  - inspections of physical structures and patrolling of security rounds;
  - routine inspections for maintenance of buoys, pipes, docks, breakwater cranes, roof ships, and other structures that are conventionally difficult to access;
  - stock measurement to calculate (bulk) volumetric mass inventory. Inventory volume tracking of outdoor bulk material storage areas;
  - detection of irregular situations, leaks, or abnormalities through (thermal and gas) sensors, as supplemental emergency support without the need to expose people to the affected areas;
  - measurement and control of environmental aspects, detection of contamination, and tracking and monitoring those responsible for the environmental breaches. Environmental and ecological monitoring of both on-site processes and surrounding areas.

Figure 4. *Examples of vehicular applications in smart maritime port environments.*

Figure 5. *Fleet of AGVs as part of port operations automation (Credits: wikiwand.com).*

3.2 Levels of intelligence in vehicular applications in SMP

The different types of vehicles used in the port yard have the objective of facilitating the port operation through the optimization of processing time, safety of goods and personnel, and achieving cost and energy efficiency. However, higher levels of optimization are only reached through the automation of different operations, even also through the operation of vehicles in maritime ports. Therefore, we talk about autonomous or intelligent vehicles. The transition from traditional/classical vehicles towards autonomous port vehicles cannot be done in one transition. Thus, there are different levels of autonomy of vehicles, as illustrated in Figure 6. These intermediate levels allow a smooth transition, where the research and development works can be accompanied with the proof of concepts in the real life of maritime ports. This will allow the engineers get feedback from the practice to adapt their systems to any existing or new challenges in such a harsh and critical infrastructure. Higher autonomy signifies higher intelligence that requires the collection of larger amount of data, which in its turn must be treated and/or transmitted in very short delay and in high reliable manner. In other words, these vehicles will need ultra-low latency and ultra-reliable communications links.

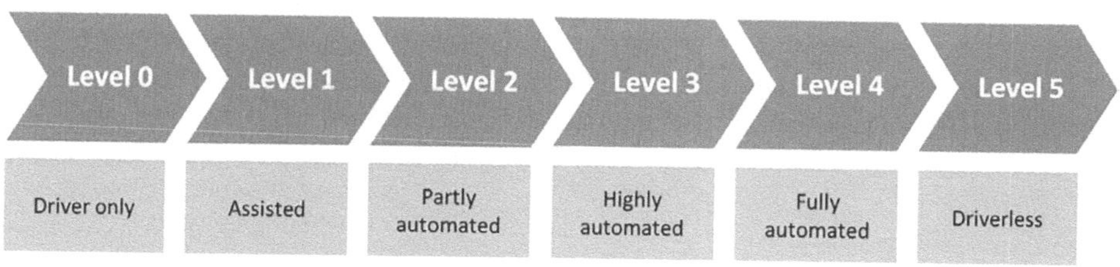

Figure 6. *Generic model for autonomy levels of vehicles in port operations [5].*

In a general context, the introduction of autonomous vehicles has three main objectives [5]:

- increased efficiency of transport, which brings alongside better capacity;
- utilization;
- less negative environmental impact;
- increased safety.

4. VEHICULAR COMMUNICATIONS IN MARITIME PORTS

4.1 General hierarchical architecture/infrastructure in the smart port

From the economic point of view, it is worthy to select communication technology that can support all (or at least the maximum of) customer services. However, in practice, this is not possible. On the one hand, the diversity in services and the generated traffic have different quality-of-service requirements. On the other hand, each developed technology has its strengths and weaknesses, as can be seen in Figure 7, where only two performance criteria are plotted for different transmission technologies, namely, coverage and throughput. In reality, operators and engineers of the network must take into consideration further performance criteria such as delay, reliability, deployment as well as operational costs, etc.

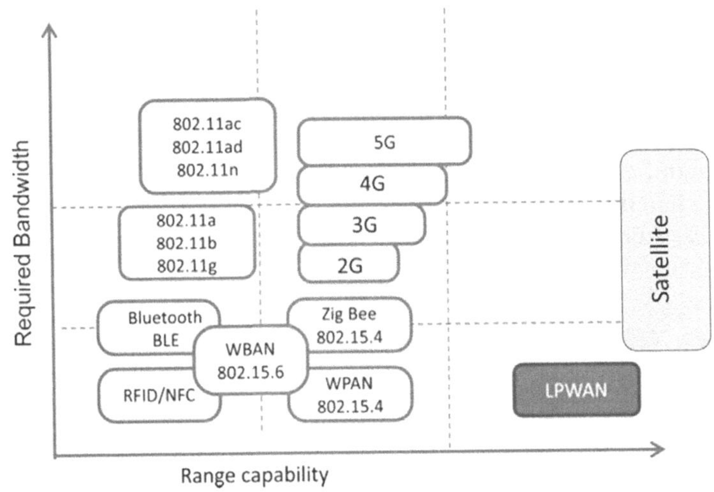

Figure 7. *Classification of wireless communications technologies in a two-dimensional plane (coverage and throughput).*

An optimal design of the networking layer consists in having a networking infrastructure, which can support a variety of services from different vertical smart domains. A simplified topology of communication infrastructure in the area of a smart maritime port is illustrated in Figure 8. In such architecture, we recognize three levels, as discussed first in [6]:

Level of low band or low bandwidth: This connects to the end-devices of the sensor layer (built through sensors and actuators). These nodes are dispatched in wide areas and transmit, in general, short and few parquets. Therefore, these nodes are not very requiring concerning the quality of services. The networking technologies to build this level must be very energy efficiency (low energy consumption) while offering low bit rates. Generally, wireless technologies such as Bluetooth, Bluetooth energy (BLE), ZigBee, Z-wave, low-power wide area (LPWA) networks (SigFox or LoRa or WightLess). As we remark, the wireless solutions are advantageous in this level, as they are easy to deploy and to maintain.

Level of medium bandwidth: This level must transmit the traffic aggregated from the nodes of the lower level. Normally, for each node pool (or cluster of nodes), there is a gateway that collects and aggregates the traffic outgoing from these nodes. Because of that, the demand of the transmission capacity is relatively higher. The determination of the size of the clusters is subject to optimization under different constraints. But, we can consider that a bit rate of 2 Mbps a good compromise.

Level of broadband technologies: In its turn, the level of medium bandwidth must be aggregated and connected to the (national) wide area networks. For most of the cases in the IoT, wireless solutions are more advantageous. Furthermore, mobility plays an important role here. Therefore, we have greater requirement for mobile broadband communications in this level. This is the mobile backbone of the networking layer in the smart port area. In addition, this level supplies direct connections to end-devices with higher demand of bandwidth, like HD camera, mobile PC/tables/device of the field force, normal Internet services everywhere in the port area, etc. The notion of mobile broadband has appeared in the engineering field related to third generation (3G) of mobile communications, technically known as universal mobile telecommunications system (UMTS). However, this was not able to offer transmission capacity more than 2 Mbps. Therefore, this has been extended to offer higher capacities and its derivation was known as high-speed downlink packet access (HSDPA) that focused on the improvement of downlink connection. This was followed by high-speed uplink packet access (HSUPA) that focused on the improvement of uplink connection. The last two variants have converged in a relative advanced technology called evolved high-speed packet access (eHSPA or HSPA+), which adopted advanced antenna technology as well as advanced modulation and coding schemes to reach up to 40 Mbps. The major (re)evolution in mobile broadband communications was realized by the fourth generation (4G) or long-term evolution (LTE), which can achieve bit rates over 300 Mbps. This milestone was rapidly overruled by the releases of long-term evolution-advanced pro (LTE-A Pro, called also 4.5G) reaching the magic threshold of 1 Gbps. This generation has offered different performance milestones in form of capacity of the bandwidth, shorter transmission delay, and service coverage for massive IoT systems. Nowadays, the fifth generation (5G) is struggling to find field where it can beat the performance of 4.5G. This seems to be a hard task, because different technical aspects as well as economic factors are putting lot of pressure on the deployment roadmaps of 5G. This has resulted in a delay in the widespread deployment of 5G technology.

4.2 Qualitative comparison of broadband technologies for vehicular applications

When we consider the evaluation of communication technologies for vehicular applications in smart port environment, a qualitative comparison is used, according to different criteria, which are throughput, network costs, coverage, adequacy to support massive IoT and the delay/latency. Here, the short-range communications technologies or short-range networks (SRNs) are excluded, because the main challenge of the net-

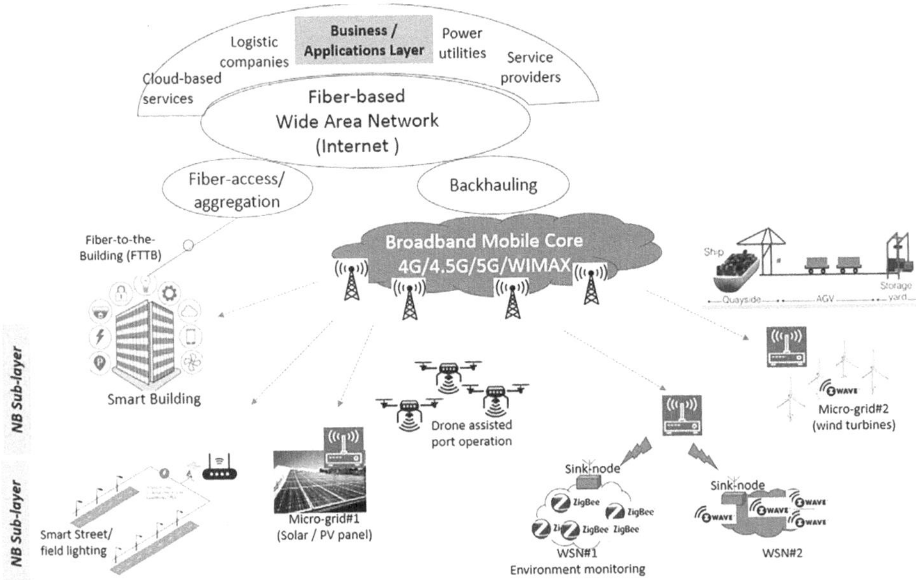

Figure 8. *Sublayers of communication architecture in a smart port.*

work communications in port application lies in the set of (large coverage, short delay, and sometimes broadband). Furthermore, we have also taken into consideration the costs, which are very important for the network operator, the reliability as main challenge for sensitive application, and finally the IoT adaptability. This latter one is very important because any smart environment has a sensor or IoT layer as basis for any smart thing. Therefore, the communication layer should be adequately chosen for such environment.

Long-term evolution/advanced (LTE/LTE-A) is the first real mobile broadband technology. Having a real success worldwide, this is the backbone for the current mobile network and mobile Internet. Its major strength is its characteristic of "long-term" evolution; thus, it was designed to be able to be extended anytime for any new services. Indeed, the first LTE releases (Release 8 and Release 9) have been quickly evolved to Release 10 and Release 11, etc. Major transitions have realized in a short time, so LTE overrides just usual broadband Internet to support machine-type communication (MTC), machine-to-machine (M2M), vehicle-to-vehicle (V2V), Internet-of-Things (IoT) communication/services/applications. This lets this technology show higher performance in all criteria compared with other technologies. Of course, it has a unique weakness, which is the coverage compared to the satellite. A very large number of research works have investigated the performance of the LTE in vehicular communications either in the form of V2V, vehicle-to-infrastructure (V2I), vehicle-to-everything (V2X), including Pedestrian, or Internet of Vehicles (IoV) [7, 8, 9, 10, 11]. Typical types of utilization of vehicular communications are depicted in (Figure 9) for general use as well as for use in the case of maritime port environment by mobile work force.

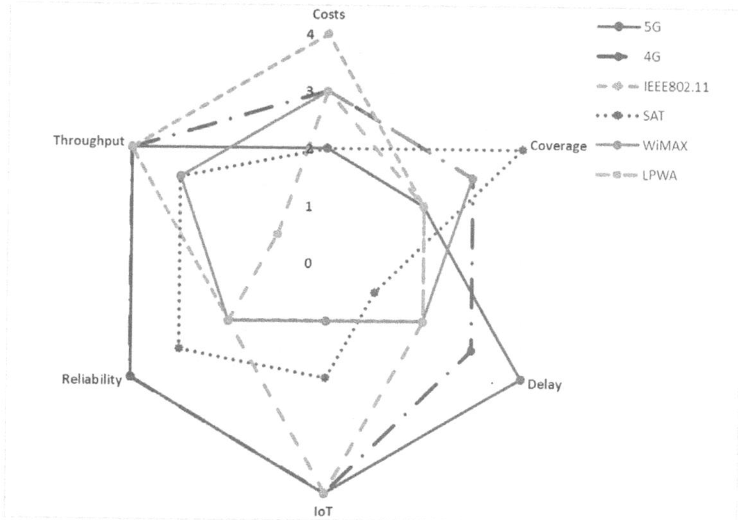

Figure 9. *Types of V2X in LTE standards (left, [12]) and its applications in maritime port environment (right, [13]).*

5G is the newest mobile technology generation. It should bring higher performance than 4G, especially in three dimensions: enhanced broadband throughput, ultrareliable communication, and support of massive IoT devices. The enhanced broadband should support new applications requiring very high capacities, such as virtual reality/augmented reality (VR/AR), which are not expected to be required in smart port environments. A major weakness is relative high cost to be invested. This is because this generation requires the deployment of new infrastructure. In the opposite of the evolved releases of 4G, which the upgrade of the networks is made with less costs. Different works evaluated the use of 5G for vehicular applications, and some of them are even talking about 6G [9, 14, 15, 16]. An example of application of 5G for vehicular applications in maritime port environment is given in [17], where higher capacity and lower latency are the objectives of the adopted architecture (Figure 10). The competition between 4G/LTE and 5G to build network layer for smart maritime ports will be intensive, because they are on the same level of performance in most of the considered criteria, as illustrated in Figure 11. However, 4G has the main advantage of network costs, because 4G network are already massively deployed and proofed in the field, while 5G massive rollout suffer from delay, viability of its business model and other different economic circumstances.

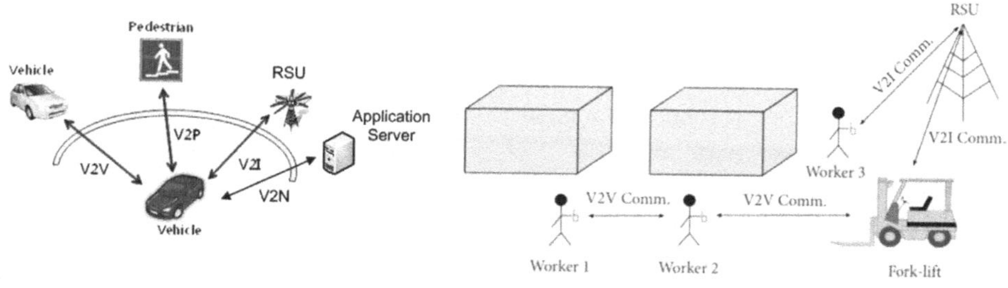

Figure 10. *Example from the practice for the use of 5G in maritime port vehicular applications [17].*

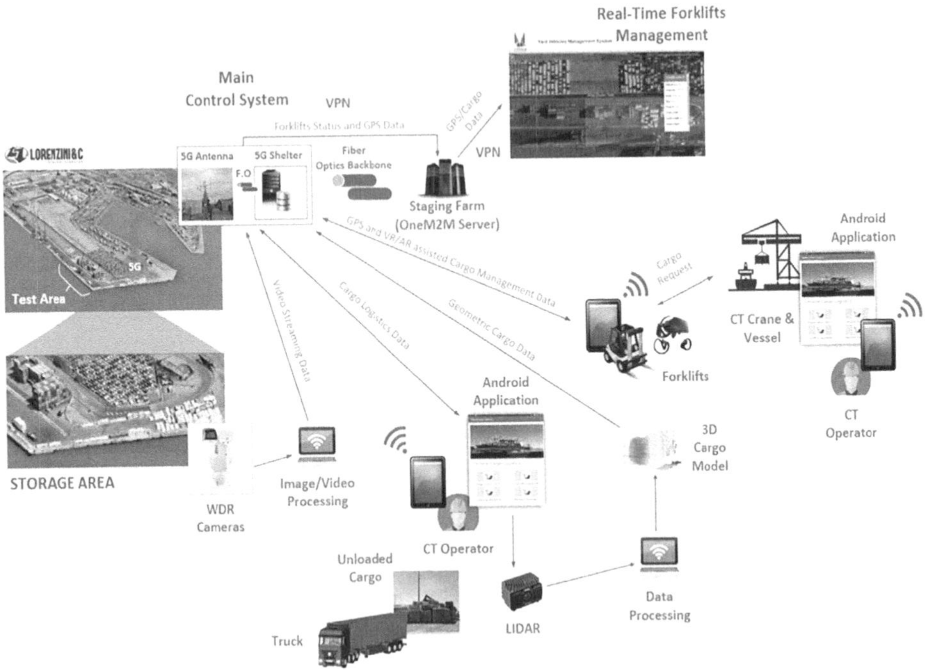

Figure 11. *Qualitative performance of communications technology (relative to each other) for vehicular applications (scores meaning: 4—very good, 3—good, 2—average/acceptable, 1—bad, and 0—very bad).*

The satellite communication is the excellent solution for the large coverage (everywhere and anytime). This technology did not cease to evolve over the years. However, two major weak points are inevitable: the high costs for broadband communications and the long delay, which is caused by the signal propagation to the sky and back. In spite of this, different mechanisms have been investigated to improve its performance [18, 19, 20, 21].

To realize the objective of long range or long coverage of IoT applications and devices, without the use of satellite, low-power wide area (LPWA) networks have been developed especially for the narrowband IoT. Thus, high throughput and implicitly the delay are the main drawbacks of this technology. Examples of the utilization of LPWA in vehicular applications can be found in [22, 23, 24, 25].

WiMAX technology has been reserved for some industrial applications, after it lose the battle of 4G, and this is the adoption of the usual broadband network for the public mobile network. This worked well for a while, however, with the apparition of massive IoT applications in industrial fields under the name of Industrie 4.0; WiMAX was not able to support the modern services, because its standardization has been frozen more than a decade ago. Therefore, it cannot be a good favorite or alternative for such modern smart environment like smart maritime ports.

The success of the WLAN standard (IEEE 802.11) has led to its extension beyond residential and enterprise areas. Special applications have been developed, including for smart cities (IEEE 802.11h) and smart transportation (IEEE 802.11p). While studies have investigated the use of IEEE 802.11p for vehicular applications, such as [26, 27, 28, 29, 30], there is a lack of research considering its use in the specific context of a smart maritime port environment.

5. CONCLUSIONS

The maritime transport networks are playing a unique and major role in an always more globalized world economy. In such networks, the maritime ports play the role of hubs. Any disturbances in these hubs will negatively influence the worldwide economy Therefore, economy players are transforming the ports through an evolving process to become smart maritime ports. These smart ports are built through an ensemble of smart domains that are adopting sensing, data transmission, and data intelligence to support intelligent decision-making processes. Smart grid/microgrids, smart container management, smart/automated terminal operations, etc. are examples of such smart domains.

In different smart domains, vehicular applications are present. Therefore, the vehicular communications must be taken into consideration while developing the roadmap for smart maritime port. In this chapter, we focused on five communication technologies through the presentation of a qualitative performance evaluation of these technologies. The introduction of autonomous imposes hard requirements for the quality of services to be guaranteed by the communication networks, such as ultra-low latency, ultra-high reliable communication links.

Acknowledgments

The authors would like to thank our colleagues in the department for their lecture, corrections and valuable feedbacks that helped to improve the scientific quality of this chapter.

REFERENCES

1. Bodewig K. Shaping the Future Policy of the European Maritime Space Motorways of the Sea Detailed Implementation Plan of the European Coordinator: Report EU Mobility and Transport. Brussels, Belgium: European Commission; 2020
2. Miller G. Record shattered: 73 container ships stuck waiting off California, FREIGHTS WAVES News on Thursday, September 16, 2021. Available from: https://www.freightwaves.com/news/record-shattered-61-container-ships-stuck-waiting-off-california
3. Spangler W, Kreulen J, Chen Y, Proctor L, Alba A, Lelescu A, et al. A smarter process for sensing the information space. IBM Journal of Research and Development. 2010;54:1-13. DOI: 10.1147/JRD.2010.2050541
4. Bexiga V. Blue Innovation—Drones in Port Operations, Escola Europa Intermodal Transport. 2021. Available from: https://escolaeuropea.eu/odiseo/issue-30-spring-2019/blue-innovation-drones-in-port-operations [Accessed: June 23, 2021]
5. Fiedler R, Bosse C, Gehlken D, Brümmerstedt K, Burmeister H-C. Autonomous Vehicles' Impact on Port Infrastructure Requirements. Hamburg: Fraunhofer Center for Maritime Logistics and Services CML; 2019

6. Haidine A, Ait-Allal A, Aqqal A, Dahbi A. Networking layered for the evolution of maritime ports into a smart environment. In: The International Archives of Photogrammetry, Remote Sensing and Spatial Information Sciences; Gottingen. Gottingen: Copernicus GmbH; 2021. pp. 251-257. DOI: 10.5194/isprs-archives-XLVI-4-W5-2021-251-2021
7. Bazzi A, Cecchini G, Zanella A, Masini BM. Study of the impact of PHY and MAC parameters in 3GPP C-V2V mode 4. IEEE Access. 2018;6:71685-71698. DOI: 10.1109/ACCESS.2018.2883401
8. Izydorczyk T, Tavares FML, Berardinelli G, Bucur M, Mogensen P. Performance evaluation of multi-antenna receivers for vehicular communications in live LTE networks. In: 2019 IEEE 89th Vehicular Technology Conference (VTC2019-Spring). Kuala Lumpur, Malaysia. 2019. pp. 1-6. DOI: 10.1109/VTCSpring.2019.8746609
9. Muhammad M, Safdar GA. Survey on existing authentication issues for cellular-assisted V2X communication. Vehicular Communications. 2018;12:50-65. DOI: 10.1016/j.vehcom.2018.01.008. Available from: https://www.sciencedirect.com/science/article/pii/S2214209617302267
10. Sliwa B, Patchou M, Wietfeld C. Lightweight simulation of hybrid aerial- and ground-based vehicular communication networks. In: 2019 IEEE 90th Vehicular Technology Conference (VTC2019-Fall). Honolulu, Hawaii. 2019. pp. 1-7. DOI: 10.1109/VTCFall.2019.8891340
11. Gyawali S, Xu S, Qian Y, Hu R. Challenges and solutions for cellular based V2X communications. IEEE Communications Surveys & Tutorials. 2020;2020:1-1. DOI: 10.1109/COMST.2020.3029723
12. European Telecommunications Standardization Institute (ETSI). LTE, Service Requirements for V2X Services. ETSI Technical Specification for (3GPP TS 22.185 version 14.3.0 Release 14). Available from: http://www.etsi.org/standards-search
13. Sanja B, Jose Angel L, Schmeink A, Azam S, Mathar R. On deploying vehicular communications at the developing seaport and related innovation success impediment. Journal Transport. 2019;34:126-134. DOI: 10.3846/transport.2019.7987
14. Molina-Masegosa R, Gozalvez J. LTE-V for sidelink 5G V2X vehicular communications: A new 5G technology for short-range vehicle-to-everything communications. IEEE Vehicular Technology Magazine. 2017;12(4):30-39. DOI: 10.1109/MVT.2017.2752798
15. Zeadally S, Javed MA, Hamida EB. Vehicular communications for ITS: Standardization and challenges. IEEE Communications Standards Magazine. 2020;4(1):11-17. DOI: 10.1109/MCOMSTD.001.1900044
16. Cheng X, Huang Z, Chen S. Vehicular communication channel measurement, modelling, and application for beyond 5G and 6G. IET Communications. 2020;14:3303-3311. DOI: 10.1049/iet-com.2020.0531/*Fgiure port
17. Tesei A, Lattuca D, Tardo A, Di Mauro L, Pagano P, Luise M, et al. Securing seaport logistic vehicles using a distributed ledger-based credential management system. IEEE Open Journal of Vehicular Technology. 2021;2021:2. DOI: 10.1109/OJVT.2021.3067209
18. Cui G, Long Y, Xu L, Wang W. Joint offloading and resource allocation for satellite assisted vehicle-to-vehicle communication. IEEE Systems Journal. 2020;2020:1-12. DOI: 10.1109/JSYST.2020.3017710
19. Yin Z, Jia M, Cheng N, Wang W, Lyu F, Guo Q, et al. UAV-assisted physical layer security in multi-beam satellite-enabled vehicle communications. IEEE Transactions on Intelligent Transportation Systems. 2021;23(3):2739-2751
20. Lei A, Ogah C, Asuquo P, Cruickshank H, Sun Z. A secure key management scheme for heterogeneous secure vehicular communication systems. ZTE Communications. 2019;14(S0):21-31

21. Yuhan R, Rui Z, Yongzhao L, et al. Spectral-energy efficiency tradeoff in cognitive satellite-vehicular networks towards beyond 5G. In: 2019 IEEE Wireless Communications and Networking Conference (WCNC). 2019. pp. 1-6

22. Yuke L, Shuangshuang H, Linyao Y, et al. LoRa on the move: Performance evaluation of LoRa in V2X communications. In: IEEE Intelligent Vehicles Symposium (IV). Suzhou, China. 2018. pp. 1107-1111

23. Akira S, Yoshitaka S, Toshihiro T. Evaluation of performance on lpwa network realizes for multi-wavelength cognitive v2x wireless system. In: 2019 IEEE 10th International Conference on Awareness Science and Technology (iCAST). Morioka, Japan: IEEE; 2019. pp. 1-6

24. Jinhyung O, Dong-Woo L, Kyu-Min K. Unmanned aerial vehicle identification success probability with LoRa communication approach. In: IEEE 31st Annual International Symposium on Personal, Indoor and Mobile Radio Communications. London, United Kingdom. 2020. pp. 1-6

25. Fei G, Jianwei N, Landu J, et al. Survey of the low power wide area network technologies. Journal of Network and Computer Applications. 2020;149:102459

26. Singh A, Singh B. A study of the IEEE802.11p (WAVE) and LTE-V2V technologies for vehicular communication. In: International Conference on Computation, Automation and Knowledge Management (ICCAKM). 2020. pp. 157-160. DOI: 10.1109/ICCAKM46823.2020.9051468

27. Singh PK, Nandi SK, Nandi S. A tutorial survey on vehicular communication state of the art, and future research directions. Vehicular Communications. 2019;18:2214-2096. DOI: 10.1016/j.vehcom.2019.100164

28. Trujillo-Flores JI, Torrealba-Meléndez R, Muñoz-Pacheco JM, Vásquez-Agustín MA, Tamariz-Flores EI, Colín-Beltrán E, et al. CPW-fed transparent antenna for vehicle communications. Applied Sciences. 2020;10:6001. DOI: 10.3390/app10176001

29. Arena F, Pau G, Severino A. A review on IEEE 802.11p for intelligent transportation systems. Journal of Sensor and Actuator Networks. 2020;9:22. DOI: 10.3390/jsan9020022

30. Klapez M, Grazia CA, Casoni M. Minimization of IEEE 802.11p packet collision interference through transmission time shifting. Journal of Sensor and Actuator Networks. 2020;9:17. DOI: 10.3390/jsan9020017

Chapter 4
A Real-Time Ship Detector via a Common Camera

Penghui Zhao[1], Xiaoyuan Yu[1], Zongren Chen[1,2] and Yangyan Liang[1]

[1] *Faculty of Innovation Engineering, Macau University of Science and Technology, Taipa, Macau 999078, China*

[2] *Computer Engineering Technical College (Artificial Intelligence College), Guangdong Polytechnic of Science and Technology, Zhuhai 519090, China*

ABSTRACT

Advanced radars and satellites, suitable for remote monitoring, inappropriately reach the economical requirements of short-range detection. Compared with far-sightedness skills, common visible-light sensors offer more ample features conducive to distinguishing the classes. Therefore, ship detection based on visible-light cameras should cooperate with remote detection technologies. However, compared with detectors applied in inland transportation, the lack of fast ship detectors, detecting multiple ship classes, is non-negligible. To fill this gap, we propose a real-time ship detector based on fast U-Net and remapping attention (FRSD) via a common camera. The fast U-Net offered compresses features in the channel dimension to decrease the number of training parameters. The remapping attention introduced boosts the performance in various rain–fog weather conditions while maintaining the real-time speed. The ship dataset proposed contains more than 20,000 samples, alleviating the lack of ship datasets containing various classes. Data augmentation of the cross-background is especially proposed to further promote the diversity of the detecting background. In addition, the rain–fog dataset proposed, containing more than 500 rain–fog images, simulates various marine rain–fog scenarios and soaks the testing image to validate the robustness of ship detectors. Experiments demonstrate that FRSD performs relatively robustly and detects 9 classes with an mAP of more than 83%, reaching a state-of-the-art level.

Keywords: *convolutional neural network; remapping attention; rain–fog dataset; fast U-Net; ship dataset; ship detector; frames per second (FPS)*

1. INTRODUCTION

Considering the marine environment and the detecting sphere, various radars or optical satellites are the primary resources for ship detection. Regardless of the classification, advanced radars and satellites detect ships by extraordinary resolution images containing ample information, which is helpful to detect ships in a remote range. Nevertheless, advanced military radars, suitable for far detection, hardly economically maintain real-time performance in the middle-range or below detection, considering the suitability of remote detection and material deformation [1,2] raised by long-term operations. Although synthetic aperture radars (SAR) undertake all-day detection tasks and the locating accuracy is practical, classifying the ship classes is in a weak position due to the relatively single features offered [3,4]. Optic satellites can offer ample features to distinguish the classes, while ultra long-distance detection is susceptible to cloud and fog. In addition, satellites do not meet the requirement of real-time speed in an economical way [5,6] due to high maintenance and replacement costs. For civilian radar and infrared imaging technology, the limited detection capacity raised by unideal weather conditions is non-negligible [7,8]. The images offered from the civilian radars and infrared sensors have relatively single features compared with the visible-light sensors. Therefore, it is challenging to meet the real-time and appropriate requirements of ship detection with a single skill. Compared with the detection skills above, although ship detection utilising the standard camera of visible-light sensors is susceptible to rain and fog, visible-light sensors economically meet the the relatively close range detection requirement under ideal weather conditions. In addition, the images obtained by visible-light sensors are far more abundant, which is suitable to distinguish the ship classes [9]. Comprehensively, ship detection based on visible-light sensors can cooperate with remote detection technologies via compensation.

The common target detectors contain conventional and modern detectors. Conventional detectors tend to be dependent on manual interference and have relatively little generalisation in various conditions or a large redundancy in window calculation [10]. Modern detectors, mainly consisting of CNNs, are increasingly mounting, owing to the impressive performance in the computer version. Most modern target detectors are categorised into two kinds, two-stage and one-stage detectors. The main classic two-stage detectors, whose workflow is divided into classifying and localizing regression, contain the R-CNN series [11,12,13], RepPoints [14], etc. R-CNN series detectors are based on predesigned anchors, in contrast, RepPoints is the anchor-free detector. The representative one-stage detectors, whose workflow unifies the classifying processing and localizing regression into one regression, contain SSD [15], YOLO series [16,17,18,19,20], EfficientDet [21], etc. The one-stage detectors tend to better balance inference and accuracy than two-stage detectors. In particular, YOLOv5s [20] (the light model of YOLOv5) has fewer parameters and relatively fast inference, which becomes one basement of potential ship detectors. However, there is a relative lack of ship detectors that detect multiple ship classes with real-time inference by visible cameras. Common target detectors are not directly applicable to ship detection, owing to the uniqueness of marine transportation. There is a serious lack of open datasets for intelligent maritime transport compared with KITTI [22], Torontocity [23], and RobotCar [24], which are suitable for inland transportation. Although an enhanced ship detector [9] based on YOLOv3 detects multi-class ships, the detecting background lacks the consideration of aerial scenes, and the corresponding application is suitable for shore-based detection. In addition, the testing datasets of COCO [25] and VOC [26] tend to validate detectors in an ideal state, leading to the fact that the robustness of ship detectors is rarely validated in actual application scenarios.

This paper considers the common visible-light camera to establish the low–high cooperation with radars or satellites suitable for far-sightedness. To pursue the ship detector detecting various ship classes and validate algorithms quantitatively in several rain–fog soaking degrees, our contributions are summarised as follows:

- We propose a real-time ship detector which is built on fast U-Net and remapping attention proposed. The fast U-Net offered adopts pixel period insertion to prompt the inference when multiple testing batches are set and decrease the number of training parameters. Remapping attention is specifically introduced to remap global features to local calculations, which is conducive to increasing the robustness in actual rain and fog scenes;
- We offer a ship dataset containing more than 14,000 images and data augmentation for the diversity of detecting backgrounds. The ship dataset provided alleviates the dataset's lack of multiple ship classes. The data augmentation of the crossing-background randomly selects the targets copied to the new background, promoting the value of ship datasets and background diversity;
- We develop a rain–fog dataset containing 500 samples of pure rain–fog backgrounds. The pure images are collected from non-artificially synthesised scenes and the real world. This dataset is suitable for quantitatively validating ship detectors and testing the practical effects in marine environments.

2. RELATED WORKS

The ship detection of modern detectors has been introduced recently, while several conventional approaches have also been proposed for ship detection. According to the networks of conventional or modern detectors, ship detectors consist of traditional and modern ship detectors.

2.1. Traditional Ship Detectors

The traditional ship detectors, which focus on the background, mainly initialise the specified remote sensing or ship-borne video to establish the background model and utilise the difference to detect marine targets. Wang Mingfan et al. propose a Gaussian mixture model [27] to combine the contrast between local and global backgrounds, constraining the influence of phosphorescence under intense light. Xu Fang et al. establish a background model [28] based on the feature of remote sensing on the sea surface, which combines with the multiple modules to gain the target features according to the visual significance, and then unifies the thick and fine segmentation to detect the ship targets. Borghgraef et al. resort to the spatiotemporal correlation of dynamic background and introduce a surface floating-object algorithm [29] based on ViBe [30] and Behaviour Subtraction [31], which detects potentially dangerous objects by updating the background in a complex dynamic scene. Hu et al. propose a robust background-iterative algorithm [32], eliminating the influence of waves on the background to some extent, and a fast four-link module to accelerate the detection speed.

The traditional ship detectors, which focus on the foreground, build the ship detection model based on salient features or textures. According to the attention mechanism of human psychology, Itti et al. simulate the bottom-up visual selection processing of human beings and adopt Gaussian filtering and difference subtraction of feature expression [33] to obtain the detection model. Arshad et al. consider ship boundaries to localise ships by "thickening", "expanding" and "bridging" operations [34], whose operations also combine background features to establish a real-time ship detector by morphological operations and differential framing methods. Fefilatyev et al. propose a sea-level monitoring algorithm [35] that can be installed on a fast-moving ship-borne platform according to the texture of the skyline, which effectively detects ships in a parallel perspective.

However, the target detectors designed on manual features have a large window calculation redundancy and instability, especially for various surrounding changes [10]. Unavoidably, while inheriting the advantages of interpretability, traditional ship detectors also gain the corresponding disadvantages of instability. Therefore, traditional ship detectors based on conventional detectors meet the plateau effect.

2.2. Modern Ship Detectors

Representative one-stage detectors, whose workflow unifies classifying processes and localizing regression into one regression, contain SSD [15], EfficientDet [21], YOLO series [16,17,18,19,20], etc. In particular, the YOLO series make non-negligible efforts to improve the better balance between inference and performance. YOLOv1 is an initial version and has a real-time speed, while the set anchors are initially fixed, leading to an unpractical and low accuracy [16]. To some extent, YOLOv2 improves the detection effects based on the Batch Normalization (BN) and adjustable anchors [17]. YOLOv3 adopts multiple output layers and DarkNet53 to unify the classification and location regression [18], achieving impressive improvement results. Recently, YOLOv4 and YOLOv5 are proposed and maintained based on the pre-version of the YOLO series [19,20]. YOLOv4 takes the "Bag of freebies" and "Bag of specials" to ensure a further impressive balance between inference and accuracy. Although YOLOv4 has a parallel performance or even outperforms the YOLOv5, YOLOv5s(the light model of YOLOv5) has fewer parameters and faster inference, which leaves more optimisation space for the post-adjustments, which becomes one basement of potential ship detectors.

Along with the mounting development of CNNs, modern ship detectors are gradually appearing, which utilise CNNs to automatically extract the target features and calculate the classification and position of the ship. Modern ship detectors are mainly built on two-stage and one-stage common detectors [16,17,18,19,36].

Modern ship detectors, built on two-stage common detectors, tend to achieve impressive performance without consideration of computational costs. R-DFPN [36] is proposed to improve the recall of ships in complex port scenarios by adding the rotation dimension based on the feature pyramid network. However, R-DFPN, belonging to two-stage networks with a complex network structure, and speeds less than 15 fps, makes it challenging to meet the real-time requirements. An autonomous ship-oriented method [37] is introduced by a novel hybrid algorithm that combines GAN and CNN. Nonetheless, the ship classes classified and inference are limited.

Modern ship detectors, constructed on one-stage common detectors, tend to better balance inference and accuracy compared with two-stage modern ship detectors. An enhanced ship detector [9] based on YOLOv3 detects multi-class ships with real-time inference speed, while the background and target collected are single to some extent. Although meeting the basic real-time speed(more than 25 fps), the inference speed hovers at approximately 30 fps, leaving less room for subsequent enhancements. Compared with the application of YOLOv3 in ship detection, YOLOv4 and YOLOv5 are platforms of potential ship detectors, owing to improved performance and relatively fast inference. Although YOLOv4 has a parallel performance or even outperforms the YOLOv5, YOLOv5s (the light model of YOLOv5) has fewer parameters and faster inference [20], which leaves more optimisation space for the post-adjustments. However, YOLOv5s lacks application and improvements corresponding to the marine surroundings. In addition, the lack of a ship dataset containing various scenes and a rain–fog dataset suitable for validating ship detectors is a non-negligible barrier.

2.3. Ship Datasets

In terms of the datasets used for ship detection, there is a serious lack of open datasets for intelligent maritime transport, compared with KITTI [22], Torontocity [23], and RobotCar [24] properly used for inland transportation. Public datasets, such as COCO [25] and VOC [26], have nearly 9000 ship samples, but the types of targets and ratios are single, and the scenes are simple. In addition, the corresponding public datasets tend to validate detectors in an ideal state, leading to the fact that the practical effects of ship detectors are rarely validated in actual application scenarios. The ship dataset [9] proposed by R.W Liu has relatively considerable targets compared with public datasets, while the most frequently detected backgrounds are the horizontal field of vision, lacking various angles, multi-dense states, multiple classes or target scenes, etc. In the absence of sufficient datasets related, it is difficult to give full play to the advantages of CNNs or the algorithms involved. Therefore, the establishment of ship and rain–fog datasets is essential in the field of ship detection.

3. METHODOLOGIES

Although YOLOv4-tiny (the tiny version of YOLOv4) decreases the training parameters, the average precision of YOLOV4-tiny is just 21.7% [20], far less than the YOLOv5s. YOLOv4 and YOLOv5 belong to the YOLO series and have fewer differences compared with the gap between YOLOv2 and YOLOv3. Furthermore, YOLOv5s better balance the performance and inference. Therefore, we decide to choose YOLOv5s as the optimization platform and construct our ship detector. This chapter proposes fast U-Net and remapping attention to improve the accuracy and robustness of the real-time ship detector under marine meteorological weather. The ship dataset is developed to increase the diversity of ship targets, and data augmentation of cross-background is mainly proposed for producing new detection backgrounds. The rain–fog dataset is introduced to validate the practical effects of detectors comprehensively.

3.1. Remapping Attention and Fast U-Net

Attention modules have been utilised to recommend CNN models "what" or "where" solutions, which are non-negligible methods to further prompt the accuracy of detectors. The popular attention modules, utilizing both space and channel features, contain CBAM [38] and Coordattention [39]. In particular Coordattention has been validated and approved recently by comparing various attention methods, such as SE attention [40], CBAM [38], and "X-Y" Attention [39]. However, the generalization ability is limited to some extent when validating situations are changing. Considering, especially, the improving of the robustness of YOLOv5s and retaining the balance of inference and accuracy in various marine weather conditions, remapping attention is proposed as shown in Figure 1. Remapping attention of space exploits the positional information by "Maxpool" and "AVGpool", and then remaps the global space information to the local features by "softplus" activation. Remapping attention of the channel exploits the similar information remapped. Compared with the CBAM [38] and Coordattention [39], which both utilise the space and interchannel information, remapping attentions maintain local and global features of YOLOv5s (the light-version model of YOLOv5 [20]) by mapping the attention parameters to the probability adjustment. The probability adjustment based on local and global features is a reasonable assumption to boost the robustness in adverse weather conditions. Considering the cost of computation and GPU memory, the insertion position of remapping attention models is shown with CBL+ in Figure 2. The following experiments demonstrate that remapping attention obtains a relative peak compared to the CBAM [38] and Coordattention [39] and promotes robustness in fog–rain soaking conditions. The detailed workflow of remapping attention can be calculated as:

$$\begin{cases} C_{atten} = F\left(\frac{K(APavg2d_{(1)}((f))+APmax2d_{(1)}((f)))}{Sum_{dim=1}(K((f)))}\right) \\ S_{atten} = F\left(\frac{K(MAX_{dim=1}((f)) \cup AVG_{dim=1}((f)))}{APavg2d(1)((f))}\right) \\ Feature_{pro} = S_{atten} \times C_{atten} \times (f) \end{cases} \quad (1)$$

where Catten and Satten are the remapping attention of space and interchannel, respectively. F is the activation function of "softplus". K denotes the processing of kernels. APavg2d(1)) and APmax2d(1) are the operations of "nn.AdaptiveAvgPool2d" and "nn. AdaptiveMaxPool2d" in the channel dimension. f and Featurepro are the original features and enhanced features after the processing of Satten and Catten. MAXdim=1and AVG-dim=1 are the "torch.mean" and "torch.max" functions in the dimension of the channel. "nn" and "torch", imported from PyTorch, provide module tools for creating and training neural networks. During the utilisation of "nn." and "torch.", "Module" is an abstract concept representing either a layer in a neural network or a neural network with multiple layers.

The head of YOLOv5s is the "U" tube network, constructed to upsample the features by the two upsampling layers of "nn.Upsampling" operation, which is conducive to merging the semantic and detailed features. However, the "nn.Upsampling" operation is inefficient compared with indexing operations of tensors. The inference comparison of popular upsampling methods is shown in Figure 3. We utilise a laptop with RTX 2060 as the inference testing platform and collect the logs of inference speeds based on various testing batch sizes. The deconvolution spends the most time in the upsampling processing, which is the primary reason to abandon "Deconv" in the following experiments. "Pixelshuffle" is an efficient upsampling method, while the requirement of four times between channels

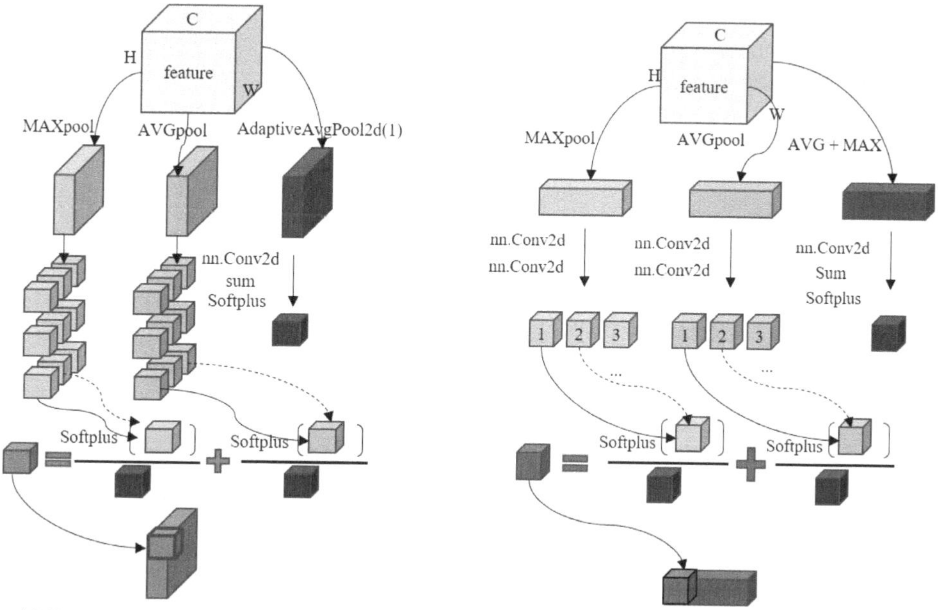

(**a**) Remapping attention of the space. (**b**) Remapping attention of the channel.

Figure 1. *Remapping attention. The remapping attention of space and channel is a unified operation of probability adjustment, which exploits the local features by "Maxpool" and "AVGpool", and remaps the global information to the local features by "softplus" activation. The probability adjustment based on local and global features is a reasonable assumption to boost the robustness in adverse weather.*

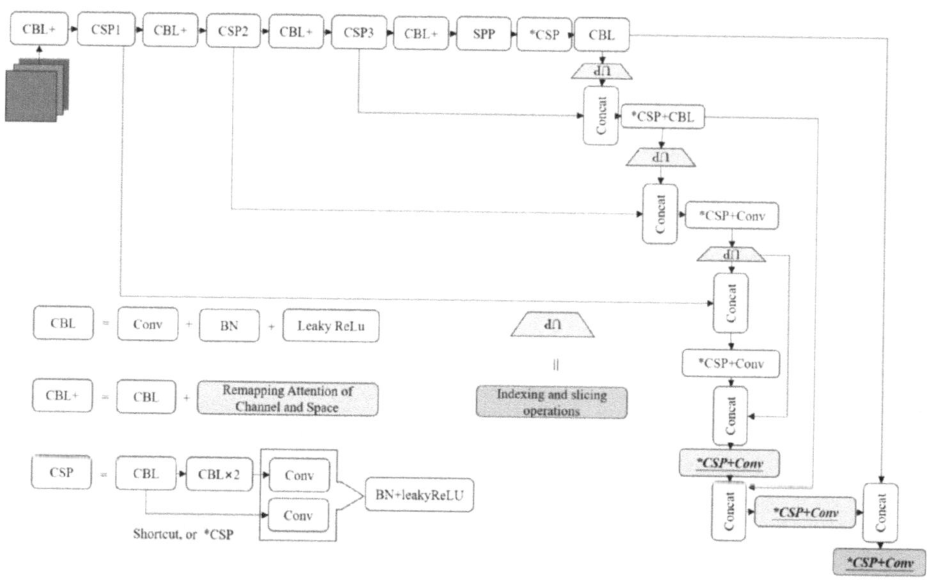

Figure 2. *The structure of FRSD. CBL, CBL+, and CSPi are the CNN modules, and Conv is the convolution operation. The head of FRSD is the fast U-Net, mainly constructed by reversing "Up", denoting the novel upsampling by indexing and slicing operations. "×" represents the number of consecutive accumulations of CBL. *CSP means that the light version of CSP, eliminating the shoutout.*

of adjacent layers is incompatible with the heads of the network that is built. "nn.Upsampling" is a common upsampling method with the attribute of the broadest utilisation, which is the major reason to add "nn.Upsampling" to the following experiments. The fast U-Net is introduced by the indexing operations to further prompt the inference and robustness because of the appropriate decrease of parameters, whose structure is shown in the head of Figure 2. The reversing "Up" represents a fast U-Net which is most efficient compared with the others in Figure 2. The fast U-Net offered is shown in Figure 4. The tensors are indexed and sliced to double the resolutions by period distribution, while the channels are decreased which is conducive to indirectly compressing features and saving computing costs. Compared with the "nn.Pixelshuffle", the RandomCopy of Figure 5 ensures avoidance of not being divisible [41] and feature loss.

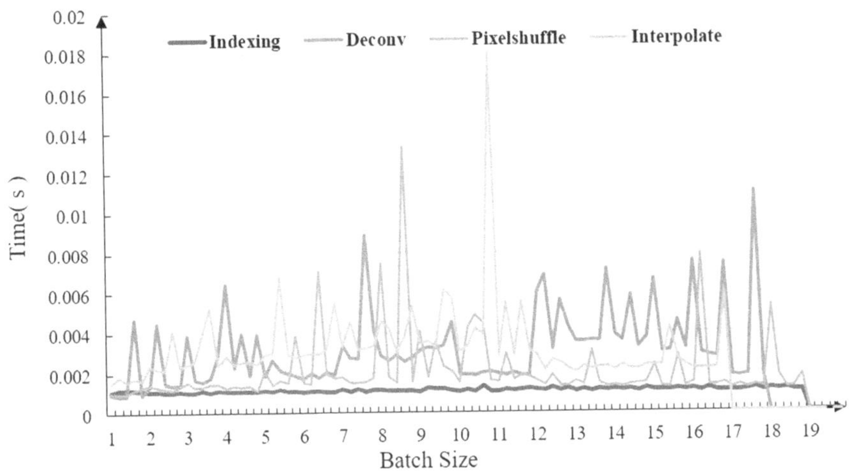

Figure 3. *The inference comparison of the upsampling method in the same experimental platform (Laptop NVIDIA GeForce RTX 2060). "Batch Size" represents the testing batch size set, and "Time" denotes the total processing time of the upsampling methods. "Indexing" represents the upsampling utilised in "Up" of* **Figure 2** *and* **Figure 4**, *"Deconv" denotes the deconvolution with a 3×33×3 kernel, and "Pixelshuffle" means "nn.PixelSshuffle" operation of PyTorch, "Interpolate" is the common upsampling method using "nn.interpolate" operation of PyTorch. The curve goes directly to "0", indicating that the maximum memory of the GPU is exceeded.*

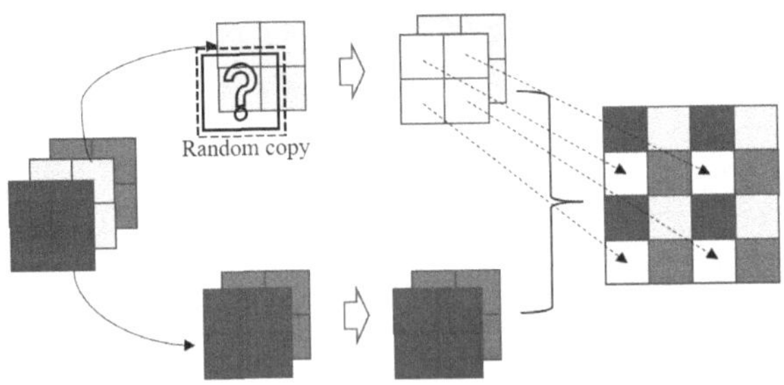

Figure 4. *The upsampling method of FRSD. "Random Copy" denotes a random copy from the original features, ensuring a fourfold relation between features which is more compatible than "nn. PixelShuffle".*

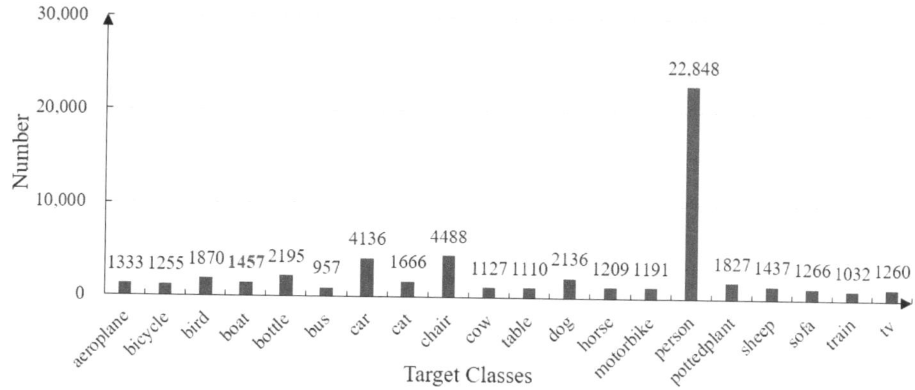

Figure 5. *The number and classes of targets in VOC [**26**]. All ship targets are labelled as "boat", and the number of "boat" is impressively small.*

3.2. The Ship Dataset and Data Augmentation

VOC [26] and COCO [25] are public datasets for detection or segmentation. VOC contains about 20 classes of daily life, however all ships are labelled as "boat". COCO is relatively more challenging. However, by converting and visualising COCO, we found that labels have more than one human error, a box which carelessly contains multiple targets and is resistant against training and testing to some extent when the number of samples is insufficient. The number of categories and labels of VOC used only for training are shown in Figure 5. The targets of the top three largest numbers are "person", "chair", and "car", respectively. There are more than 20,000 samples of "person", while the amount of "boat" is just 1457. The size distribution of the "boat" in VOC is shown in Figure 6. The absolute values of the horizontal and vertical axes are the width and height of the images and targets. Red and blue dots represent an image and a target, respectively. The distribution in the upper right corner shows the size distribution of images, those in the lower-left corner show the size distribution of the "boat", and the colour depth indicates the degree of concentration. The length or width of the images is mainly constrained to 500, which is evenly distributed in a line, while the corresponding "boat" is excessively concentrated in the area of "200 × 200". Although VOC meets the requirements of ship detection to some extent, the distribution is single with only 1457 targets. The target distribution is too concentrated in one location, lacking diversity distribution. If 1457 targets and labels are used as the training dataset of ship detectors, the training process is prone to overfitting. Therefore, it is necessary to establish a dataset containing substantial ship targets and ample classes, which brings into full play the advantages of CNN.

Due to the requirement to visualise producing the ship dataset, this paper further develops the datasets [42] based on the format of VOC [26]. The ship dataset proposed contains 21,000 samples, subdivided into eight classes and one unidentified class. Eight ship classes are eight specific categories which can be distinguished when we label the targets, while unidentified ships are categorised into a class in which the specific class is indistinguishable and we label unidentified ships with "ship". Our ship dataset has two components, "VOC-ship" and "VOC-ship-test", utilised for the training and testing, respectively. Both VOC-ship and VOC-ship-test consider the weather factors, such as detection angles, light, and dense occlusion. The specific classes and numbers of our ship dataset are shown in Fig-

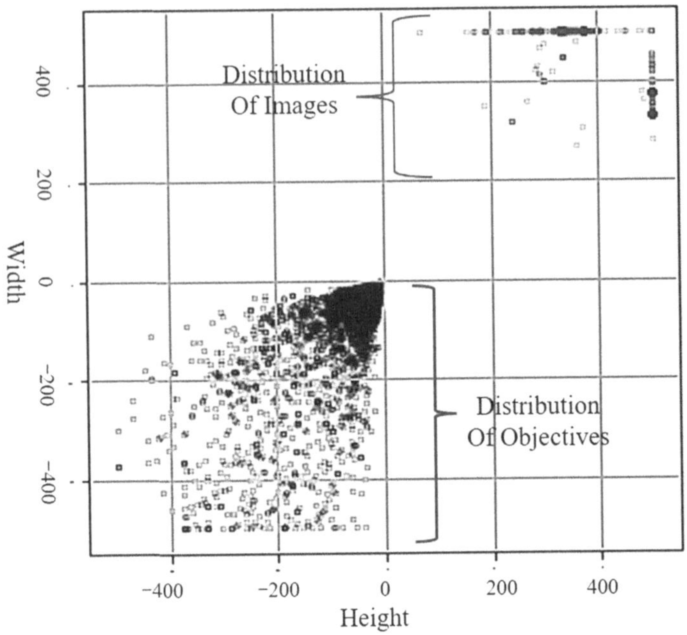

Figure 6. *The boat distribution of VOC [26]. Each light-red and light-blue dot denotes one image and one objective, respectively. VOC is a public dataset for the detection and segmentation of common targets, while all ships are labelled as "boat" and the number of boats hardly meets the requirement of ship detection.*

ure 7. The ship classes contain "engineering_ship, freighter, passenger_ship, public_service_vessel, sailboat, speedboat, submarine, warship," and the numbers of training and testing targets are "1179, 2604, 802, 2311, 2749, 3145, 2625, 1163" and "515, 638, 241, 499, 650, 782, 685, 320", respectively. The steps to make the ship datasets are as follows: (1) we use cameras to capture multiple scenes while utilising Google Chrome and a plug-in to crawl ship pictures. (2) We delete pictures with low quality through manual screening, classify ships filtered, and name pictures in batches with Arabic numerals. (3) We utilise the "labelImg" downloaded from GitHub to make labels, and label files are XML format which is conducive for visualisation. The code of "labelImg" is https://github.com/tzutalin/labelImg (accessed on 26 July 2022).

The target distribution of our ship dataset is shown in Figure 8. The denotation of light-red and light-blue dots is the same as the presentations in Figure 6. The images of our dataset are concentrated in the area near "500×500500×500". The corresponding targets are focused on the room with a length or width of "0–500". In addition, there are also enough middle or large targets in our ship dataset, shown in the area of "500–800", which meets the requirements of medium-range detection utilising common cameras. Compared with Figure 6, the targets of our ship dataset are more evenly distributed in the range of "500×500500×500" while increasing the number and aspect ratios of targets. Comprehensively, the ship dataset proposed has the ratio diversity of targets and potentially exploits

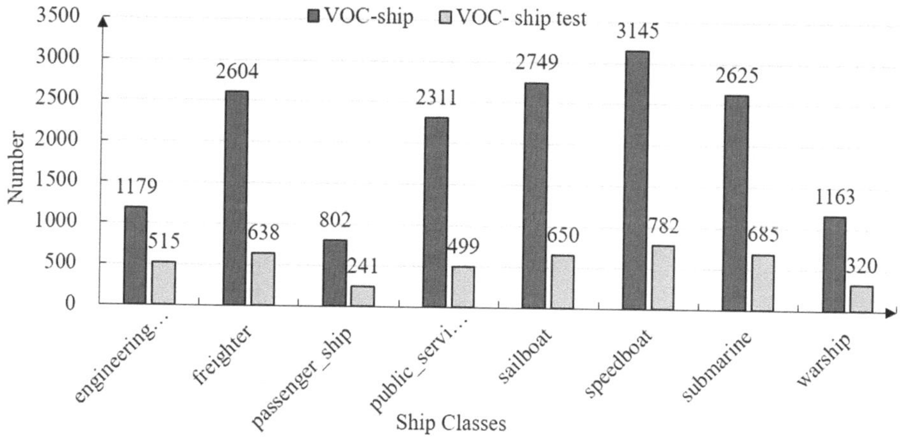

Figure 7. *The numberand classes of targets in our ship dataset. The ship datasets have eight specific classes and contain two components, VOC-ship and VOC-ship-test, which are used to train and test models, respectively.*

the advantages of visible-light cameras to form a match with the remote detection technology. Part samples of our ship datasets are shown in Figure 9, which also demonstrates the factors considered to build the ship dataset, containing multiple detection angles. Different detection angles show that the detectors, trained by VOC-ship, can be used in static shore-based detection scenes, as well as shipborne or airborne detection scenes.

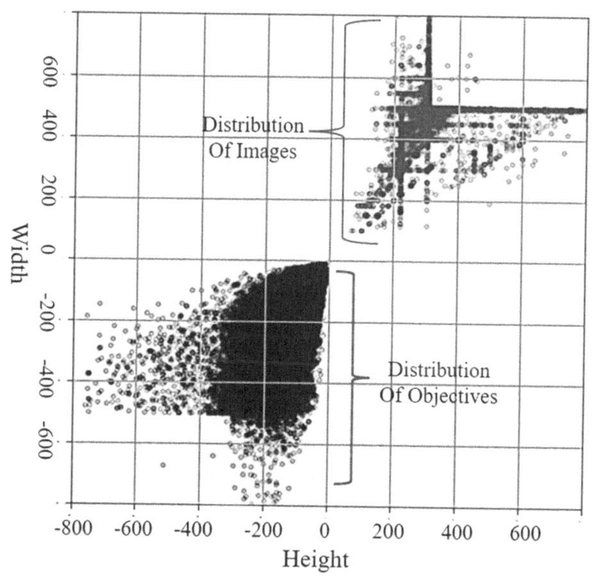

Figure 8. *The distribution of the ship dataset offered. The denotation of dots is the same as in* **Figure 6**.

Figure 9. *The samples labelled from our ship dataset show various ship classes. Various detecting angles are considered. The bounding boxes and the labels corresponding denote the object of interest and the ship classes. To enhance the robustness of the algorithm, some targets are not marked.*

The data augmentations of YOLOv5 are practical to enhance the value of samples, while the stability of ship navigation is relatively complicated, especially in this case that the shipboard camera detects other ships under the influence of rains and waves, leading to a complexing background. To increase the diversity of detecting surroundings, data augmentation of the cross-background is introduced to increase the diversity of detecting backgrounds, as shown in Figure 10. A larger merged image contains 4 middle merged images. The copied targets have a location in the new background, which releases the relative lack of background diversity in ship datasets and enhances the utilisation of the remaining grey area.

3.3. The Rain-Fog Dataset

VOC [26] and COCO [25] are public datasets for the detection and segmentation of common targets. However, the corresponding testing datasets tend to validate detectors in an ideal state, leading to the fact that the robustness of ship detectors is hardly validated in an adverse environment. Therefore, collecting a rain–fog dataset suitable for the marine environment is necessary to validate the practicability of ship detectors. Considering adverse impacts from rainy or foggy weather, the SPA [43], Rain100H [44], and Rain1400 [45] have been offered to minimize the influence of bad weather. However, the rain datasets above are mainly utilized for the rain removal of inland

Figure 10. *Data augmentation of the cross-background. This is a merged image containing 4 middle merged images, which are also merged by 4 images. The two middle images above have four different backgrounds merged. The two middle images below have several copied targets labelled by red arrows, enhancing the usage of the remaining grey area and the diversity of surroundings.*

environments. In addition, the public rain datasets have a few pure images only having little rain or drops, which hardly meet the requirements of soaking ship images. Most backgrounds of rain datasets are cut or synthetically made from entire images, and the backgrounds have buildings and streets, hardly representing the marine surroundings, such as the rain–fog crossing scenes and raindrops created by rain or waves. Our rain–fog dataset contains 550 images with pure weather backgrounds, which consist of pure rainy or foggy scenes, rain–fog crossing scenes, and raindrops wetting lenses. The steps for collecting pure rain–fog images are similar to the ship dataset, except for the labelling processing. The images of the rain–fog dataset are entire images collected from the real world without artificial data interference. To enhance the value of the rain–fog dataset, we can randomly choose rain–fog images to paste or soak the testing images in staggered repetition. Part of the images soaked by different degrees are shown in Figure 11, and classic scenes of the rain–fog dataset are shown in Figure 12. The soaking operation is based on the Python Imaging Library (PIL), and the different soaking degrees are set from 0.4 to 0.8. The adverse marine scenes are more complicated than the major scenes of public rain–fog datasets, which are suitable for the rain removal task of the inland environment. Our rain–fog dataset is built by considering various adverse weather conditions and scenes, which appropriately simulate the marine environment. Compared to rain datasets containing buildings and streets, our rain–fog dataset is more suitable for quantitatively testing the practical effects of ship detectors based on deep learning.

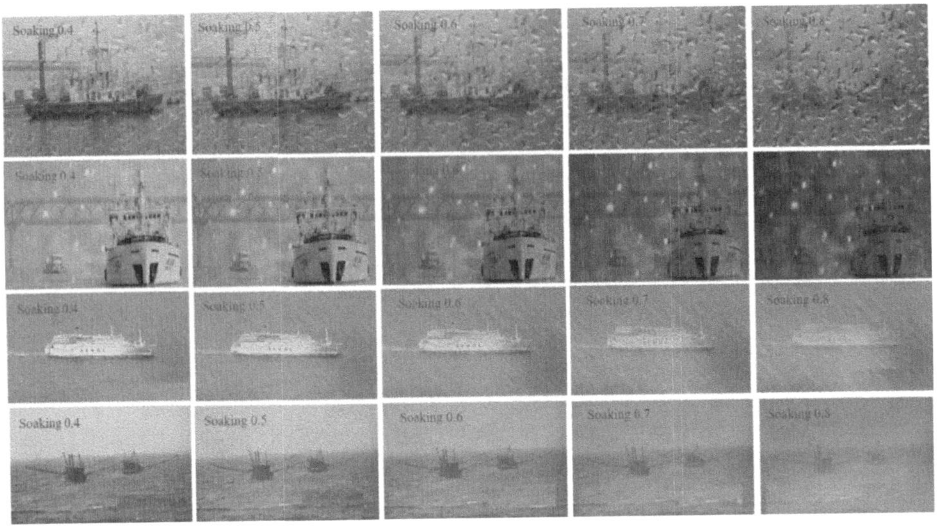

Figure 11. *Part soaking scenes. Classical rain–fog scenes contain lenses soaked by raindrops, the pure fog or rain, and rain–fog intertwined scenes. Images containing the ships above are stemmed from our ship dataset and soaked by the rain–fog dataset, and the values of each row above are the soaking degree from 0.4 to 0.8. More severe soaking cases are excluded in the following experiments, considering visible light sensors' maximum and practical applicability.*

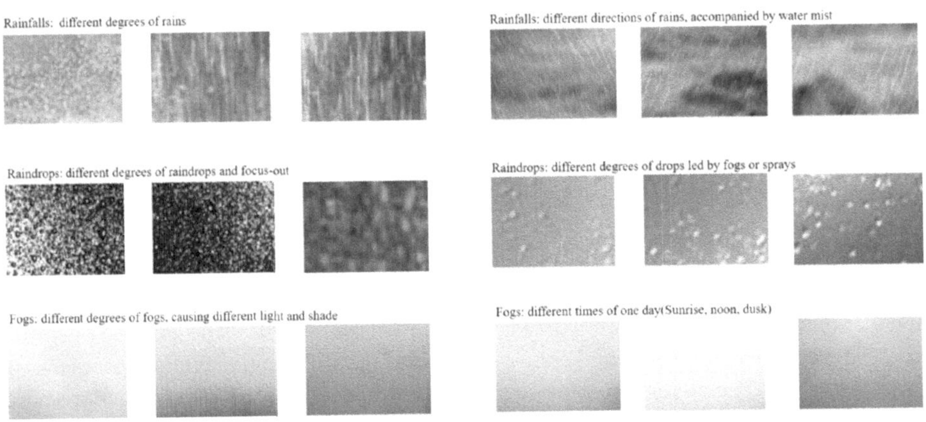

Figure 12. *Part displays of the rain–fog dataset proposed. The images of the rain–fog dataset are classic rain–fog scenes without other targets, which is more suitable to validate the robustness of ship detectors, compared with Rain100 [**44**], Rain1400 [**45**], SPA [**43**], and Raindrops [**46**].*

4. EXPERIMENTS AND RESULTS

4.1. Experimental Setting

4.1.1. Experimental Conditions

The experimental software contains Ubuntu 16.04.4, PyTorch 1.7, and CUDA10.2. TITAN V and RTX 2080ti are GPUs utilised. If not specifically mentioned, TITAN V is the default device.

4.1.2. Datasets

The datasets, utilised in our manuscript, contain COCO [25], VOC-ship, VOC-ship-test, and rain–fog datasets. COCO [25] is a public dataset containing training and testing components, which can be used to, respectively, train and test the common detectors. Correspondingly, the ship dataset proposed consists of VOC-ship and VOC-ship-test. VOC-ship is utilised for the training ship detectors that participated. VOC-ship and VOC-ship-test are the ship datasets we proposed. VOC-ship is used to train ship detectors and VOC-ship-test is utilised to test ship detectors. The training samples of our ship dataset are shown in Figure A1. The rain–fog dataset proposed is a pure rain–fog image collection which can validate the practical generalisation ability of ship detectors.

4.1.3. Parameter Setting

With regard to the FRSD training in COCO [25], the detailed settings are as follows: Epochs: 300; Batch Size: 32; Image Size: 640×640640×640; Initial Learning Rate: 1×1×10−310−3; Optimizer: Adam; Adam Beta1: (0.30.30.60.60.980.98); and Data Augmentation: Mosaic augmentation. With regard to the FRSD training in VOC-ship, the detailed settings are following: Epochs: 300; Batch Size: 64; Image Size: 640×640640×640; Initial Learning Rate: e^{-2}; Optimizer: Adam; Adam Beta1: (0.30.3 0.60.6 0.980.98); and Data Augmentation: Data Augmentation of Cross-background (at the time of mention). Detectors who participated take their own published parameters. Concerning the testing in the VOC-ship-test, the FPS is contained with a batch size of 1, and inference augmentation is unset. Rain–fog dataset randomly soaks the testing images to the degree of "0, 0.40.4, 0.50.5, 0.60.6, 0.70.7, 0.80.8". The loss function used in the training step of FRSD relevant is as follows:

$$\begin{cases} L_{class} = -\sum_{n=i}^{N}(x_i^* log(\frac{1}{1+e^{-x_i}}) + (1-x_i^*)log(\frac{e^{-x_i}}{1+e^{-x_i}})) \\ L_{location} = 1 - (IOU - (\frac{D_{angles}}{D_{centers}})^2 - \frac{\alpha^2}{1-IOU+\alpha}) \\ L_{confi} = (1 - if_{gt}) + if_{gt} * IOU_{score} \\ Total_{loss} = L_{class} + L_{location} + L_{confi} \end{cases} \tag{2}$$

where x∗i and xi are the probability of the category predicted and Ground Truth, respectively. Dangles/Dcenters is the contrast value of Distance of opposite angles and centers. α is a parameter that measures the consistency of aspect ratio. IOU is the union ratio of the prediction box to the GT box (Ground Truth). ifgt denotes that there is or is not a target in the noticing area. The square root of α is set as $2/\pi \times (arctan(W_{gt}/H_{gt}) - arctan(W_{pred}/H_{pred}))$

4.1.4. Data Augmentation

The data augmentation of the cross-background is controlled by Thresh, Prob, Copy_times, and Epochs, which are set to 50×5050×50, 0.50.5, 3, and 30. Thresh is the threshold of the object copied whose H×W<thresh, Prob is the probability of copy occurring, Copy_timesis the times to be copied, and Epochsrepresents that the data augmentation proposed is efficient until the Epochsset.

4.1.5. Metrics

The metrics utilised to validate the performance of detectors contain average precision (AP) and other metrics based on AP. Frames per second (FPS) is the popular measurement to represent the running or processing time. The higher the FPS value, the better the real-time performance of the algorithm. The detailed concepts are the following:

$$AP = \int_0^1 p_{smooth}(r)dr \Leftrightarrow \sum_{r=0}^{1}(r_{n+1} - r_n) \max_{\tilde{r}:\tilde{r} \geq r_{n+1}} p(\tilde{r}) \tag{3}$$

where psmooth(r) and p(r˜) are the max value of precision recall(PR) after smoothing the curve of PR. r˜:r˜≥rn+1 represent the split points of precision recall.

AP0.5:0.95: The threshold of intersection over union (IOU) is adjusted from a fixed 0.5 to calculate the multi-AP values at intervals of 0.5 to 0.95, and the average of all results is taken as AP0.5:0.95.

AP50 and AP75: The AP is gained when the IOU is set to 0.5 and 0.75, respectively.

APS, APM, and APL: The AP is gained when pixel areas of targets set are 0:3220:322, 322:962322:962, and 962:end962:, respectively.

FPS: Frames per second(FPS) denotes the number of images processed per second and directly the real-time effect of algorithms.

4.2. Results

4.2.1. Influence of the Remapping and Fast U-Net

Comparing YOLOv5s_ReMap_FastUnet (another denotation of FRSD) with other public detectors in COCO [25] is the first step to constructing robust ship detectors, which is necessary to validate the remapping attention and fast U-Net in the public dataset. The comparison of popular detectors in the public dataset is shown in Table 1. YOLOv5s_Unet is the detector utilizing three "nn.Upsampling" layers in the head network of YOLOv5s. YOLOv5s_CBAM_Unet, YOLOv5s_Coord_Unet, and YOLOv5s_ReMap_Unet are the detectors using the CBAM [38], Coord [39], and remapping methods, respectively. CBAM and Coord are inserted into the backbone, whose position is the same as the remapping attention in the FRSD. Compared with the results of YOLOv5s_Unet and detectors promoted by three attention insertions, attention methods reliably prompt the accuracy. The AP50 of YOLOv5s_ReMap_Unet has increased by 3.5%, 3%, and 0.2%, respectively, compared with YOLOv5s_Unet,

YOLOv5s_Coord_Unet, and YOLOv5s_CBAM_Unet, demonstrating the rationality of the remapping method. Especially, the most gap of the AP50 between YOLOv5s_ReMap_Unet and YOLOv5s_CBAM_Unet reaches approximately 1%, as shown in Figure 13, indicating that the remapping attention proposed has the potential advantage of promoting the robustness in rainy and foggy weather, because of the relatively high performance in the case of undertraining conditions. YOLOv5s_ReMap_FastUnet maintains almost the same speed when the batch size is set to 1 compared with YOLOv5s_ReMap_Unet. Owing to the compression of the channel by the fast U-net, as shown in Figure 4, the inference speed increases by 15 fps to 278.4 fps when multiple batches are taken in the testing stage, which is suitable for the multi-task model and suggests the rationality of the fast U-Net. YOLOv5s_ReMap_FastUnet has a slightly lower speed than YOLOv5s_ReMap_Unet when the testing batch is set to 1, which the seemingly weird result is led by the preprepared production of double-size features. The advantages of fast U-Net are more visible when the testing batch size is set to multiple batches. Although the remapping attention and fast U-Net have an unimpressive influence on YOLOv5s_ReMap_FastUnet when adopting the testing images of COCO, the practical effect is visible in the rain–fog soaking situation, as shown in Table 2. The immediate influence of the fast U-Net on features is shown in Figure 14. Features of fast U-Net have more visible sparsity compared with the upsampling method of "nn.Upsampling", which suggests why fast U-Net is evident when the channels are indirectly compressed.

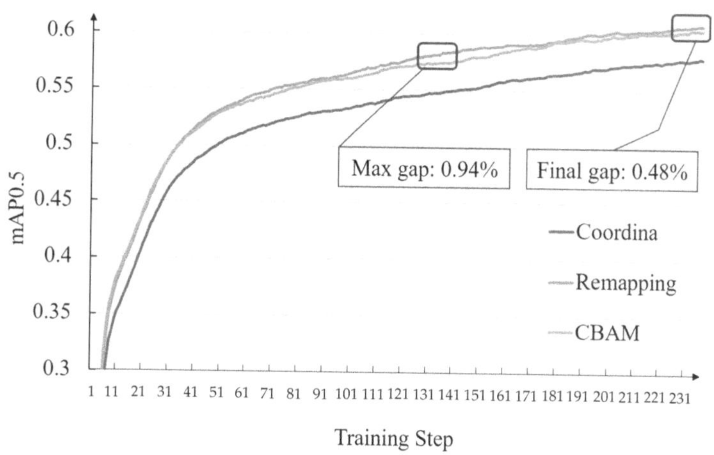

Figure 13. *The influence of Coord, CBAM, and Remapping attention to YOLOv5s. The validating dataset used is "COCO val2017"* [**25**]. *The "Max gap" reaches approximately 1%, suggesting the method of remapping attention has the potential advantage of prompting the performance of the ship detector proposed, because actually collected ship datasets cannot completely represent the actual detection environment, the corresponding model unfits the actual detection scenes.*

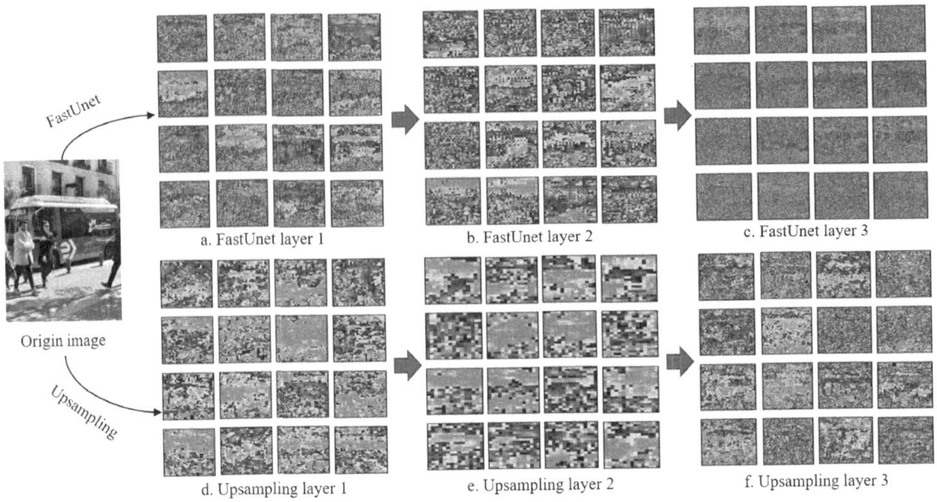

Figure 14. *The sparsity visualization of fast U-Net versus "nn.Upsampling"(a common operation of PyTorch). "FastUnet" and "Upsampling" denote the upsampling processing of fast U-Net and "nn.Upsampling". The subfigures represent the visualization of each upsampling layer, and three layers of images per row indicate the features produced by three upsampling layers, respectively. The features of "Fast U-Net" have smaller areas of the same values, suggesting the practical effects from another angle.*

Table 1. Comparison of detectors in COCO (devtest2019) [**25**]. To distinguish algorithms, FRSD is denoted by "YOLOv5s_ReMap_FastUnet". The batch sizes of all detectors are set to 1 in the testing stage. To display the inference boosting of fast U-Net, the batch sizes of YOLOv5s_ReMap_Unet and YOLOv5s_ReMap_FastUnet are set to 1 and 10, respectively, and the values before and after "/" correspondingly represent the inference speed.

Detectors	**Backbone**	**ImgSize**	**FPS**	$AP_{0.5:0.95}$	AP_{50}	AP_{75}	AP_S	AP_M	AP_L
SSD300 [15]	VGG16	300	43.2	0.251	0.431	0.258	0.066	0.259	0.414
Faster RCNN [12]	ResNet-50	1000	9.4	0.398	0.592	0.435	0.218	0.426	0.507
RetinaNet [47]	ResNet-101	800	5.1	0.378	0.575	0.408	0.202	0.411	0.492
CenterNet [48]	Hourglass-104	512	4.2	0.449	0.624	0.481	0.256	0.474	0.574
EfficientDet [21]	EfficientDet-B0	512	62.3	0.338	0.525	0.358	0.12	0.383	0.512
YOLOv4s [19]	CSPDarknet-53	416	38.0	0.412	0.628	0.443	0.204	0.444	0.56
YOLOv5s [20]	CSPDarknet-53	640	**69.9**	0.369	0.561	0.4	0.196	0.413	0.457
YOLOv5s_Unet	CSPDarknet-53	640	59.0	0.393	0.582	0.431	0.23	0.432	0.466
YOLOv5s_Coord_Unet	CSPDarknet-53	640	48.7	0.396	0.587	0.433	0.232	0.436	0.468
YOLOv5s_CBAM_Unet	CSPDarknet-53	640	45.87	0.424	0.615	0.465	0.254	0.462	0.512
YOLOv5s_ReMap_Unet	CSPDarknet-53	640	43.5/263.2	0.425	0.617	0.465	0.251	0.463	0.513
YOLOv5s_ReMap_FastUnet	CSPDarknet-53	**640**	**40.3/278.4**	**0.425**	**0.617**	**0.465**	**0.251**	**0.463**	**0.513**

4.2.2. The Comparison of Ship Detectors

The robustness of ship detectors in the VOC-ship-test is shown in Table 2, in which all ship detectors are validated on the same software and hardware, and the GPU used is 2080ti. Soaking 0.4 to 0.8 represent the mean average precision of ship detectors under

Table 2. Comparison of ship detectors in the VOC-ship-test. FRSD_ship_dataaug is the final model which utilizes the data augmentation of the cross-background. Soaking 0.0–0.8 denote the mean average precision under the different levels of influence corresponding to the rain–fog dataset. FRSD_CBAM_FU and FRSD_FU are the ship detectors utilizing CBAM and remapping methods proposed in the backbone, respectively. "U" and "FU" denote the upsampling methods of "nn.Upsampling/nn.interpolate" and fast U-Net in heads, respectively. The positive impacts of fast U-Net and remapping attentions proposed are more visible under the most severe rain–fog wetting conditions (Soaking 0.8).

Detectors	Imgsize	Soaking 0.0	Soaking 0.4	Soaking 0.5	Soaking 0.6	Soaking 0.7	Soaking 0.8	MEAN	DEV	FPS
Faster RCNN [12]	800	70.59%	66.75%	62.65%	55.13%	46.16%	26.27%	54.592%	16.38%	25.10
RetinaNet [47]	600	76.90%	73.92%	70.98%	66.92%	58.98%	41.49%	64.865%	13.03%	34.61
SSD300 [15]	300	78.30%	74.40%	70.90%	64.50%	54.40%	36.70%	63.20%	15.45%	66.02
CenterNet [48]	512	78.70%	75.20%	71.90%	66.90%	58.300%	39.80%	65.13%	14.30%	42.30
EfficientDetd0 [21]	512	74.00%	70.10%	66.80%	61.50%	52.80%	35.50%	60.12%	14.13%	36.40
YOLOV4_ship [19]	416	74.26%	69.19%	65.81%	61.17%	52.22%	36.37%	59.87%	13.72%	34.20
YOLOV5s_ship [20]	640	78.79%	74.90%	71.60%	66.90%	57.80%	40.00%	65.16%	14.41%	**68.01**
FRSD_U	640	80.10%	76.70%	73.90%	69.10%	60.90%	41.80%	67.08%	14.07%	43.61
FRSD_CBAM_FU	640	80.19%	76.90%	74.10%	69.78%	61.60%	43.30%	67.65%	13.55%	43.31
FRSD_FU	640	80.39%	76.90%	74.20%	69.80%	62.22%	44.50%	68.01%	13.12%	42.56
FRSD_ship_dataaug	640	**83.15%**	**80.70%**	**78.40%**	**74.50%**	**66.60%**	**48.40%**	**71.958%**	**12.91%**	42.56

different rain–fog soaking, in which the higher the number at the end, the more intrusive soaking degree. Although a better performance without considering the inference speed is achieved compared with YOLOv5s, YOLOv4 has a unimpressive practical detection results in multiple rain–fog scenes. YOLOV5s_ship and YOLOv5s_ReMap_ship are the ship detectors directly based on YOLOV5s and YOLOv5s_ReMap_Unet, and YOLOv5s_ReMap_Unet is the detector adopting only the remapping method in the YOLOV5s. FRSD_CBAM_FU and FRSD_FU are the ship detectors utilizing CBAM and remapping methods in the backbone, respectively, and "FU" denotes the fast U-Net in the head network. FRSD_FU maintains a real-time inference speed, reaching 42.56 fps, which meets the real-time ship detection. After the prompts of data augmentation proposed, FRSD_ship_dataaug have an impressive performance. The mean average precision of FRSD_ship_dataaug reaches the best under different soaking levels, reaching 83.30%, 80.50%, 78.00%, 74.40%, 67.40%, and 51.10%, respectively. Compared with FRSD_U, FRSD_FU achieves a lower deviation and better accuracy, suggesting that the fast U-Net also prompts robustness in the fog–rain conditions except for the inference speed accelerating. Meanwhile, FRSD_FU obtains better accuracy than FRSD_CBAM_FU, suggesting that remapping attention has the advantage of promoting robustness in rain–fog weather, and the probability adjustment based on the remapping attention proposed is a reasonable assumption to boost the robustness in adverse weather. FRSD_ship_dataaug has the best accuracy and lowest deviation, demonstrating that the data augmentation of the cross-background is effective. A more impressive comparison of ship detectors' robustness is shown in Figure 15. The average precision of specific ship classes is shown in Table 3. The "ship" class represents the ship target is detected while the specific class is indistinguishable. The detection results of FRSD are shown in Figure 16, Figure 17 and Figure 18.

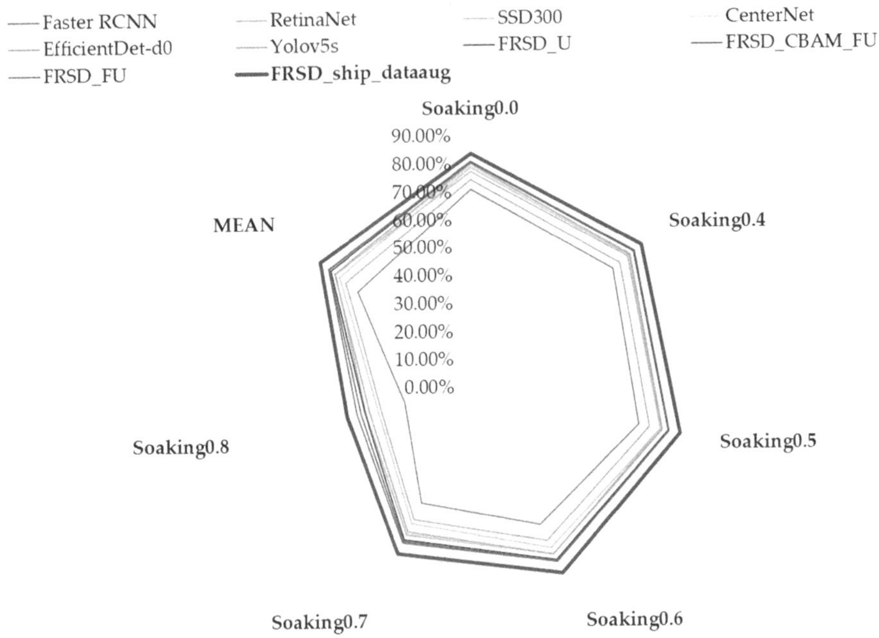

Figure 15. *The heptagon of performance comparison. Soaking 0.40.4 to 0.80.8 represent the mean average precision of ship detectors under different levels of rain–fog soaking. MEAN denotes the mean accuracy of each ship detector in all soaking situations.*

Figure 16. *The detection results of FRSD corresponding to* **Figure 9**. *The images above are the same as those in* **Figure 9**, *and several targets unlabelled are still detected accurately, showing that our ship detector is practical.*

Figure 17. *The performance of FRSD at different soaking degrees. Each column of pictures shows the detection results under different rain and fog scenes.*

5. DISCUSSION

Ship detectors are mainly dependent on radars or remote sensing satellites, partially, but significantly, because of the relatively complex hydrology and the advantages of remote monitoring. Advanced radars detect not just ships with intense precision and even real-time speed, while satellites obtain extraordinary resolution images in the sky, containing ample information. Nevertheless, it is not economical and is wasteful to utilize remote monitoring systems in ship detection over short or medium ranges. Advanced radars are impressively money-consuming even after adopting products to reduce costs, without consideration of follow-up tedious maintenance, satellites are scarce in terms of expensive launching costs and limited service life. Compared with the detection technologies above, although ship detection utilizing visible-light cameras is susceptible to rain and fog, the observation distance reaches more than 10 km under ideal weather conditions. More importantly, the information obtained by visible-light sensors is more abundant.

Therefore, this paper considers that ship detection based on standard cameras should form a low and high collocation with radars or other technologies suitable for far-sightedness. We propose a ship detector based on fast U-Net and remapping attention (FRSD). With the datasets proposed and the novel methods, FRSD performs relatively robustly and detects 9 classes with an mAP of more than 83%, and the practical robustness is fully validated, reaching a state-of-the-art level. The practical performance in rain-fog conditions is further displayed in Figure A2. Therefore, the FRSD proposed has an impressive capacity to build a low and high collocation with radars or other technologies suitable for far-sightedness. We believe that the cooperation of FRSD and remote detection skills should improve the management of maritime security and transport.

Figure 18. *The detection visualization of FRSD in the original VOC-ship-test. There are 24 results whose original images are derived from the VOC-ship-test unsoaked. The images above also demonstrate the factors considered to build the ship dataset, containing high density or occlusion scenes, which is more practical than the dataset proposed by R.W. Liu [9].*

Although our ship detector has impressive performance in the related field, this paper also has limitations. The rain–fog dataset proposed is only suitable for validating the practical effects of ship detectors and hardly meets the requirements of model training. Although the ship dataset proposed has considerable samples, the original fog–rain scenes of the ship dataset are fewer compared with the practical marine environments. The ship detector proposed has the initial anti-adverse weather ability. How-

Table 3. Comparison of ship detectors for specific ship detection. FRSD_ship_dataaug is the final model which utilizes the data augmentation of the cross-background during the training period. "ship" is categorised into a class in which the location is detected while the specific class is indistinguishable.

Detectors	Imgsize	Engi_ship	Freighter	Passe_ship	Public_ser	Sailboat	Speedboat	Submarine	Warship	Ship
Faster RCNN [12]	800	70.70%	85.55%	59.68%	86.36%	79.67%	85.81%	87.78%	65.08%	14.72%
RetinaNet [47]	600	81.58%	93.31%	69.60%	90.35%	86.35%	91.79%	87.58%	77.72%	13.80%
SSD300 [15]	300	84.37%	88.77%	78.08%	88.61%	82.22%	87.31%	90.21%	85.10%	20.47%
CenterNet [48]	512	77.60%	91.70%	75.00%	92.50%	85.70%	90.50%	92.50%	81.90%	20.05%
EfficientDetd0 [21]	512	74.10%	90.40%	66.40%	89.10%	87.30%	89.80%	88.10%	72.70%	11.30%
YOLOV5s_ship [20]	640	71.10%	92.00%	72.20%	91.79%	88.10%	91.40%	94.20%	84.40%	23.90%
FRSD_ship_dataaug	640	**84.70%**	**94.20%**	**80.60%**	**93.70%**	**89.20%**	**93.90%**	**95.01%**	**85.80%**	**26.90%**

ever, the mean average precision decreases impressively in full rainy or fog weather. Suppose we enable the anti-adverse weather ability to the detection algorithm. In that case, the real-time effect of FRSD template hardly meets post-improvement, which easily leads to a visible decrease in inference speed. Overcoming the shortcomings is our next direction, we will build a ship detector of anti-adverse weather by proposing novel methods.

6. CONCLUSIONS

The remapping attention method and fast U-Net are proposed to gain a real-time ship detector suitable for various ocean weather conditions. Remapping attention combines the local and global features that are relatively compatible with YOLOv5s. Fast U-Net compresses the parameters by the index operation, accelerating the inference speed when multiple batches are set and increasing stability in various weather conditions. The ship dataset is offered to alleviate the lack of multiple ship datasets, and the data augmentation of the cross-background is proposed to increase the value of the ship dataset. The rain–fog dataset is introduced to quantitatively validate the ship detectors. Experiments demonstrate that FRSD detects 9 classes with an mAP of more than 83%, and the practical robustness is fully validated by a quantitatively soaking method, reaching relevant state-of-the-art ship detectors.

APPENDIX A

Figure A1 demonstrates the factors considered to build the ship dataset, containing detection angles. Different detection angles represent that the detectors, trained by our ship dataset, can be used in static shore-based detection scenes, as well as shipborne or airborne detection scenes. Figure A2 represents the practical detection results in various rain–fog weather, although some rain–fog scenes exceed the detection ability of FRSD.

Figure A1. *The training examples from VOC-ship. The images above, derived from the merge of random four images, also demonstrate various detection angles and reflect the application scenes, such as horizontal, and overlooking detection scenes for fixed and moving cameras.*

Figure A2. *The performance of FRSD in multiple rain–fog conditions.*

REFERENCES

1. Xiao, R.; Jianga, S.; Zhongb, J. Status and Development Trend of Active Sub-arrays Structure Design in Active Phased Array Antenna. *IOP Conf. Ser. Mater. Sci. Eng.* **2020**, *914*, 012038.
2. Wang, X.; Li, G.; Zhang, X.P.; He, Y. A Fast CFAR Algorithm Based on Density-Censoring Operation for Ship Detection in SAR Images. *IEEE Signal Process. Lett.* **2021**, *28*, 1085–1089.
3. Gao, F.; He, Y.; Wang, J.; Hussain, A.; Zhou, H. Anchor-free convolutional network with dense attention feature aggregation for ship detection in SAR images. *Remote Sens.* **2020**, *12*, 2619.
4. Cui, Z.; Li, Q.; Cao, Z.; Liu, N. Dense attention pyramid networks for multi-scale ship detection in SAR images. *IEEE Trans. Geosci. Remote Sens.* **2019**, *57*, 8983–8997.
5. Caon, M.; Ros, P.M.; Martina, M.; Bianchi, T.; Magli, E.; Membibre, F.; Ramos, A.; Latorre, A.; Kerr, M.; Wiehle, S.; et al. Very low latency architecture for earth observation satellite onboard data handling, compression, and encryption. In Proceedings of the 2021 IEEE International Geoscience and Remote Sensing Symposium IGARSS, Brussels, Belgium, 11–16 July 2021; pp. 7791–7794.
6. Solari, L.; Del Soldato, M.; Raspini, F.; Barra, A.; Bianchini, S.; Confuorto, P.; Casagli, N.; Crosetto, M. Review of satellite interferometry for landslide detection in Italy. *Remote Sens.* **2020**, *12*, 1351.
7. Krišto, M.; Ivasic-Kos, M.; Pobar, M. Thermal object detection in difficult weather conditions using YOLO. *IEEE Access* **2020**, *8*, 125459–125476.
8. Wang, N.; Li, B.; Wei, X.; Wang, Y.; Yan, H. Ship Detection in Spaceborne Infrared Image Based on Lightweight CNN and Multisource Feature Cascade Decision. *IEEE Trans. Geosci. Remote Sens.* **2020**, *59*, 4324–4339.
9. Liu, R.W.; Yuan, W.; Chen, X.; Lu, Y. An enhanced CNN-enabled learning method for promoting ship detection in maritime surveillance system. *Ocean Eng.* **2021**, *235*, 109435.
10. Pang, G.; Shen, C.; Cao, L.; Hengel, A.V.D. Deep learning for anomaly detection: A review. *ACM Comput. Surv. (CSUR)* **2021**, *54*, 1–38.
11. Girshick, R. Fast R-CNN. In Proceedings of the IEEE International Conference on Computer Vision, Santiago, Chile, 11–18 December 2015; pp. 1440–1448.
12. Ren, S.; He, K.; Girshick, R.; Sun, J. Faster R-CNN: Towards real-time object detection with region proposal networks. *Adv. Neural Inf. Process. Syst.* **2015**, *28*, 91–99.
13. Girshick, R.; Donahue, J.; Darrell, T.; Malik, J. Rich feature hierarchies for accurate object detection and semantic segmentation. In Proceedings of the 27th IEEE Conference on Computer Vision and Pattern Recognition (CVPR), Columbus, OH, USA, 24–27 June 2014; pp. 580–587.
14. Yang, Z.; Liu, S.H.; Hu, H.; Wang, L.; Lin, S. RepPoints: Point Set Representation for Object Detection. In Proceedings of the IEEE/CVF International Conference on Computer Vision (ICCV), Seoul, Korea, 27–28 October 2019; pp. 9656–9665.
15. Liu, W.; Anguelov, D.; Erhan, D.; Szegedy, C.; Reed, S.; Fu, C.Y.; Berg, A.C. SSD: Single Shot MultiBox Detector. In Proceedings of the 14th European Conference on Computer Vision (ECCV), Amsterdam, The Netherlands, 11–14 October 2016; Volume 9905, pp. 21–37.
16. Redmon, J.; Divvala, S.; Girshick, R.; Farhadi, A. You Only Look Once: Unified, Real-Time Object Detection. In Proceedings of the 2016 IEEE Conference on Computer Vision and Pattern Recognition (CVPR), Las Vegas, NV, USA, 27–30 June 2016; pp. 779–788.
17. Redmon, J.; Farhadi, A. YOLO9000: Better, Faster, Stronger. In Proceedings of the 30th IEEE/CVF Conference on Computer Vision and Pattern Recognition (CVPR), Honolulu, HI, USA, 21–26 July 2017; pp. 6517–6525.

18. Redmon, J.; Farhadi, A. Yolov3: An incremental improvement. *arXiv* **2018**, arXiv:1804.02767.

19. Wang, C.Y.; Bochkovskiy, A.; Liao, H.Y.M. Scaled-YOLOv4: Scaling Cross Stage Partial Network. In Proceedings of the IEEE/CVF Conference on Computer Vision and Pattern Recognition (CVPR), Nashville, TN, USA, 20–25 June 2021; pp. 13024–13033.

20. Ge, Z.; Liu, S.; Wang, F.; Li, Z.; Sun, J. Yolox: Exceeding yolo series in 2021. *arXiv* **2021**, arXiv:2107.08430.

21. Naddaf-Sh, S.; Naddaf-Sh, M.M.; Kashani, A.R.; Zargarzadeh, H. An Efficient and Scalable Deep Learning Approach for Road Damage Detection. In Proceedings of the 8th IEEE International Conference on Big Data (Big Data), Atlanta, GA, USA, 10–13 December 2020; pp. 5602–5608.

22. Geiger, A.; Lenz, P.; Urtasun, R. Are we ready for Autonomous Driving? The KITTI Vision Benchmark Suite. In Proceedings of the IEEE Conference on Computer Vision and Pattern Recognition (CVPR), Providence, RI, USA, 16–21 June 2012; pp. 3354–3361.

23. Wang, S.L.; Bai, M.; Mattyus, G.; Chu, H.; Luo, W.J.; Yang, B.; Liang, J.; Cheverie, J.; Fidler, S.; Urtasun, R.; et al. TorontoCity: Seeing the World with a Million Eyes. In Proceedings of the 16th IEEE International Conference on Computer Vision (ICCV), Venice, Italy, 22–29 October 2017; pp. 3028–3036.

24. Maddern, W.; Pascoe, G.; Linegar, C.; Newman, P. 1 year, 1000 km: The Oxford RobotCar dataset. *Int. J. Robot. Res.* **2017**, *36*, 3–15.

25. Lin, T.Y.; Maire, M.; Belongie, S.; Hays, J.; Perona, P.; Ramanan, D.; Dollar, P.; Zitnick, C.L. Microsoft COCO: Common Objects in Context. In Proceedings of the 13th European Conference on Computer Vision (ECCV), Zurich, Switzerland, 6–12 September 2014; Volume 8693, pp. 740–755.

26. Everingham, M.; Van Gool, L.; Williams, C.K.; Winn, J.; Zisserman, A. The pascal visual object classes (voc) challenge. *Int. J. Comput. Vis.* **2010**, *88*, 303–338.

27. Wang, M.; Li, C.; Yu, Q. Ship Detection Based on Neighboring Information Fusion. *J.-Xiamen Univ. Nat. Sci.* **2007**, *46*, 645.

28. Xu, F.; Liu, J.; Zeng, D.; Wang, X. Detection and identification of unsupervised ships and warships on sea surface based on visual saliency. *Opt. Precis. Eng.* **2017**, *25*, 1300–1311.

29. Borghgraef, A.; Barnich, O.; Lapierre, F.; Van Droogenbroeck, M.; Philips, W.; Acheroy, M. An evaluation of pixel-based methods for the detection of floating objects on the sea surface. *EURASIP J. Adv. Signal Process.* **2010**, *2010*, 1–11.

30. Barnich, O.; Van Droogenbroeck, M. ViBe: A universal background subtraction algorithm for video sequences. *IEEE Trans. Image Process.* **2010**, *20*, 1709–1724.

31. Jodoin, P.M.; Konrad, J.; Saligrama, V. Modeling background activity for behavior subtraction. In Proceedings of the 2nd ACM/IEEE International Conference on Distributed Smart Cameras, Palo Alto, CA, USA, 7–11 September 2008; pp. 1–10.

32. Hu, W.C.; Yang, C.Y.; Huang, D.Y. Robust real-time ship detection and tracking for visual surveillance of cage aquaculture. *J. Vis. Commun. Image Represent.* **2011**, *22*, 543–556.

33. Itti, L.; Koch, C.; Niebur, E. A model of saliency-based visual attention for rapid scene analysis. *IEEE Trans. Pattern Anal. Mach. Intell.* **1998**, *20*, 1254–1259.

34. Arshad, N.; Moon, K.S.; Kim, J.N. An Adaptive Moving Ship Detection and Tracking Based on Edge Information & Morphological Operations. In Proceedings of the International Conference on Graphic and Image Processing (ICGIP), Cairo, Egypt, 1–3 October 2011; Volume 8285.

35. Fefilatyev, S.; Goldgof, D.; Shreve, M.; Lembke, C. Detection and tracking of ships in open sea with rapidly moving buoy-mounted camera system. *Ocean Eng.* **2012**, *54*, 1–12.

36. Yang, X.; Sun, H.; Fu, K.; Yang, J.; Sun, X.; Yan, M.; Guo, Z. Automatic ship detection in remote sensing images from google earth of complex scenes based on multiscale rotation dense feature pyramid networks. *Remote Sens.* **2018**, *10*, 132.

37. Chen, Z.; Chen, D.; Zhang, Y.; Cheng, X.; Zhang, M.; Wu, C. Deep learning for autonomous ship-oriented small ship detection. *Saf. Sci.* **2020**, *130*, 104812.

38. Woo, S.H.; Park, J.; Lee, J.Y.; Kweon, I.S. CBAM: Convolutional Block Attention Module. In Proceedings of the 15th European Conference on Computer Vision (ECCV), Munich, Germany, 8–14 September 2018; Volume 11211, pp. 3–19.

39. Hou, Q.B.; Zhou, D.Q.; Feng, J.S. Coordinate Attention for Efficient Mobile Network Design. In Proceedings of the IEEE/CVF Conference on Computer Vision and Pattern Recognition (CVPR), Nashville, TN, USA, 20–25 June 2021; pp. 13708–13717.

40. Hu, J.; Shen, L.; Albanie, S.; Sun, G.; Wu, E.H. Squeeze-and-Excitation Networks. *IEEE Trans. Pattern Anal. Mach. Intell.* **2020**, *42*, 2011–2023.

41. Shi, W.Z.; Caballero, J.; Huszar, F.; Totz, J.; Aitken, A.P.; Bishop, R.; Rueckert, D.; Wang, Z.H. Real-Time Single Image and Video Super-Resolution Using an Efficient Sub-Pixel Convolutional Neural Network. In Proceedings of the 2016 IEEE Conference on Computer Vision and Pattern Recognition (CVPR), Las Vegas, NV, USA, 27–30 June 2016; pp. 1874–1883.

42. Zhao, P.; Meng, C.; Chang, S. Single stage ship detection algorithm based on improved VGG network. *J. Optoelectron Laser* **2019**, *30*, 719–730.

43. Wang, T.Y.; Yang, X.; Xu, K.; Chen, S.Z.; Zhang, Q.; Lau, R.W.H.; Soc, I.C. Spatial Attentive Single-Image Deraining with a High Quality Real Rain Dataset. In Proceedings of the 32nd IEEE/CVF Conference on Computer Vision and Pattern Recognition (CVPR), Long Beach, CA, USA, 15–20 June 2019; pp. 12262–12271.

44. Yang, W.; Tan, R.T.; Feng, J.; Guo, Z.; Yan, S.; Liu, J. Joint rain detection and removal from a single image with contextualized deep networks. *IEEE Trans. Pattern Anal. Mach. Intell.* **2019**, *42*, 1377–1393.

45. Fu, X.Y.; Huang, J.B.; Zeng, D.L.; Huang, Y.; Ding, X.H.; Paisley, J. Removing rain from single images via a deep detail network. In Proceedings of the 30th IEEE/CVF Conference on Computer Vision and Pattern Recognition (CVPR), Honolulu, HI, USA, 21–26 July 2017; pp. 1715–1723.

46. Qian, R.; Tan, R.T.; Yang, W.H.; Su, J.J.; Liu, J.Y. Attentive Generative Adversarial Network for Raindrop Removal from A Single Image. In Proceedings of the 31st IEEE/CVF Conference on Computer Vision and Pattern Recognition (CVPR), Salt Lake City, UT, USA, 18–23 June 2018; pp. 2482–2491.

47. Lin, T.Y.; Goyal, P.; Girshick, R.; He, K.M.; Dollar, P. Focal Loss for Dense Object Detection. *IEEE Trans. Pattern Anal. Mach. Intell.* **2020**, *42*, 318–327.

48. Zhou, X.; Wang, D.; Krähenbühl, P. Objects as points. *arXiv* **2019**, arXiv:1904.07850.

Chapter 5

Rescheduling Strategy for Berth Planning in Container Terminals: An Empirical Study from Korea

Armi Kim[1], Hyun-Ji Park[2], Jin-Hyoung Park[2] and Sung-Won Cho[2]

[1] *Department of Transportation and Logistics Engineering, Hanyang University, 55, Hanyangdaehak-ro, Sangnok-gu, Ansan-si 15588, Gyeonggi-do, Korea*

[2] *Maritime Safety and Environmental Research Division, Korea Research Institute of Ships and Ocean Engineering, 32, Yuseong-daero 1312 beon-gil, Yuseong-gu, Daejeon 34103, Korea*

ABSTRACT

The rapid increase in international trade volume has caused frequent fluctuation of the vessels' arrival time in container terminals. In order to solve this problem, this study proposes a methodology for rescheduling berth and quay cranes caused by updated information on the arrival time of vessels. A mixed-integer linear programming model was formulated for the berth allocation and crane assignment problem, and we solved the problem using a rolling-horizon approach. Numerical experiments were conducted under three scenarios with empirical data from a container terminal located in Busan, Korea. The experiments revealed that the proposed model reduces penalty costs and overall delayed departure time compared to the results of the terminal planner.

Keywords: *container terminals; berth allocation problem; quay crane assignment problem; dynamic rescheduling; mixed-integer linear programming; rolling-horizon approach; empirical study*

1. INTRODUCTION

As the global economy and supply chain has become more complex, maritime transportation faced the challenge to improve operational reliability and flexibility against supply chain risks. For port container terminals, such operational capabilities are directly related to attracting customers and staying ahead of the competition, as well as sustaining profitability. This is because the reputation for their service can be achieved by catching up with the required departure times of the calling vessels, even if the arrivals of the shipping companies are delayed [1]. Considering the frequent occurrence of deviations from the scheduled arrival times of the vessels [2,3], the development of sophisticated optimization models for berth operations is of utmost importance to deal with the uncertain environment in container terminals.

The berth operations are most relevant to the allocation of berths and Quay Cranes (QCs), which are scarce and expensive resources in container terminals. The Berth Allocation Problem (BAP) is to determine the berthing position and the berthing time of the calling vessels. As depicted in Figure 1, the solution of the BAP (i.e., berth plan) can be represented as a non-overlapped rectangular packing on a two-dimensional time-space diagram which is expressed by the planning time horizon and the length of quay [4]. The Quay Crane Assignment Problem (QCAP) is to determine the number of QCs that are engaged to the vessels at each time. Since the handling times of vessels depend on the number of QCs that are assigned to the vessels, the QCAP is also considered for establishing the berth plan. In this regard, Figure 2 shows an example of the solution for the integrated Berth Allocation and Crane Assignment Problem (BACAP) where the QCs are assigned to the vessels in a time-variant manner. The QC assignment can inherently accelerate or decelerate the handling times of the vessels.

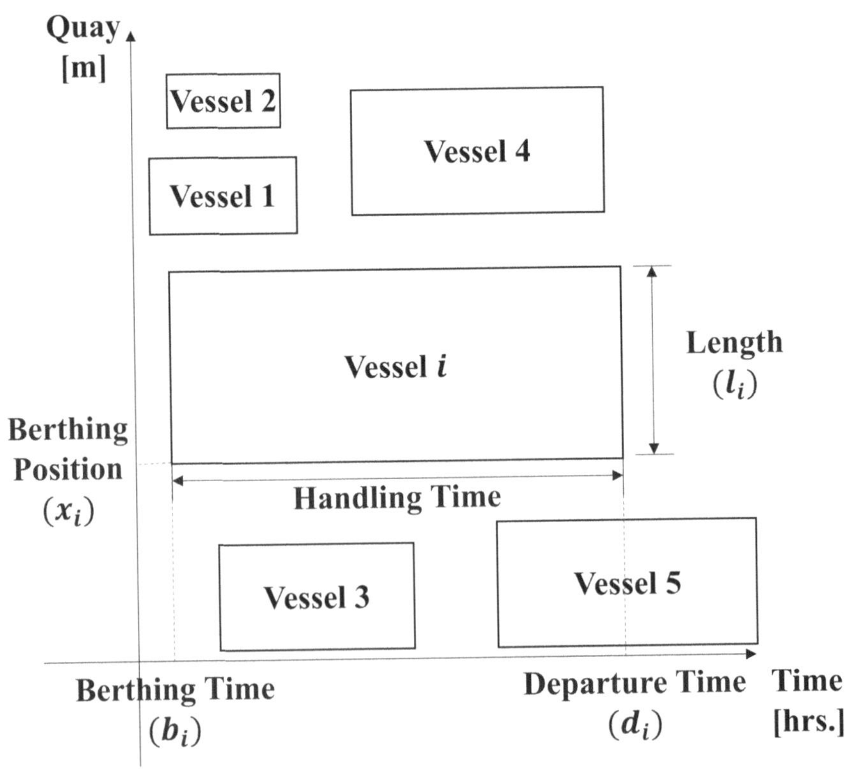

Figure 1. *Time-space diagram.*

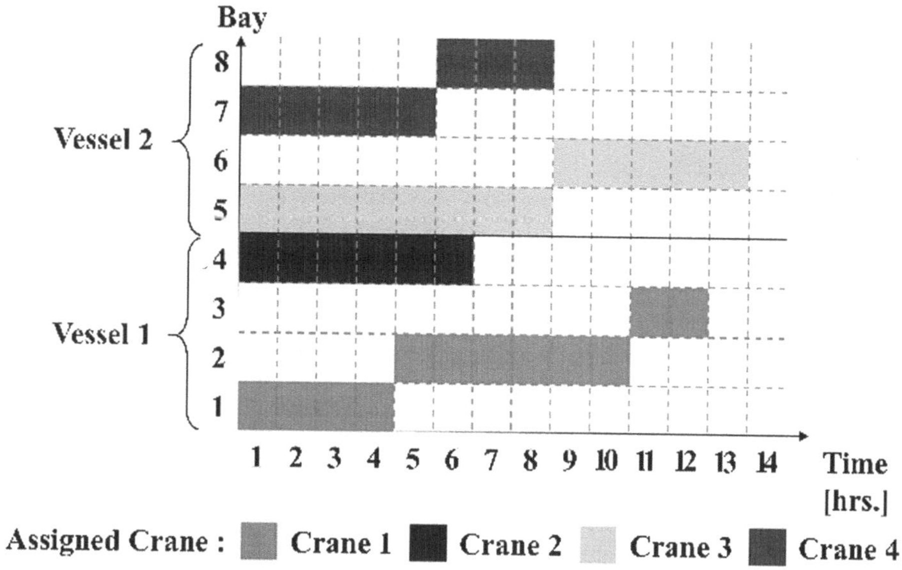

Figure 2. *An example of QCAP.*

The uncertain arrivals of the calling vessels due to unforeseen events, such as weather conditions or a delay from the previous container terminal, often lead to disruption in the berth operations. The disruption may incur substantial costs to the terminal because the berth and yard operations must be re-planned and relevant resources such as container handling equipment must be re-allocated. Moreover, it may induce delays of the other consecutive vessels due to the chain-effect which severely declines in terminal service reputation for shipping companies. Therefore, one of the main tasks of terminal berth operators is to figure out the optimal reallocation of berths and QCs for incoming vessels to respond to the real-time schedule disruption. In that case, if the actual time of departure (ATD) exceeds the estimated time of departure (ETD) of the vessels, it may result in inevitable modifications to the operational plan of the vessel for the next terminal of destination established by the shipping company. For the shipping companies, such a disruption causes loss of reliability of their services and logistic costs [5]. In order to provide high-quality service to the shipping companies, the berth rescheduling is made in such a way as to minimize the difference between the ETD and the ATD. However, there is a gap between actual operations and optimized operations in the berth rescheduling because the rescheduling in practice is manually conducted by the berth operators according to experience [1].

It is worth mentioning that the smart port initiatives and high-quality terminal operating systems have provided operational improvement by automation and optimization with significantly mitigating manual work. The berth rescheduling operations have the potential for these attempts to take effect with the following two reasons: First, the loss in optimality for the rescheduling operation may cause substantial costs including declined terminal service reputations for shipping companies. Second, co-incidences of vessel delays can make it difficult to execute the berth rescheduling in manual and accelerate the operational inefficiency. It is noted that vessel delays occur frequently [2,3] and further, the more complex global supply chain may provide more uncertainties in vessel arrivals. Accordingly, more efforts of rescheduling will be required to cope with the uncertainties and attract the shipping companies. Therefore, as one of the prerequisites for smart port initiatives, it is imperative to gauge the effect that can be achieved through automation and optimization. To this end, we intend to quantify and identify the gap between research and practice for the berth rescheduling.

This paper provides the methodology to reschedule berths and QCs simultaneously to respond to the updated information where the initial berth plan is disrupted due to the uncertainty of vessels' arrival. We formulate a mixed-integer linear programming (MILP) model for the dynamic berth rescheduling problem and propose a rolling horizon approach to solve the model. It is especially notable that we conducted the empirical study with the real-world berth operational data in a typical container terminal in Korea. We designed three scenarios that restrict the optimization to reflect the possible fluctuation while terminal operating. The impact of the proposed methodology was assessed in measure of the delayed departure time of vessels, and it was compared with the empirical berth plan conducted by the terminal operator.

The remainder of this paper is organized as follows. A review of previous research is given in Section 2. In Section 3, the problem description and mathematical model that minimizes operation cost associated with reallocation of berths and QCs are presented. As a solution approach, the rolling horizon approach is addressed in Section 4. Then, a numerical experiment is conducted with empirical data in Section 5. Finally, a conclusion is given in Section 6.

2. LITERATURE REVIEW

Many researchers have studied container terminals from various perspectives: economic [6,7,8,9], operational [10,11,12,13], and political [14,15]. In particular, the study of the various operational problems in a container terminal plays an important role in improving efficiency and advancing to a smart container terminal. In this study, we focus on seaside operations among several operational problems.

The BAP has received tremendous attention from the research community in the past decades. As modern container terminals varied in layout and equipment, terminal operation problems became more complex, and different variants of the models and solution methods have been developed in the literature [16]. The berth allocation models are typically classified into discrete and continuous models according to the berthing space [17,18,19]. Due to the complexity of the BAP, several researchers applied heuristic algorithms such as the Lagrangian relaxation procedure, tree search procedure, genetic algorithm, beam search algorithm, particle swarm optimization, etc. to their studies [20,21,22,23,24,25].

Since the QCAP is directly linked to the handling times of vessels, several researchers considered the integration of the BACAP, which causes a great increase in the complexity of the model. To solve this complexity, Guan et al. [26] considered the BAP as a multi-processor task scheduling problem where the processors and tasks correspond to consecutive QCs and vessel handling, respectively. They used minimization of total weighted completion time of tasks as an objective function, where the weight of a task is equivalent to the processing time depending on the number of serving QCs. A greedy algorithm was applied to determine the QC assignment for the arranged vessels in increasing order of the weight. Park and Kim [27] introduced a two-phase structure that makes step by step decisions for berth allocation and detailed time-variant QC assignment. For phase 1, a Lagrangian heuristic is used to solve continuous BAP with consideration of QC resources, where the handling time of a vessel is assumed to be decided by the number that is inversely proportional to the number of QCs engaged to the vessel. For phase 2, dynamic programming that minimizes QC setup cost is made to solve the QCAP. Vacca et al. [28] proposed a model which can be adapted to both BAP and BACAP. The model is based on an exponential number of variables and solved via column generation. A branch and price algorithm was carried out to obtain optimal integer solutions, and several accelerating techniques that can be generalized to other branch and price schemes were introduced. Meisel and Bierwirth [29] devised the Squeaky Wheel Optimization (SWO) heuristic based on priority list building and list insertion method. After deriving the initial priority list from the vessel list that is arranged in the order of the expected time of arrival, the SWO heuristic procedure and local refinement procedure were repeated to explore the improved solution. Iris et al. [30] proposed Generalized Set Partitioning (GSPP) formulations for the BACAP considering both time-variant and time-invariant QC assignment policies. They solved the problem introduced in [29] and applied a set of column reduction techniques to obtain the improved upper and lower bounds, which would be measures of evaluation of new heuristics. Iris et al. [31] proposed an Adaptive Large Neighborhood Search (ALNS)-based heuristic to improve the QC assignment on the mathematical model introduced in [29]. They showed the solution efficiency of the proposed heuristic comparing with the state-of-the-art heuristics.

For implementing the established berth schedule, all input parameters considered at the planning phase must be unchanged. However, in practice, a variety of factors change the plan differently from the information expected at the planning phase, thereby preventing the implementation of the original berth schedule. To cope with the uncertainties, a proactive approach and a reactive approach have been discussed. The proactive approach aims to absorb the uncertainties at the planning level [2,32,33,34,35,36], while the reactive approach aims to handle the real-time disruptions due to the uncertainties at the operation level [1,37,38,39,40]. We focus on the review of the literature using the reactive approach which is close to this study. Zeng et al. [37] devised two strategies (QC rescheduling strategy and berth reallocation strategy) to solve the disruption recovery problem through local rescheduling and tabu search. The numerical experiment showed a comparison of additional costs resulting from QC rescheduling and berth reallocation. Umang et al. [38] studied the BAP with dynamic updates on the ETA of the calling vessels. They proposed a smart greedy method by taking advantage of the estimates of the arrival and handling times from a probability distribution. Xiang et al. [39] introduced the fuzzy decision-making process for the disruption recovery in the BACAP and applied a rolling-horizon optimization algorithm. Li et al. [1], Lv et al. [40] developed an SWO-based construction heuristic to solve the real-time rescheduling for the BACAP.

As discussed above, the number of studies that concern the reactive approach for the BAP or BACAP is limited. The focus of existing studies is mainly on how to reallocate the berths and QCs, and none of them attempts the empirical study to quantify and identify the gap between research and practice for the berth rescheduling. In this paper, the practical restrictions in optimization for the berth rescheduling are considered to estimate the contribution of automation and optimization in real-world terminal environments. Our empirical analysis confirms that the rescheduling strategy for berth planning does have a cost-effective effect on automated terminals and provides managerial insights to berth planners. Furthermore, the results of our research can help understand the effect to be achieved through automation and optimization techniques, and accelerate the smart port initiatives.

3. PROBLEM DESCRIPTION

This study concerns the dynamic berth rescheduling problem. As shown in Figure 3, the conceptual framework to model the problem works as follows. We assume that the initial baseline schedule is given, and the terminal receives a dynamic update on the arriving vessel information. The baseline schedule is continuously executed until the vessel information is updated. The controller identifies the impact of the updated information on the feasibility of the plan during the execution of the baseline schedule. If it turns out to be a cause of the schedule disruption, this triggers the solver to modify the current schedule. The solver creates the re-optimized schedule to cope with the disruption. Then, the baseline schedule is considered to be replaced with the newly obtained re-optimized schedule by the solver.

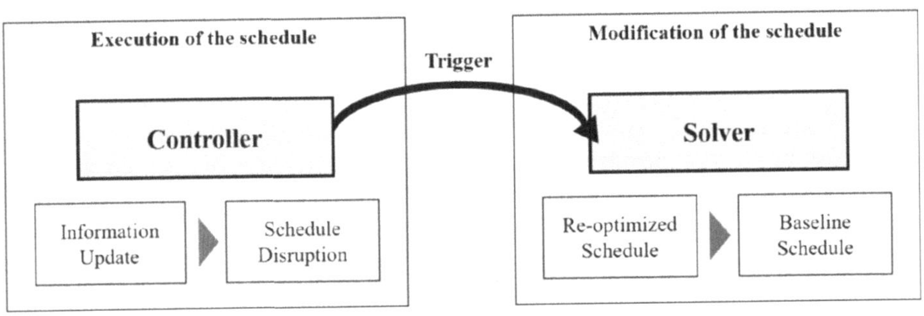

Figure 3. *Conceptual framework to model a dynamic berth rescheduling problem.*

To create the initial baseline schedule at the planning stage, a variety of service costs such as berthing delay from ETA, departure delay from ETD, deviation from the ideal berthing position are considered in literature [19,20,27,29]. In contrast, for the recovery schedule at the operation stage, catching up with the required departure times (i.e., ETDs) of the calling vessels is of utmost importance to achieve the service reputation, even if the arrivals of the shipping companies are delayed [1]. As mentioned in Section 1, the service reputation is essential to stay ahead of the competition, as well as sustain profitability. Accordingly, in the conceptual framework for rescheduling, the solver model explicitly considers the service reputation and aims to minimize the departure delay from ETD.

The solver yields the recovery schedule based on re-optimization of the berth scheduling in the event of the schedule disruption. A common approach for dynamic rescheduling is to apply an optimization model on a rolling planning horizon [38]. If the solver is triggered at time epoch τ, the baseline schedule (i.e., determination of the berthing times and positions, assignment of the dedicated QCs for the vessels), which was planned before time τ, remains unchanged, and the schedule after time τ is planned completely new. The following subsections detail the re-optimization model to recover the berth schedule.

3.1. Assumption

Quay layout is assumed as a continuous space.

Each vessel has a maximum and minimum number of QCs that can be assigned. It is decided in advance through an agreement between the terminal and the shipping company to ensure minimum productivity. For some feeder vessels, if the minimum number of QCs is not given, one is assumed.

When vessels berth and departure at the terminal, tugboat and pilot schedule between seaport to inland ports and line handling take an hour. That is, it is ensured a minimum of 2 h between vessels.

QCs have constant productivity and start the loading/discharging process the moment the vessel is berthed.

The number of QCs working on the vessel may change during loading/discharging operation. However, the minimum number of QCs that can be assigned must be met to ensure continuous work. Namely, it is assumed that the vessel which has started operation will not be stopped.

3.2. Notation

3.2.1. Parameters

$V(\tau)$	Set of vessels to be scheduled at time τ, $V(\tau) = V^1(\tau) \cup V^2(\tau)$
$V^1(\tau)$	Subset of vessels which are berthed or have not yet departed the port at time τ
$V^2(\tau)$	Subset of vessels which are not yet berthed at time τ
T	Length of planning period (unit: an hour)
L	Total length of quay
Q	Total number of QCs
q_i^{min}	Minimum number of QCs that can be assigned per hour to vessel i
q_i^{max}	Maximum number of QCs that can be assigned per hour to vessel i
n_i	Possible number of QCs assigned to vessel i at a particular point in time ($q_i^{min} \le n_i \le q_i^{max}$)
ETA_i	The most recently updated arrival time of vessel i
ETD_i	Departure time of vessel i in proforma schedule
b_i^1	Determined berthing time of specific vessel i in $V^1(\tau)$
x_i^1	Determined berthing position of specific vessel i in $V^1(\tau)$
l_i	Length of vessel i
r_i	Total number of containers to be discharged or loaded for vessel i
r_i^1	Remained number of containers to be discharged or loaded for specific vessel i in $V^1(\tau)$ at time τ
p	Productivity of one QC
m_i	Total number of QCs required for vessel i, $m_i = \left\lceil \frac{r_i}{p} \right\rceil$
m_i^1	Remained number of QCs required for specific vessel i in $V^1(\tau)$ at time τ, $m_i^1 = \left\lceil \frac{r_i^1}{p} \right\rceil$
gap^x	Minimum time interval needed between vessels
gap^y	Minimum distance needed between vessels
c_i	Penalty cost corresponding to the departure delay of vessel i
M	A large number

3.2.2. Decision Variables

b_i	Berthing time of vessel i
d_i	Departure time of vessel i
x_i	Berthing position of vessel i
v_{ij}^x	1 if vessel i is completely on the left side of vessel j on x-axis (time) in the time-space diagram, 0 otherwise (where $i \neq j$)
v_{ij}^y	1 if vessel i is completely below vessel j on y-axis (space) in the time-space diagram, 0 otherwise (where $i \neq j$)
q_{itn_i}	1 if n_i QCs are assigned to vessel i at time t, 0 otherwise

3.3. Model Formulation

In this section we describe the MILP model for the dynamic berth rescheduling problem. A list of notations, i.e. parameters, decision variables of the model, can be found in Section 3.2.

$$Min \sum_{i=1}^{N} \left[c_i(d_i - ETD_i)^+\right] \tag{1}$$

$$b_i = b_i^1,\ for\ i \in V^1(\tau) \tag{2}$$

$$x_i = x_i^1,\ for\ i \in V^1(\tau) \tag{3}$$

$$b_i \geq ETA_i,\ for\ i \in V^2(\tau) \tag{4}$$

$$x_i \leq L - l_i,\ for\ i \in V^2(\tau) \tag{5}$$

$$b_j \geq d_i + gap^x + M\left(v_{ij}^x - 1\right),\ for\ i,j \in V(\tau),\ i \neq j \tag{6}$$

$$x_j \geq x_i + l_i + gap^y + M\left(v_{ij}^y - 1\right),\ for\ i,j \in V(\tau),\ i \neq j \tag{7}$$

$$v_{ij}^x + v_{ji}^x + v_{ij}^y + v_{ji}^y \geq 1,\ for\ i,j \in V(\tau),\ i \neq j \tag{8}$$

$$b_i \leq t \times q_{itn_i} + M(1 - q_{itn_i}),\ for\ i \in V^2(\tau),\ max(ETA_i,\ \tau) \leq t \leq T,\ q_i^{min} \leq n_i \leq q_i^{max} \tag{9}$$

$$d_i \geq\ t \times q_{itn_i} + 1,\ for\ i \in V^1(\tau),\ max(ETA_i,\ \tau) \leq t < T,\ q_i^{min} \leq n_i \leq q_i^{max} \tag{10}$$

$$d_i - b_i \leq \left\lceil \frac{m_i}{q_i^{min}} \right\rceil,\ for\ i \in V(\tau) \tag{11}$$

$$\sum_{t=\tau}^{T} \sum_{n_i=q_i^{min}}^{q_i^{max}} n_i \times q_{itn_i} \geq m_i^1,\ for\ i \in V^1(\tau) \tag{12}$$

$$\sum_{t=\tau}^{T} \sum_{n_i=q_i^{min}}^{q_i^{max}} n_i \times q_{itn_i} \geq m_i,\ for\ i \in V^2(\tau) \tag{13}$$

$$\sum_{i \in V(\tau)} \sum_{n_i = q_i^{min}}^{q_i^{max}} n_i \times q_{itn_i} \leq Q, \; for \; t = \tau, \cdots, T \tag{14}$$

$$\sum_{n_i = q_i^{min}}^{q_i^{max}} q_{itn_i} \leq 1, \; for \; i \in V(\tau), \; t = \tau, \cdots, T \tag{15}$$

$$\sum_{n_i = q_i^{min}}^{q_i^{max}} q_{itn_i} \geq \sum_{n_i = q_i^{min}}^{q_i^{max}} \left(q_{i(t-1)n_i} + q_{it'n_i} \right) - 1, \; for \; i \in V(\tau), \; \tau \leq t < t' \leq T \tag{16}$$

$$x_i \geq 0, \; b_i \geq 0, \; d_i \geq 0 \tag{17}$$

$$q_{itn_i}, \; v_{ij}^x, v_{ij}^y \; : binary \tag{18}$$

The rescheduling of the berth plan aims to minimize the delays from the ETD of the vessels. The objective function (1) defines a weighted penalty costs for delays from ETD, taking into account the service priorities of each vessel. Constraints (2) and (3) ensure to sustain the already determined berthing time and position for the vessels which are berthed or have not yet departed the port at time epoch τ. Constraints (4)–(8) ensure the feasible berthing position, berthing time, and departure time of all vessels so that they do not overlap with each other within the time-space diagram. Constraints (9)–(16) are related to QC assignment to vessels. Constraint (12) indicates only the required QC capacities to handle the remained containers are considered for the vessels which are berthed or have not yet departed the port at time epoch τ. Constraint (16) satisfies the assumption of the model to ensure continuous work. Lastly, Constraints (17) and (18) define decision variable type.

4. SOLUTION APPROACH

In this study, we apply a rolling-horizon approach for the rescheduling method to accommodate the framework presented in Section 3. In general, dynamic rescheduling is to apply an optimization model on a rolling planning horizon [38]. A rolling-horizon optimization algorithm has been widely employed in research on rescheduling problems [39]. The main idea of the rolling-horizon approach is to divide the planning horizon into sub-horizons, and then iteratively solve sub-problems for each sub-horizon. The rolling-horizon approach allows to plan for the planning horizon, and then reflects the results of the previous sub-horizon on the planning for the next sub-horizon. Through the rolling-horizon approach, we can reflect the highly dynamic and fluctuating environment generated during the berth planning and derive the flexible plan within a reasonable computation time by updating the plan every sub-horizon.

For the description of the rolling-horizon approach, the following notations are introduced: length of the planning horizon (*T*), length of the rolling sub horizon (*S*), and length of the frozen time window (*F*). Then, the total number of iterations (*N*) for the rolling-horizon method is equivalent to T divided by S. The rolling-horizon rescheduling procedure is terminated if the number of iterations N is reached.

More specifically, Figure 4 shows a snapshot in the rescheduling framework. In Figure 4, the leftmost one presents the initial situation where the baseline schedule is executed in grey-shaded time period and at the specific time epoch, the berth rescheduling is required. The controller component triggers the solver component. Then, for the time period to be rescheduled, the sub-problem for each sub-horizon is iteratively solved by the solver component. Here, the exact optimization algorithm is implemented based on the mathematical model in Section 3, where the length of planning period T is reduced to the length of rolling sub-horizon S. The result of the solver is kept until the vessel information is updated. The next one shows the updated situation again, and the controller and solver works for the reallocations of the berth and QCs in the same way described.

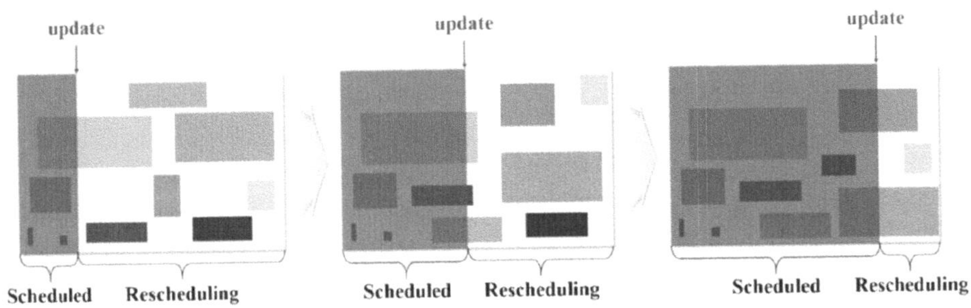

Figure 4. *Rolling-horizon with reallocation of berth and QCs.*

5. NUMERICAL EXPERIMENTS

5.1. Experiment Environment

To find out the effectiveness of the proposed method in real-life environment, the berth plan generated by the proposed model and the result of the berth planner in the terminal were compared. The experiment environment was set to a berth length of 1200 m and a total of 12 QCs, which is the actual environment from a terminal in Pusan Newport, Republic of Korea. In addition, the data of vessels that berthed at the container terminal in the actual berth schedule, including ETA, ETD, vessel length, number of containers to be handled, minimum/maximum number of QCs that can be assigned, and the actual time of berthing/departure (ATB/ATD), were used. For the penalty costs corresponding to the departure delay of the vessels, the costs resulting from reassigning a vessel to another terminal were used [41].

In real-life environments, berth planning is conducted without knowing the information of all vessels. In the experiments, this factor is considered by using an actual time of arrival (ATA), ATB, and ATD gathered from vessels in the terminal. For the periodic rescheduling strategy, information from each vessel is known 24 h prior to arrival, and the proposed model is triggered to reschedule every 6 h. That is, the length of the planning horizon T, rolling sub-horizon S, and rescheduling period are 1 week, 24 h, and 6 h, respectively. The experiments are carried out on a computer with Intel Core i7-7700HQ 2.80GHz and 16.0 GB of RAM working under Windows 10 Server operating system. We ran commercial solver Gurobi 8.0.1 to solve the subproblems for rolling-horizon optimization. For implementing the rolling-horizon optimization, the average computation time is in the range of 500 and 1000 s.

5.2. Experiment Scenarios

As mentioned in Section 4, in numerical experiments, we derived the result of rescheduling using a rolling-horizon approach. However, there are not only the dynamic nature of the information but also other various factors that make efficient berth operations difficult for planners to determine berth allocation at the actual terminal. For example, there are specific situations where it is difficult to change the berth plan due to the request of the consignor, the tight schedule to the next destination port, the relationship between terminal and shipping company, and the relationship between the mother ship and feeder ship, and so on. Moreover, because of the complexity of the terminal's operations, the loading/discharging processing time of the vessel may be delayed longer than expected. Even though such factors do not appear on the data, they can be considered an inherent factor considering the actual operating environment of the terminal.

When comparing the results of the rescheduling of the berth plan with those of the terminal planner, to eliminate these inherent factors as much as possible and analyze the results, three scenarios were assumed as shown in Table 1. Scenario 1 is a case where intrinsic factors are not considered. In this scenario, all vessels can berth immediately after their ATA instead of ETA. Scenario 2 assumed that berthing time is determined by the terminal planner due to exceptional circumstances, and it is best for modifying the berth plan. In this scenario, every vessel can berth only after the ATB in place of ATA. In addition to the assumption of Scenario 2, Scenario 3 assumed the processing time of the vessel was delayed. In this scenario, the productivity of a QC, p is reduced from 30 to 25 containers per hour to derive the berth plan. Numerical experiments were conducted on each of the three scenarios, and the results were analyzed.

Table 1. Summary of the experiment scenarios.

Scenario 1	$\Omega = \{\omega \mid ETA_i = ATA_i \; for\ all\ vessels\ i, \;\; p = 30 \; for\ all\ QCs\}$
Scenario 2	$\Omega = \{\omega \mid ETA_i = ATB_i \; for\ all\ vessels\ i, \;\; p = 30 \; for\ all\ QCs\}$
Scenario 3	$\Omega = \{\omega \mid ETA_i = ATB_i \; for\ all\ vessels\ i, \;\; p = 25 \; for\ all\ QCs\}$

5.3. Result of Experiments under Three Scenarios

5.3.1. Scenario 1: A Case Where Vessels Can Berth after Their ATA

For the actual planner's result, as shown in Figure 5, a total of 12 vessels had delayed ETD, which accumulates to 126 h of delay. For the proposed model under Scenario 1, a total of 3 vessels had delays, which accumulate to 6 h of delay. This scenario showed that the penalty cost, the expected cost of reassigning the vessel to other terminals, can be reduced by about 2.23 million dollars.

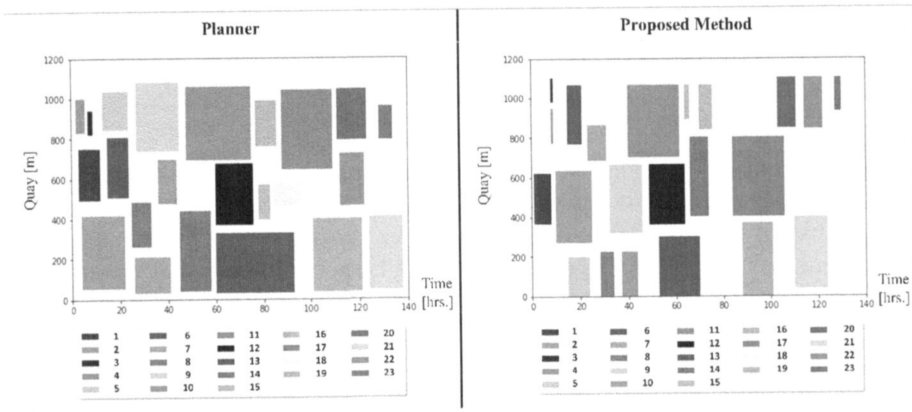

Figure 5. *The result of Scenario 1.*

As can be seen in Figure 6, the proposed method in this study reduced departure delay time on most vessels from the results of the planner. For Vessel 14, a delay did not occur in the result of the planner, but it occurred in the proposed methodology. As the vessel is berthed more greedily than the proposed methodology, vessels arriving at similar times are delayed and it incurs significant costs to the terminal.

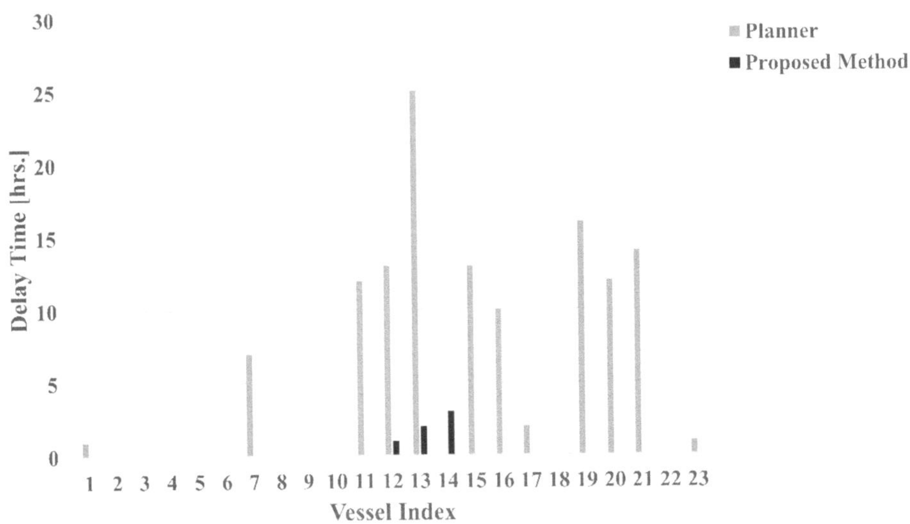

Figure 6. *Departure delay time by vessel in Scenario 1.*

5.3.2. Scenario 2: A Case Where Vessels Can Berth after Their ATB

As the result of Scenario 2, the model had a delay of only 8 vessels, which accumulates to 74 h of delay, thus reducing the objective value by 44.67% compared to the planner's berth allocation results. The expected amount of saved penalty cost by Scenario 2 is about 1.11 million dollars. In Scenario 2, although the vessel's berthing time had to be determined disadvantageously, the objective value and delay time were generally reduced in comparison with the result of the planner. Significantly, it can be seen in Figure 7 that the handling time for vessels 9, 11, 13, and 19 that were affected by QC allocation and productivity had shorter times compared to the result of a planner.

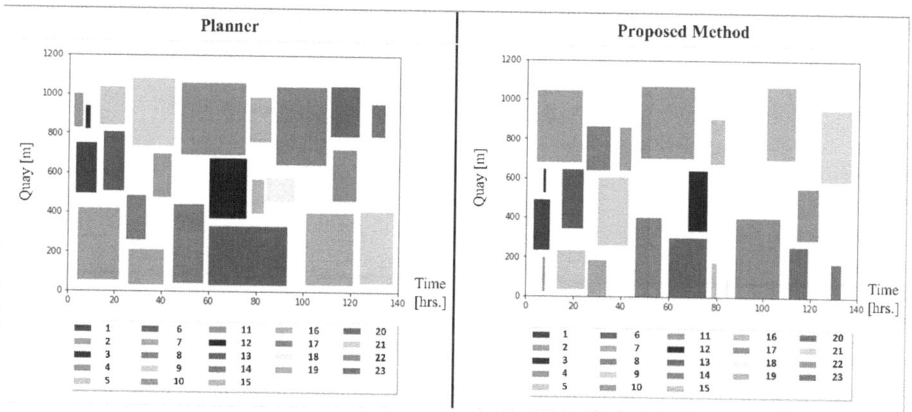

Figure 7. *The result of Scenario 2.*

Figure 8 shows that the proposed method had a lower departure delay on most vessels than the planner's results. However, compared to Scenario 1, the results of the model also show that there were delays in the departure of vessels.

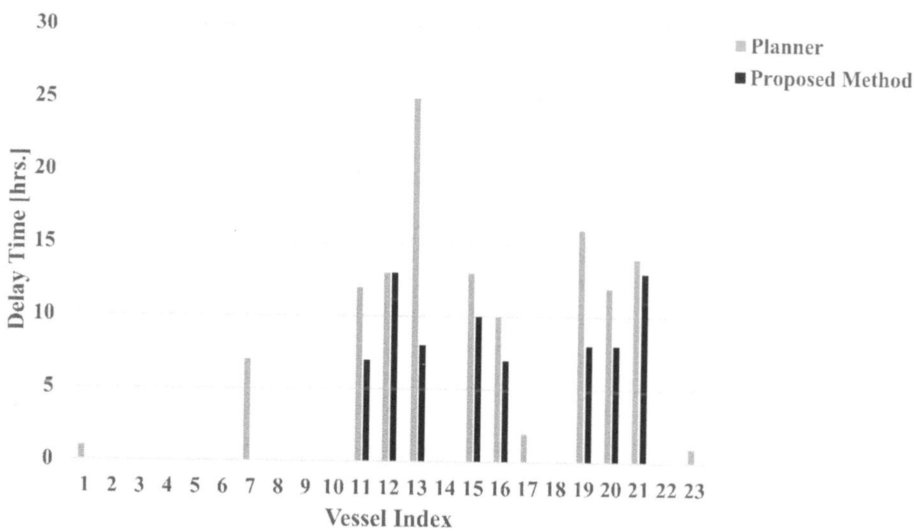

Figure 8. *Departure delay time by vessel in Scenario 2.*

5.3.3. Scenario 3: A Case Where Productivity of QCs Is Reduced

As can be seen in Figure 9, only 9 vessels were delayed, and the total delay time was 103 h, decreasing the objective value by 21.82%. By Scenario 3, the penalty cost corresponding to about 0.55 million dollars can be reduced. Despite the case, where the minimum productivity of the QCs is assumed and takes a longer time to process the vessel, in general, the objective value and delay time was reduced compared to the planner. It means that the strategic method of determining QC allocation based on ETD and penalty cost worked effectively.

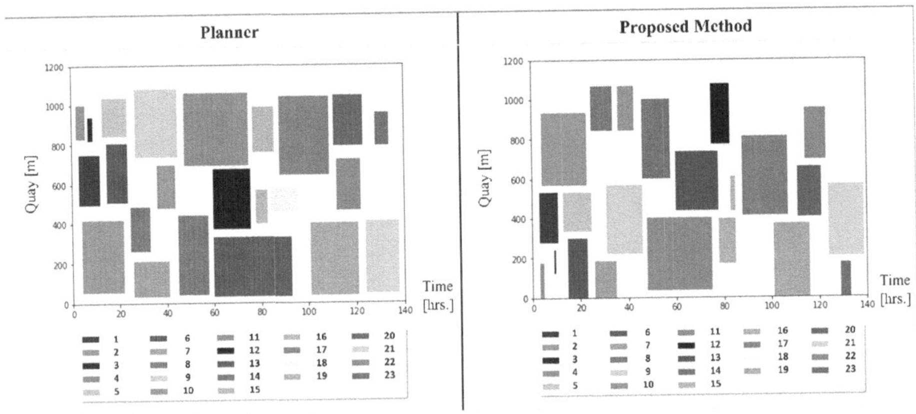

Figure 9. *The result of Scenario 3.*

In Figure 10, the berth schedule planned by the model resulted in more delay for a few vessels, while the overall cost was reduced by decreasing the delay for other vessels. In particular, vessel 12, which was delayed more in the result of the proposed model than the result of the planner, costs only half as much as vessel 13. It indicates that vessel 12 was strategically delayed to reduce the total cost by allocating QCs to vessel 13 intensively.

6. CONCLUSIONS

Since a vessel arrives late due to various uncertainties, the baseline berth plan can no longer be executed, which causes huge financial losses to terminals. To prevent this, we proposed a berth and QC rescheduling strategy that minimizes costs when disruptions occur in the initial plan. A MILP model was formulated and a rolling-horizon approach was suggested to solve the mathematical model effectively. Numerical experiments based on the empirical data from the terminal in Busan were conducted to compare the results of the proposed method with those of a planner. Experimental results show that the delay tendencies of the planner and the proposed method are similar, but overall delay time is reduced in the results of the proposed method. As setting the scenario conservatively, the delay time of the proposed method is gradually increased. Nevertheless, the dynamic rescheduling results in fewer penalty costs than the planner scheduling manually.

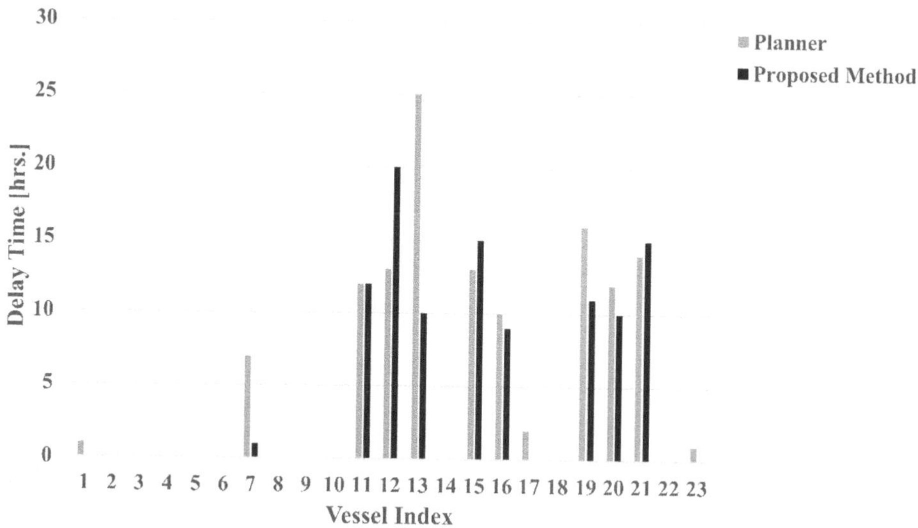

Figure 10. *Departure delay time by vessel in Scenario 3.*

It is noted that the optimization for berth rescheduling can be enhanced with taking the information on the uncertainties into account. This is because it can adapt to uncertainties in a robust way and further enrich the proactive approaches for berth planning. If the vessel's arrival time is predictable through analysis of the vessel's delay data, it can help predict the expected losses of the terminal due to the possible disruptions and berth rescheduling operation at the planning stage. Then, this anticipation enables to proactively create the berth plan with schedule robustness by minimizing the expected losses in the terminal or reducing the possible occurrence of disruptions. Predictive-based operational logic can realize the global optimization of the terminal through the smart port technology, as collecting real-time information on oceans and terminal, and predicting various port management systems (e.g., berth planning, yard planning) comprehensively.

Furthermore, this study can be extended by taking into account the energy efficiency aspect for next-generation green ports [42]. The integration of energy management and operational improvements can attain energy savings and emission reductions [43].

REFERENCES

1. Li, M.Z.; Jin, J.G.; Lu, C.X. Real-time disruption recovery for integrated berth allocation and crane assignment in container terminals. *Transp. Res. Rec.* **2015**, *2479*, 49–59.
2. Moorthy, R.; Teo, C.P. Berth management in container terminal: The template design problem. *OR Spectr.* **2006**, *28*, 495–518.
3. Yu, J.; Tang, G.; Song, X.; Yu, X.; Qi, Y.; Li, D.; Zhang, Y. Ship arrival prediction and its value on daily container terminal operation. *Ocean. Eng.* **2018**, *157*, 73–86.
4. Kim, K.H.; Park, Y.M. A Crane Scheduling Method for Port Container Terminals. *Eur. J. Oper. Res.* **2004**, *156*, 752–768.

5. Notteboom, T.E. The time factor in liner shipping services. *Marit. Econ. Logist.* **2006**, *8*, 19–39.
6. Sanchez, R.J.; Hoffmann, J.; Micco, A.; Pizzolitto, G.V.; Sgut, M.; Wilmsmeier, G. Port efficiency and international trade: Port efficiency as a determinant of maritime transport costs. *Marit. Econ. Logist.* **2003**, *5*, 199–218.
7. Notteboom, T.; Rodrigue, J.P. Containerisation, box logistics and global supply chains: The integration of ports and liner shipping networks. *Marit. Econ. Logist.* **2008**, *10*, 152–174.
8. Acciaro, M. A critical review of port pricing literature: What role for academic research? *Asian J. Shipp. Logist.* **2013**, *29*, 207–228.
9. Notteboom, T.; Pallis, T.; Rodrigue, J.P. Disruptions and resilience in global container shipping and ports: The COVID-19 pandemic versus the 2008–2009 financial crisis. *Marit. Econ. Logist.* **2021**, *1*, 1–32.
10. Stahlbock, R.; Voß, S. Operations research at container terminals: A literature update. *OR Spectr.* **2008**, *30*, 1–52.
11. Carlo, H.J.; Vis, I.F.; Roodbergen, K.J. Storage yard operations in container terminals: Literature overview, trends, and research directions. *Eur. J. Oper. Res.* **2014**, *235*, 412–430.
12. Vis, I.F. Survey of research in the design and control of automated guided vehicle systems. *Eur. J. Oper. Res.* **2006**, *170*, 677–709.
13. Vis, I.F.; De Koster, R. Transshipment of containers at a container terminal: An overview. *Eur. J. Oper. Res.* **2003**, *147*, 1–16.
14. Monios, J.; Wilmsmeier, G.; Ng, A.K. Port system evolution—The emergence of second-tier hubs. *Marit. Policy Manag.* **2019**, *46*, 61–73.
15. Martin, J.; Thomas, B.J. The container terminal community. *Marit. Policy Manag.* **2001**, *28*, 279–292.
16. Lai, K.K.; Shih, K. A Study of Container Berth Allocation. *J. Adv. Transp.* **1992**, *26*, 45–60.
17. Imai, A.; Nagaiwa, K.I.; Tat, C.W. Efficient Planning of Berth Allocation for Container Terminals in Asia. *J. Adv. Transp.* **1997**, *31*, 75–94.
18. Kim, K.H.; Moon, K.C. Berth Scheduling by Simulated Annealing. *Transp. Res. Part B Methodol.* **2003**, *37*, 541–560.
19. Bierwirth, C.; Meisel, F. A Survey of Berth Allocation and Quay Crane Scheduling Problems in Container Terminals. *Eur. J. Oper. Res.* **2010**, *202*, 615–627.
20. Bierwirth, C.; Meisel, F. A Follow-up Survey of Berth Allocation and Quay Crane Scheduling Problems in Container Terminals. *Eur. J. Oper. Res.* **2015**, *244*, 675–689.
21. Imai, A.; Nishimura, E.; Papadimitriou, S. The Dynamic Berth Allocation Problem for a Container Port. *Transp. Res. Part. B Methodol.* **2001**, *35*, 401–417.
22. Guan, Y.; Cheung, R.K. The Berth Allocation Problem: Models and Solution Methods. *OR Spectr.* **2004**, *26*, 75–92.
23. Imai, A.; Nishimura, E.; Hattori, M.; Papadimitriou, S. Berth Allocation at Indented Berths for Mega-containerships. *Eur. J. Oper. Res.* **2007**, *179*, 579–593.
24. Wang, F.; Lim, A. A Stochastic Beam Search for the Berth Allocation Problem. *Decis. Support Syst.* **2007**, *42*, 2186–2196.
25. Ting, C.J.; Wu, K.C.; Chou, H. Particle Swarm Optimization Algorithm for the Berth Allocation Problem. *Expert Syst. Appl.* **2014**, *41*, 1543–1550.
26. Guan, Y.; Xiao, W.Q.; Cheung, R.K.; Li, C.L. A Multiprocessor Task Scheduling Model for berth Allocation: Heuristic and Worst-case Analysis. *Oper. Res. Lett.* **2002**, *30*, 343–350.

27. Park, Y.M.; Kim, K.H. A scheduling method for Berth and Quay cranes. *OR Spectr.* **2003**, *25*, 1–23.
28. Vacca, I.; Salani, M.; Bierlaire, M. An exact algorithm for the integrated planning of berth allocation and quay crane assignment. *Transp. Sci.* **2013**, *47*, 148–161.
29. Meisel, F.; Bierwirth, C. Heuristics for the Integration of Crane Productivity in the Berth Allocation Problem. *Transp. Res. Part E: Logist. Transp. Rev.* **2009**, *45*, 196–209.
30. Iris, Ç.; Pacino, D.; Ropke, S.; Larsen, A. Integrated berth allocation and quay crane assignment problem: Set partitioning models and computational results. *Transp. Res. Part E Logist. Transp. Rev.* **2015**, *81*, 75–97.
31. Iris, Ç.; Pacino, D.; Ropke, S. Improved Formulations and an Adaptive Large Neighborhood Search Heuristic for the Integrated Berth Allocation and Quay Crane Assignment Problem. *Transp. Res. Part E Logist. Transp. Rev.* **2017**, *105*, 123–147.
32. Zhou, P.F.; Kang, H.G. Study on Berth and Quay-crane Allocation under Stochastic Environments in Container Terminal. *Syst. Eng. Theory Pract.* **2008**, *28*, 161–169.
33. Zhen, L.; Lee, L.H.; Chew, E.P. A decision model for berth allocation under uncertainty. *Eur. J. Oper. Res.* **2011**, *212*, 54–68.
34. Xu, Y.; Chen, Q.; Quan, X. Robust berth scheduling with uncertain vessel delay and handling time. *Ann. Oper. Res.* **2012**, *192*, 123–140.
35. Zhen, L. Tactical berth allocation under uncertainty. *Eur. J. Oper. Res.* **2015**, *247*, 928–944.
36. Iris, Ç.; Lam, J.S.L. Recoverable robustness in weekly berth and quay crane planning. *Transp. Res. Part B Methodol.* **2019**, *122*, 365–389.
37. Zeng, Q.; Yang, Z.; Hu, X. Disruption Recovery Model for Berth and Quay Crane Scheduling in Container Terminals. *Eng. Optim.* **2011**, *43*, 967–983.
38. Umang, N.; Bierlaire, M.; Erera, A.L. Real-time management of berth allocation with stochastic arrival and handling times. *J. Sched.* **2017**, *20*, 67–83.
39. Xiang, X.; Liu, C.; Miao, L. Reactive strategy for discrete berth allocation and quay crane assignment problems under uncertainty. *Comput. Ind. Eng.* **2018**, *126*, 196–216.
40. Lv, X.; Jin, J.G.; Hu, H. Berth allocation recovery for container transshipment terminals. *Marit. Policy Manag.* **2020**, *47*, 558–574.
41. Cho, S.W.; Park, H.J.; Lee, C. An integrated method for berth allocation and quay crane assignment to allow for reassignment of vessels to other terminals. *Marit. Econ. Logist.* **2021**, *23*, 123–153.
42. Iris, Ç.; Lam, J.S.L. Optimal energy management and operations planning in seaports with smart grid while harnessing renewable energy under uncertainty. *Omega* **2021**, 102445.
43. Iris, Ç.; Lam, J.S.L. A review of energy efficiency in ports: Operational strategies, technologies and energy management systems. *Renew. Sustain. Energy Rev.* **2019**, *112*, 170–182.

Chapter 6

M3C: Multimodel-and-Multicue-Based Tracking by Detection of Surrounding Vessels in Maritime Environment for USV

Dalei Qiao[1,2], Guangzhong Liu[1], Jun Zhang[2], Qiangyong Zhang[2], Gongxing Wu[3] and Feng Dong[1,4]

[1] *College of Information Engineering, Shanghai Maritime University, Shanghai 201306, China*

[2] *Jiangsu Maritime Institute, Nanjing 211100, China*

[3] *College of Ocean Science and Engineering, Shanghai Maritime University, Shanghai 201306, China*

[4] *College of Information Engineering, Shaoyang University, Shaoyang 422000, China*

ABSTRACT

It is crucial for unmanned surface vessels (USVs) to detect and track surrounding vessels in real time to avoid collisions at sea. However, the harsh maritime environment poses great challenges to multitarget tracking (MTT). In this paper, a novel tracking by detection framework that integrates the multimodel and multicue (M^3C) pipeline is proposed, which aims at improving the detection and tracking performance. Regarding the multimodel, we predicted the maneuver probability of a target vessel via the gated recurrent unit (GRU) model with an attention mechanism, and fused their respective outputs as the output of a kinematic filter. We developed a hybrid affinity model based on multi cues, such as the motion, appearance, and attitude of the ego vessel in the data association stage. By using the proposed ship re-identification approach, the tracker had the capability of appearance matching via metric learning. Experimental evaluation of two public maritime datasets showed that our method achieved state-of-the-art performance, not only in identity switches (IDS) but also in frame rates.

Keywords: *maritime surveillance; tracking by detection; multimodel and multicue (M3C); deep learning; unmanned surface vessels*

1. INTRODUCTION

As a sea surface agent, an unmanned surface vessel (USV) needs to independently navigate and perform specific tasks in a maritime environment. Robust and accurate awareness and understanding of the surrounding environment are required to achieve this desired outcome, especially the ability to detect and track surrounding vessels in real time. On a crewed vessel, such critical data can mainly be obtained from navigation radar, sonar, and millimeter wave radar, but these are far from sufficient for a USV, because radar and sonar have minimum detection ranges, and shipborne automatic identification system (AIS) stations are only required for the vessels that weigh over 300 tons, according to the regulations of the International Maritime Organization (IMO) [1,2]. In other words, most of the common sensors have a blind zone at a certain distance, which is precisely what the USV should be paying attention to. Using a vision sensor (visible light or infrared cameras) as auxiliary equipment can compensate for this deficiency.

Approximately 80% of ship and bridge collision accidents are related to human factors, according to the statistical evidence [3]. As deep learning and computer vision have attracted increasing attention, researchers are dedicated to using them to more intuitively performing maritime vessel detection, classification, and tracking tasks. Some maneuvering behavior may occur during collision avoidance, such as acceleration or turning, which must take into account the capricious marine environment, the swing, and yaw imposed by the ego vessel, the change of light, and possible occlusion [4]. All of these pose great challenges to vessel detection and tracking research in the maritime environment. Tracking by detection is one of the state-of-the-art frameworks for multi target tracking (MTT). It is a two-stage strategy, which includes two independent steps: detection and tracking. Detection constructs a solid base for the subsequent tracking. We focus on the latter, on the basis of a state-of-the-art detector.

The significant contributions of this paper can be summarized, as follows. Firstly, we propose a multimodel (MM) filter approach that is based on a gated recurrent unit with attention (GRU-attention) mechanism. The maneuverability of the surrounding vessel is predicted by using the GRU recurrent neural network, being driven by the time series of the historical state of the target vessels. Secondly, a multicue (MC) data association method that considers the long-term and short-term cues is presented. Thirdly, treating the tracking by detection of maritime vessels as a typical problem of monitoring a moving object from video images captured by a moving camera [5], we propose a ship re-identification (Ship-ReID) method, which uses metric learning to determine whether the same ship is in the incoming video frame. We are the first to introduce GRU-attention for surrounding vessel maneuver prediction, to the best of our knowledge. This paper is also the first to introduce Ship-ReID to solve the identity switches (IDS) problem that is caused by camera motion, appearance variation, occlusion, and even blur in the process of maritime vessel detection and tracking.

The structure of this paper is organized, as follows. The related works are introduced in Section 2. Section 3 presents the proposed MM and MC (M^3C) tracking by detection pipeline. We evaluate the performance of M^3C on the Singapore Marine Dataset (SMD) [1] and PETS 2016 maritime dataset [6] in Section 4. Finally, Section 5 discusses the main conclusions and future work.

2. RELATED WORK

2.1. Deep Learning for Generic Target Detection

The convolutional neural network (CNN) has played an increasingly important role in artificial intelligence, especially in the field of computer vision, as a new structure of neural network. The CNN-based detectors are roughly classified into two categories: two stages and single stage. For the two-stage detection method, the features are first extracted from the image, then proposal regions are generated, followed by classification task performance, and finally, the positioning regression task is completed. The typical algorithms include regions with CNN features (R-CNN), fast/faster R-CNN, R-FCN [7,8,9], and so forth. The single-stage detector omits the task of generating proposal regions in the two-stage method, and such as single shot multibox detector (SSD) [10], YOLOv1/v2/9000/v3 [11] are the typical frameworks. In recent years, anchor-free methods have become very popular, which use the corner or area as the anchor instead of the bounding box directly [12,13].

2.2. Deep Learning for Generic Target Tracking

Traditional target tracking algorithms are divided into tracking by detection and tracking before detection, according to the interaction between the detector and tracker in the tracking process [14]. Tracking before detection can effectively detect dim and small targets. It has been extensively used in radar and infrared image processing. Based on machine vision, especially in visible vision processing, more attention has been paid to joint detection and tracking [15], rather than tracking before detection. Gordon et al. proposed a real-time generic tracker by incorporating temporal information that is founded on a recurrent neural network [16]. Bae et al. proposed an online tracking framework by combining tracklet confidence with deep appearance learning [17]. Arandjelović applied a multiview appearance to track and recognize vehicles in the aerial sequential images [18]. Re-identification is a subissue of image retrieval, which aims to identify the same target from multiple cross cameras or the same target at different moments from a single camera. In the field of using target re-identification to improve the tracking performance, most of the current achievements have focused on pedestrians [19,20] or vehicles [21,22], and there have been no reports that are related to achievements in the detection and tracking of maritime vessels.

2.3. Visible Vessel Detection and Tracking

Despite state-of-the-art object detection and tracking for generic objects having recently demonstrated impressive performances, these have not been adequate for the complicated maritime environment. It is common knowledge that, as surrounding vessels traveling from far to near enter the ego vessel's surveillance area, they always first appear near the sea antenna (horizon line). With regard to moving shipborne or even buoy-mounted cameras, ship detection and tracking while using the guidance of the sea-sky line has also been proposed in recent years [23]. Jeong et al. proposed a method to estimate the horizon line while using the region of interest [24]. Horizon line detection, dynamic background, and foreground segmentation was performed by means of discrete cosine transform in [25].

Sun et al. also proposed a robustly coarse-fine-stitched strategy to detect the horizon line for USV [26].

Bovcon et al. explored semantics segmentation assisted by an inertial measurement unit (IMU) for stereo obstacle detection [27]. Cane et al. evaluated the semantic segmentation networks on several public maritime datasets and compared their performances [28]. Kim et al. proposed a probabilistic method to detect and classify ships based on a faster R-CNN detector and improved the probability of ship detection by intersection over union (IOU) tracking [29]. Marié et al. proposed a key point tracking method to generate high-quality region proposals and then fed them into a fast R-CNN for further processing [30]. Cao et al. made use of a CNN to extract the features from the vessel image and to eventually identify the ship in a frame of a video sequence [31]. Leclerc et al. took full advantage of ship classification that is based on deep transfer learning to enhance the estimation capability of the trackers [32].

3. THE PROPOSED TRACKING BY DETECTION APPROACH

In this section, we present a novel tracking by detection framework combining MM and MC (M³C), which has the ability to re-identification the vessels that have lost their track. The overall architecture of M³C is detailed in Figure 1.

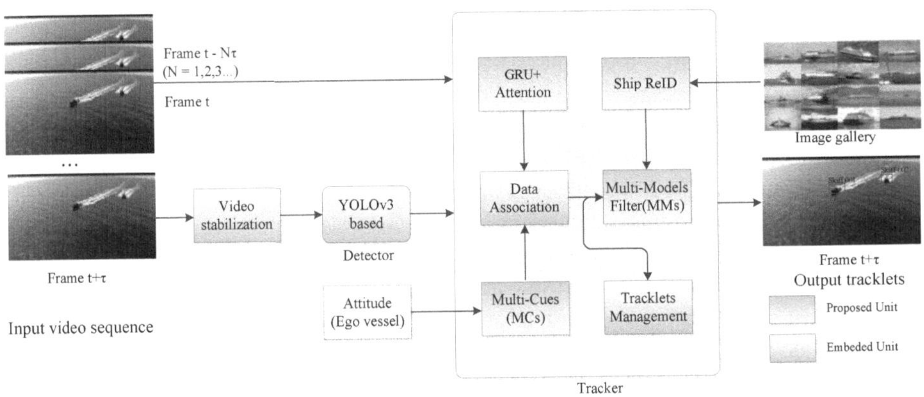

Figure 1. *The architecture of our proposed M³C tracking by detection pipeline. It comprises eight modules, among which the video stabilization unit uses the algorithm introduced by [33]. We proposed the four blue units, and their interaction with the other three green units is discussed in detail later.*

3.1. The Detector Based on YOLOv3

We devised a maritime vessel detector based on YOLOv3. YOLOv3 is one of the state-of-the-art CNN-based generic object detectors, and it treats the detection and classification of targets as a regression problem [11]. By regressing the location of the detected objects and performing classification, the bounding box and class of the objects are obtained by looking at each video frame only once. It can conduct three different scale predictions. When feeding into a 608 × 608 pixel image, the detection is performed using the scales of 19 × 19, 38 × 38, and 76 × 76, respectively. The structure of the detector is shown in the Figure 2.

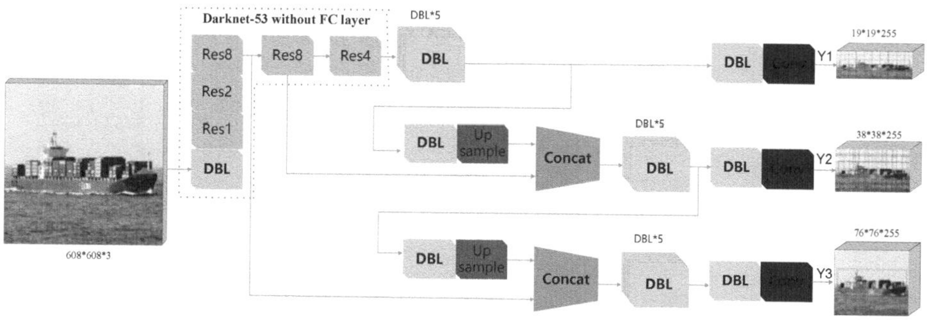

Figure 2. *Framework of the detector. Among them, Darknetconv2d_BN_Leaky (DBL) is the basic unit, which consists of a convolutional layer (conv), batch normalization, and a leaky rectified linear unit (ReLU).*

3.2. Multimodel for Tracking by Detection.

We formulated the maneuvering prediction of a surrounding vessel as the estimation of the matching probability of each candidate model at the next moment based on the vessel's historical trajectory and while considering the ego vessel's posture via GRU recurrent neural networks.

3.2.1. Maneuvering Model of Vessels

According to the kinematic behavior [34,35,36] of maritime vessels, we predefined the three most common kinds of motion maneuvers in advance as candidate models, named constant velocity (CVCV), constant acceleration (CACA), and curvilinear motion (CMCM), respectively.

The CVCV model was used to describe the target performing straight line motions at a constant velocity in the two-dimensional (2D) plane. The state vector describing the dynamic characteristics of the vessel consisted of two elements: pixel position of the target x,y and velocity x˙,y˙ (i.e., the state vector at time t can be denoted by XCV(t)=[xtx˙-tyty˙t]T, satisfying the conditional that x¨=0,y¨=0¨). The target state equation of CVCV at the current time was modeled, as follows:

$$X_{CV}(t+\tau) = F_{CV}(t)X_{CV}(t) + G_{CV}(t)v(t) \tag{1}$$

where τ is an indication of the time interval between the current and previous measurement and v(t)=[vx,vy]T is the process noise vector. The transition matrix FCV(t) and the process noise distribution matrix GCV(t) are stated, as follows, respectively:

$$F_{CV}(t) = \begin{bmatrix} 1 & 0 & \tau & 0 \\ 0 & 1 & 0 & \tau \\ 0 & 0 & 1 & 0 \\ 0 & 0 & 0 & 1 \end{bmatrix}, G_{CV}(t) = \begin{bmatrix} 0.5\tau^2 & 0 \\ 0 & 0.5\tau^2 \\ \tau & 0 \\ 0 & \tau \end{bmatrix} \tag{2}$$

A vessel modeled as CACA was usually considered to move at constant acceleration. The state vector of a vessel consisted of three components: position x,y velocity x˙,y˙, and acceleration x¨,y¨ and satisfied the conditional that x...=0. The target state equation was the same as Equation (1), but the state vector was XCA(t)=[xtx˙tx¨tyty˙ty¨t]T. The transition matrix FCA(t) and the process noise distribution matrix GCA(t) are stated, as follows, respectively:

$$F_{CA}(t)=\begin{bmatrix} 1 & \tau & 0.5\tau^2 & 0 & 0 & 0 \\ 0 & 1 & \tau & 0 & 0 & 0 \\ 0 & 0 & 1 & 0 & 0 & 0 \\ 0 & 0 & 0 & 1 & \tau & 0.5\tau^2 \\ 0 & 0 & 0 & 0 & 1 & \tau \\ 0 & 0 & 0 & 0 & 0 & 1 \end{bmatrix}, G_{CA}(t)=\begin{bmatrix} 0.5\tau^2 & 0 \\ \Delta t & 0 \\ 1 & 0 \\ 0 & 0.5\tau^2 \\ 0 & \tau \\ 0 & 1 \end{bmatrix} \tag{3}$$

The CMCM model was based on CVCV and CACA, and it depicted the parabolic trajectory of a ship in a turning mode. For this, we needed to estimate the turning rate ω in real time, and the state vector was extended to XCM(t)=[xtx˙tyty˙tω]T. The transition matrix FCM(t) and the process noise distribution matrix GCM(t) are stated, as follows, respectively:

$$F_{CM}(t)=\begin{bmatrix} 1 & \omega^{-1}\sin\omega\tau & 0 & \omega^{-1}(\cos\omega\tau-1) & 0 \\ 0 & \cos\omega\tau & 0 & -\sin\omega\tau & 0 \\ 0 & \omega^{-1}(1-\cos\omega\tau) & 1 & \omega^{-1}\sin\omega\tau & 0 \\ 0 & \sin\omega\tau & 0 & \cos\omega\tau & 0 \\ 0 & 0 & 0 & 0 & 1 \end{bmatrix}, G_{CM}(t)=\begin{bmatrix} 0.5\tau^2 & 0 & 0 \\ \tau & 0 & 0 \\ 0 & 0.5\tau^2 & 0 \\ 0 & 0 & 0 \\ 0 & 0 & 1 \end{bmatrix} \tag{4}$$

3.2.2. Encoder-Decoder Model of GRU-Attention

We utilized the multimodel weighting method to adapt the change of the target statel in our proposed M[3]C pipeline, during the kinematic filter procedure. The primary task was to estimate the match score of each predefined candidate model and the current state of the vessels. For this, the GRU-attention model was deployed to improve the prediction performance. Long short-term memory (LSTM) networks were proposed to solve the long-term dependencies problem of traditional recurrent neural networks (RNNs) with increasing time intervals [37]. Through proper training, it can remember the critical data and forget the less-important data. GRU [38] improves LSTM by reducing the number of gates and removing all of the memory units, so that only reset gates and update gates remain.

As shown in Figure 3, the formulas for forward propagation of GRU are as follows:

$$\begin{aligned} z_t &= \sigma(w_r \cdot [h_{t-1}, x_t] + b_z) \\ r_t &= \sigma(w_z \cdot [h_{t-1}, x_t] + b_r) \end{aligned} \tag{5}$$

where xt is the input vector at time *t*, σ represents the activation function, wr,wz are the weight matrix, ht−1*h* represents the hidden activation value at the previous moment, and bz,br is the deviation vector. The activation value ht*h* and the candidate activation value h˜t*h*˜ of the hidden node at time *t* are calculated, as follows, where ⊗ indicates the element-wise multiplication:

$$\widetilde{h_t} = tanh(W \cdot [r_t \otimes h_{t-1}, x_t] + b_h)$$
$$h_t = (1 - z_t) \otimes h_{t-1} + z_t\widetilde{h_t} \tag{6}$$

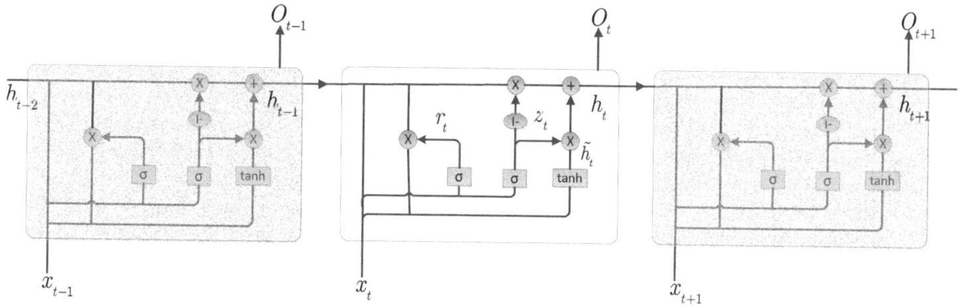

Figure 3. *Typical structure of a gated recurrent unit (GRU) neural network. Ot(ht) is the output at time t; rt denotes the reset gate, which determines the combination of current input and historical memory information; and, zt is the update gate, which determines the proportion of memory left behind.*

As stated above, the curvilinear motion model of vessels can be divided into three subpatterns, according to tangential or normal acceleration: CV,CACV,CA, and CMCM, respectively. Their coexistence and the uncertainty of switching time leads to an alignment problem. It is very difficult for a traditional LSTM/GRU-based encoder–decoder approach to solve the alignment problem. We introduced an attention model to encode each subsequence into a context vector, instead of encoding the entire sequence into an integral vector. Through training, the attention model was encouraged to selectively focus on the important part of all information while also ignoring other secondary information, so as to generate more accurate prediction results by making full use of the subpatterns.

If we were to estimate the maneuvering model of the surrounding vessel at time t, the input sequence would be zt−Nτ ,...,zt−τ,zt,zt+τ, as can be seen from Figure 4. The context vector was the weighted sum of output vector ht−Nτ ,...,ht−τ,ht,ht+τ from encoders, and weights ct−Nτ ,...,ct−τ,ct,ct+τ could be calculated by

$$c_i = \sum_{j=t-N\tau}^{t+\tau} h_j \beta_{ij} \tag{7}$$

where βij is the attention weights to be learned, as obtained by the following formula:

$$\beta_{ij} = \frac{Exp(e_{ij})}{\sum\limits_{k=t-N\tau}^{t+\tau} Exp(e_{ik})}, s.t. \sum_j \beta_{ij} = 1 \tag{8}$$

$$e_{ij} = FC(s_{i-1}, h_j) \tag{9}$$

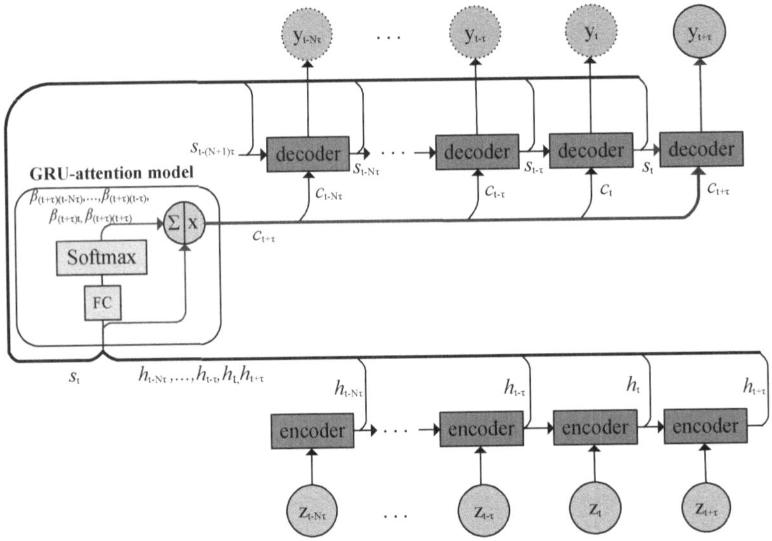

Figure 4. *The improved framework of the GRU encoder–decoder with attention. The attention model was embedded between encoders and decoders and learned the attention weights ct−Nτ ,...,ct−τ,ct,ct+τ via the fully-connected network (FC) and softmax with loss function, output vector of encoder ht−Nτ ,...,ht−τ,ht,ht+τh, and state vector of decoder st−(N+1)τ,st−Nτ,...,st−τ,st as the attention model's input sequence.*

The GRU-attention model was fed with a time-series vector, which described the short-term historical motion measurements of the surrounding vessels. It could output the probability that each candidate model matched the current motion state of the target vessels.

3.2.3. Multimodel Filter

As mentioned before, we defined a model set that meets the needs of tracking moving vessels on the sea: M={mi}i=1,2,⋯,N, consisting of a total of N candidate models. If at time t we have already obtained the historical cumulative observation vectors, the state estimation Y(t+τ)i of the *i*-th model mi can be modeled as the following formula [39], which rests on the conditional probability theory:

$$\begin{aligned} \hat{\mathcal{Y}}_i^{(t+\tau)} &= \mathbb{E}(\mathcal{Y}^{(t+\tau)}|m_i^{(t+\tau)}, Z) \\ &\triangleq \mathcal{Y}^{(t)} + K^{(t)}[X(t+\tau) - F(t+\tau)\mathcal{Y}_i^{(t)}] \end{aligned} \tag{10}$$

where K(t) represents the gainer of the Kalman filter, and m(t)≙P(m(t)|Z(0))=ψ(t)(0) is the prior probability of the initial time that satisfies the sum-to-one condition of ∑t=1Nψ(t)(0)=1.

As shown in Figure 4, Z={zt−Nτ …zt−τztzt+τ} as the input sequence, which indicates the cumulative set of measurements until time *t*. Among zj=[uj,vj,hj,γj,h˙j,γ˙j,x˙j,y˙j,ρj,ϑ]

T, the components of zj are the state's parameter of *j* time, which we separately explain below: (u,v) is the bounding box center coordinate, h*h* is the bounding box height, γ represents the bounding box ratio, h˙j,γ˙j*h*˙ are the velocities, (x˙j,y˙j) are the velocities along the *x*-axis and *y*-axis on the 2D plane, ρ represents the root-mean-square (RMS) values of the ego vessel's pitch and roll, and ϑ denotes the type of target vessel.

The sums of the multimodel were weighted according to their matched likelihood probability. The motion state of the target vessel was calculated by

$$\begin{aligned}\hat{\mathcal{Y}}^{(t+\tau)} &= \mathbb{E}[\mathcal{Y}^{(t+\tau)}|Z] \\ &= \textstyle\sum_{i=1}^{N} \hat{\mathcal{Y}}_i^{(t+\tau)} P(m_i|Z)\end{aligned} \tag{11}$$

and the estimation error of covariance matrix was:

$$\mathcal{P}(t+\tau) = \sum_{i=1}^{N} \mathbb{P}(m_i|Z)\{\mathcal{P}_i(t+\tau) + [\hat{\mathcal{Y}}_i^{(t+\tau)} - \hat{\mathcal{Y}}^{(t+\tau)}][\hat{\mathcal{Y}}_i^{(t+\tau)} - \hat{\mathcal{Y}}^{(t+\tau)}]^T\} \tag{12}$$

where P(mi|Z)ℙ is the posterior probability that can be obtained from the softmax layer of the GRU-attention model with loss and as the individual model's output:

$$\mathbb{P}(m_i|Z) = \frac{Exp(m_i)}{\sum_{k=1}^{N} Exp(m_k)} \tag{13}$$

The adaptive moment estimation (Adam) was used as the optimizer and the cross entropy for the loss function. We assumed that Θ represented the parameter set of GRU to be trained, and the loss function was written, as follows:

$$\mathcal{J}(\Theta) = -\frac{1}{N}\sum_{i=1}^{N} m_i Log(h_\Theta(m_i)) + (1-m_i)Log(1-h_\Theta(m_i)) \tag{14}$$

3.3. Multicue for Data Association

During the data association procedure, we propose a hybrid affinity model that is based on multi cues to evaluate the similarity between the detected vessels and existing tracklets, which contains both long-term cues and short-term cues. We regard appearance as a long-term cue and, meanwhile, the short-term cues consist of surrounding vessels motion measurements and the dynamic attitude of ego vessel, such as pitch and roll. We use adaptive association gate of appearance to confirm the validation of measurements before the data association algorithm is carried out.

3.3.1. Adaptive Association Gate of Appearance

The simple camera model [40] has been applied to put targets into a three-dimensional (3D) perspective considering the height of the target in the image, and it assumes that all the objects of interest rest on the ground plane. For a vessel sailing on the seaplane, which perfectly coincided with the assumptions, we additionally assumed that the ship's heading angle was consistent with its course.

Assume that camera parameters Θ consist of the following elements: focus length fΘ, height hΘ*h*, camera tilt angle ψΘ, absolute velocity υΘ, image center μc, horizon position vc, and 3D location (xΘ,zΘ). Thus, the projection f can be defined as

$$\widetilde{Z} = \begin{bmatrix} R(\psi_\Theta) & 0 \\ 0 & 1 \end{bmatrix} Z + \begin{bmatrix} x_\Theta \\ z_\Theta \\ 0 \end{bmatrix}, X = f(\widetilde{Z}) = \begin{bmatrix} \frac{f_\Theta x_Z}{z_Z} + u_c \frac{f_\Theta h_\Theta}{z_Z} + v_c \frac{f_\Theta h_Z}{z_Z} \end{bmatrix}^T \tag{15}$$

where X=[μ,v,h] represents the central coordination and the height of the vessel in the image plane, while Z˜ denotes the location in the current camera coordinates. We could easily estimate the scale and location variances of the vessel in the image based on the simple linear relationship between the vessel's image size and the *y*-axis image location [41].

When the target vessel moves closer to the camera, the bounding box of imaging area gradually becomes larger and vice versa. Suppose that the bounding box center of the current target vessel is at (x, y) coordinate of the image, x and y divide the whole image into four quadrants, and each quadrant is a candidate region. We present four possible situations in Figure 5, while considering the pitching and rolling attitude of the ego vessel at the present moment.

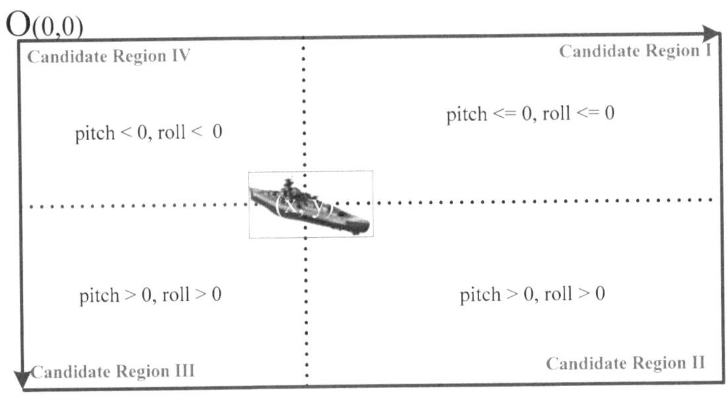

Figure 5. *Sketch map of adaptive association gate of appearance.*

3.3.2. Long-Term Cues

It is well known that the condition of a vessel on the sea is much more complex than that of vehicles and pedestrians on the ground due to the influence of environmental factors, such as waves, wind, and ocean currents. The camera also moves with the movement of the ego vessel, such as swing (surge, sway, and heave) and rotating (roll, pitch, and yaw), which cause the target to be temporarily lost from the camera scene. These dramatic changes on sea surface require the re-identification of the same vessel from adjacent frames or even across some frames according to their similarity measurements as the long-term cues.

The MDNet structure proposed in [22] was employed here, and a dataset with more than 500 vessels and nearly 150,000 images, named Ship-ReID, was constructed. We found the optimal mapping from image space to feature space after training on it. Suppose that inputs are triplet units {< Ja,Jp,Jn >}, indicating the anchor image, positive image, and negative image, respectively. < Ja,Jp > is a positive pair belonging to the same vessel, while < Ja,Jn > is the negative pair from different vessels. We used the normalized cosine similarity between the feature vector to measure the distance, such as Da,p and Da,n, and guided the network to train in the direction of pushing positive and negative samples away. We ensured that the distance between positive samples was very close by means of the improved triplet loss, as follows:

$$\mathcal{L}(\mathcal{I}_a, \mathcal{I}_p, \mathcal{I}_n) = \sum^{N}\left[\mathcal{D}_{a,p} + max(\mathcal{D}_{a,p} - \mathcal{D}_{a,n} + \lambda, 0)\right] \tag{16}$$

where λ is the pre-set threshold parameter.

3.3.3. Data Association and Fusion Method

As introduced in [42], we adopted heuristic simulated annealing to solve the assignment problem during the data association stage [43]. Assuming that we have D detectors and T tracklets waiting to match at time k, and D={d1,d2,···,dD},T={t1,t2,···,tT}, G(k) is the adaptive gate association set, and F(k) is the tracking gate set consisting of the rectangular region generated by the coordination predicted by MM filter trackers. Subsequently, the association cost matrix between the i-th tractlet's predictions t^(k)i and the j-th detectors d(k)j could be defined as

$$C^{(k)}(i,j) = \begin{cases} c^{(k)}(\hat{t}_i, d_j), if\ d_j^{(k)} \cap \mathcal{G}_i^{(k)} \neq \varnothing\ or\ d_j^{(k)} \cap \mathcal{G}_i^{(k)} \neq \varnothing \\ \infty, \quad \text{otherwise} \end{cases} \tag{17}$$

where c(t^i,dj) is the hybrid affinity metric of association at time k when the j-th detector falls into the correlation gate corresponding to the prediction coordinate of the i-th tracklet, which can be calculated as

$$c^{(k)}(\hat{t}_i, d_j) = (1 - \mathcal{A}) \cdot motion^{(k)}(\hat{t}_i^{(k)}, d_j^{(k)}) + \mathcal{A} \cdot appear^{(k)}(\hat{t}_i^{(k)}, d_j^{(k)}) \tag{18}$$

where the weight coefficient is denoted as A, which is depicted by the attitude of the ego vessel. Equation (18) demonstrates that, as the pitch and roll of the hull increase, the weight of the appearance is enhanced, and vice versa.

$$\mathcal{A} = \frac{\left[(p^{(k)})^2 + (r^{(k)})^2\right]^{1/2}}{\lambda\left[(p_{max})^2 + (r_{max})^2\right]^{1/2}} \tag{19}$$

where p,r is the dynamic pitch and roll at time *k*, which can be obtained from real-time measurement of the ego vessel's electric compass or IMU; pmax,rmax represent the maximum range of pitch and roll, which are determined by the intrinsic characteristics of the hull, which can be obtained from prior measurement; and, λ is a hyper parameter.

appear(t^i,dj) is the measurement of appearance similarity between t^i and dj, and motion(t^i,dj) represents their measurements of motion similarity that were obtained by calculating the negative logarithm of the intersection-of-union ratio:

$$motion(\hat{t}_i^{(k)}, d_j^{(k)}) = -log\left(\frac{Intersection(\hat{t}_i^{(k)}, d_j^{(k)})}{Union(\hat{t}_i^{(k)}, d_j^{(k)})}\right) \tag{20}$$

3.4. The Proposed M³C Tracking by Detection

In this section, an end-to-end and real-time tracking by detection pipeline is presented, for which the tracklet initiation and termination conditions were both determined by a continuous three-frame method, as described in Algorithm 1:

Algorithm 1: The proposed M³C tracking by detection

Inputs: The sequential frames of surrounding vessels $\mathcal{S}$, pre-trained GRU model $\mathcal{M}_{gru}$, Darknet53 model $\mathcal{M}_{d53}$, Ship-Reid model $\mathcal{M}_{reid}$, and periodic attitude data $\mathcal{A}$ of the ego vessel
Initialization: $\mathcal{T} \leftarrow \varnothing$, $\mathcal{A}_{max}$, $(v_k)_{count} = 0$, $(t_i)_{count} = 0$
Outputs: Continuous tracklets $\mathcal{T} = \{t_i\}_{i=1}^{M}$ of surrounding vessels
Procedure:

```
foreach frame at current time t
    detect all potential vessels V = {v_k}_{k=1}^N from the F_t frame using M_d53
    foreach tracklet of vessel      // Association gate generated from prior time
        Predict the kinematic track gate from MM filter from time t - 1
        & adaptive association gate of appearance
    endfor
    MC data association using simulated annealing algorithm
    foreach pair of matched detection and tracklet
        tracklet update by MM filter
    endfor
    foreach detection vessel v_k not associated with any tracklets in T
        (v_k)_count ++
        if (v_k)_count >= 3      //Three-frame tracklet initiation
            initialize a new tracklet t
            T ∪ = {t}
        end
    endfor
    foreach vessel tracklet t_i not associated with any detection in V
        (t_i)_count ++
        tracklet extrapolation
        if (v_k)_count >= λ_max      // Extrapolation times exceeding the pre-set value λ_max
            terminate the tracklet t_i
            T = T - {t_i}
    end
  endfor
endfor      //reach the end of the video sequence
```

4. EXPERIMENT AND RESULTS ANALYSIS

4.1. Experimental Setup

4.1.1. Datasets

We performed extensive experiments on SMD [1] and the PETS 2016 maritime dataset [6] to evaluate performance of the proposed M³C. The detector was pretrained on a Marvel vessel image dataset by the offline method [44].

SMD contains 51 annotated video fragments, 40 onshore videos, and 11 onboard videos. We estimated the approximate range of roll and pitch of the ego vessel from horizon ground truths (GTs). The PETS 2016 maritime dataset includes 20 RGB video fragments that were captured by four digital pan-tilt-zoom (PTZ) cameras, three of which were at the side and one at the stern. Two skiff boats and two fish boats as the supplementary target vessels performed behaviors, such as speeding up, loitering, moving around the ego vessel, and so on. The collected the Marvel vessel image dataset that consisted of more than 140,000 vessel images. It was divided into 26 superclasses, drawing from the www.shipspotting.com website of online vessel photos and trackers. We resized the image to 608 × 608 pixels and manually labeled all of the selected pictures as belonging to five superclasses, which were tug ship, container, fish boat, skiff boat, and passenger ship.

4.1.2. Implementation Details

We used an i7-8700K CPU with 32 GB of memory and dual NVIDIA Titan RTX GPU with a Pytorch deep learning framework to implement our algorithm. We devised the detector based on the state-of-the-art YOLOv3, as mentioned in the previous section. The size of prior anchors is list in Table 1.

Table 1. The feature map and prior anchors of Singapore Marine Dataset (SMD) and PETS 2016 maritime dataset. A total of nine anchors were used to generate bounding boxes, and the size of prior anchors was calculated by the k-means algorithm.

Feature Map		**19 × 19**			**38 × 38**			**76 × 76**	
Receptive Field		**Big**			**Medium**			**Small**	
Prior Anchors	110 × 84	176 × 93	267 × 166	35 × 65	62 × 57	107 × 39	43 × 19	38 × 39	68 × 29

We used GRU with 128 units for the vessel's maneuverability discrimination and trained the model while using Adam with a dropout of 0.2 for regularization and with a learning rate of 0.001. Due to the lack of adequate vessel tracklets or video galleries for training, it was pretrained by the public AIS dataset [45], and was then fine-tuned on the training set partitioned from SMD and PETS 2016.

4.2. Performance Evaluation

4.2.1. Evaluation Metric

The evaluation was carried out according to the standard root mean square error (RMSE)/mean absolute deviation (MAE) and CLEAR MOT metrics that were proposed in [46], and a brief description of each metric is listed below:

$$MOTA = 1 - \frac{\sum_t \left(N_{FN}^{(t)} + N_{FP}^{(t)} + N_{IDS}^{(t)}\right)}{\sum_t N_{GT}^{(t)}} \tag{21}$$

where N(t)FN, N(t)FP, and N(t)IDS are the number of false negatives, false positives, and IDS in the t frame index; and, N(t)GT represents the number of ground truth targets.

MOTP (↑): multi-object tracking precision, as calculated by the following formula:

$$MOTP = \frac{\sum_{i,t} d_i^{(t)}}{\sum_t \left(N_{TP}^{(t)} + N_{IDS}^{(t)}\right)} \tag{22}$$

where N(t)TP is the number of true positives in the *t* frame index and d(t)i represents the bounding box overlap of the *i*-th target.

RMSE (↓): root-mean-square error; MAE (↓): mean absolute deviation; MT (↑): mostly tracked targets; ML (↓): mostly lost targets; FN (↓): total number of false negatives; FP (↓): total number of false positives; IDS (↓): identity switches; FPS (↑): frames per second; and, mAP (↑): mean average precision. The colored arrow after each metric index indicates whether the value of increasing (↑) or decreasing (↓) is beneficial. MOTP and MOTA lie in the range of [0,100%], with the best value being 100%.

4.2.2. Qualitative Results

For the sake of fairness, performance comparison was separately carried out according to onshore or onboard datasets, because the results of some trackers can be biased due to camera motion. We discuss the comparison with state-of-the-art tracking methods, such as the framework of Markov decision process (MDP) [47], the combination of Kalman filter and Hungarian assignment algorithm (SORT) [48], the kernel correlation filter method (KCF) [49], the combination of Kalman filter and Kuhn-Munkres (POI) [50], SORT with deep association metric (DeepSORT) [19], and candidate selection combined with re-identification (MOTDT) [51].

The M³C pipeline obtained a 2.5% increase and a 33.5% decrease in MOTA and IDS as compared with the second best onshore dataset, as shown in Table 2. As far as the onboard dataset, the corresponding figures were 3% and 30.6%, and ML was also the lowest. In addition, on both onshore and onboard datasets, FPS was more than 10 in both scenarios, which indicated that real-time performance was obtained.

Table 2. Quantitative evaluation of different trackers on the two maritime datasets. We treated PETS 2016 as the onshore dataset considering a stationary large ship.

Dataset	Tracker	MOTA↑	MOTP↑	MT↑	ML↓	IDS↓	FPS↑
SMD (onshore) + PETS 2016	MDP	30.3%	71.3%	13.2%	38.4%	426	1
	SORT	59.8%	79.6%	25.4%	22.7%	631	**56**
	KCF	70.3%	80.2%	**37.8%**	22.3%	382	25
	POI	66.1%	79.5%	34.6%	**20.8%**	453	10
	M³C (Ours)	**72.8%**	80.4%	37.4%	21.2%	**254**	20
SMD (onboard)	DeepSORT	60.4%	**79.1%**	32.8%	18.2%	56	**36**
	MOTDT	57.6%	70.9%	**34.2%**	28.7%	49	23
	M³C (Ours)	**63.4%**	74.6%	26.2	**17.9%**	**34**	16

4.2.3. Visual Tracking Results

We present the visual tracking results of some videos to illustrate the performance of our proposed M³C framework intuitively.

In Figure 6 and Figure 7, we show the tracking robustness of a reappearing vessel after being lost for a short time Figure 7b, and even complete occlusion Figure 6b, for which our proposed Ship-ReID worked well as the long term cue.

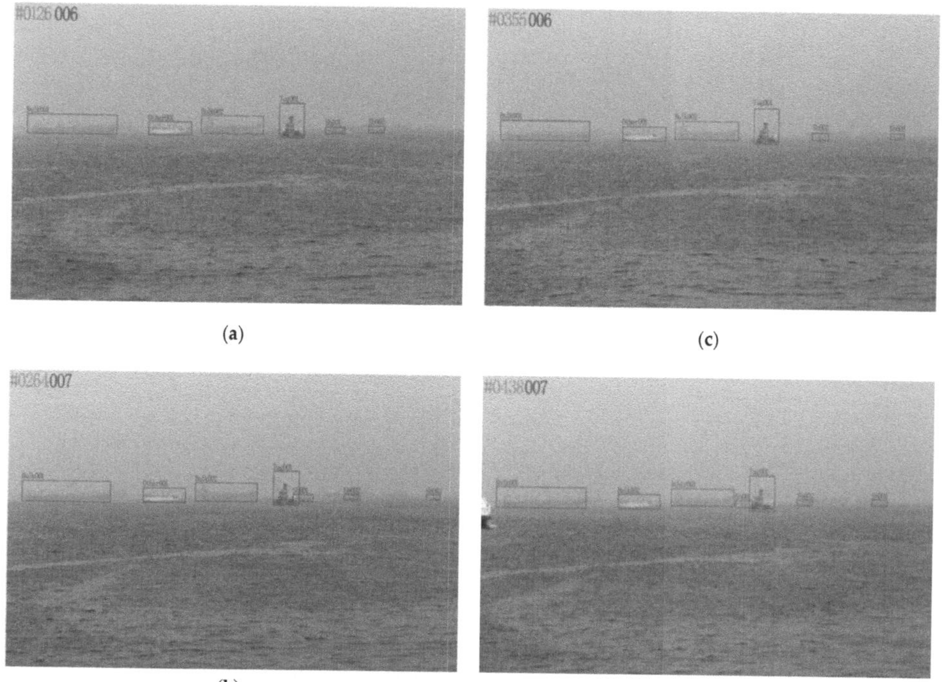

(a) (c) (b) (d)

Figure 6. *Visual tracking results of video sequence (MVI_1448_VIS_Haze) with occlusion. Six vessels have been stably tracked in frame #0126 (**a**). Starting in frame #0264 (**b**), Sb001 was occluded by Tug001 until being completely occluded in frame #0355 (**c**) and then reappeared in frame #0370 until it was completely visible in frame #0438 as shown in (**d**).*

4.3. Ablation Study and Analysis

4.3.1. Effect of Multimodel Fusion Filter

In this scheme, we considered three different motion scenarios of the target vessel, including a straight line with constant velocity motion, acceleration, and turning motion. We conducted an ablation experiment by replacing MM with a single Kalman filter (CVCV) and interactive multiple models (IMM) [52] while considering the CACA, CVCV, and CMCM model by keeping other conditions unchanged. RMSE and MAE were used as the evaluation indicators, respectively. Table 3 shows the comparison results.

IMM did not perform better than the single CVCV model with a Kalman filter when no maneuver occurred and was even worse than the latter, because of the possible delay of model transfers, as can be seen in Table 3. The proposed MM filter overcame this very well, and RMSE and MAE both increased by 30% in the three common motion scenarios.

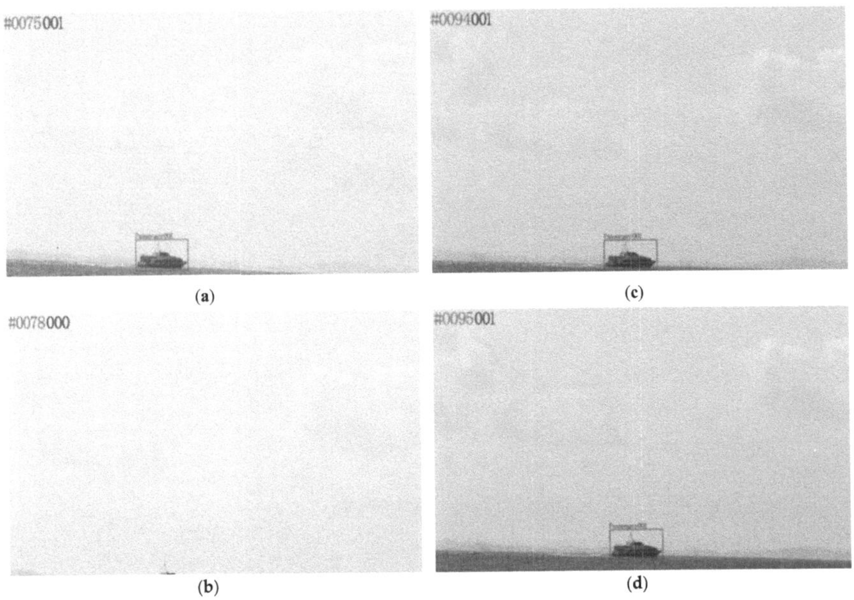

Figure 7. *Visual tracking results of video sequence (MVI_0799_VIS_OB) with reappear after lost. In frame #0075 (**a**), a vessel has been stably tracked. Due to the rolling of the ego vessel, Passenger001 gradually disappeared from the camera scene in frame #0076 until frame #0078 (**b**), when it completely disappeared. Subsequently, it began to reappear in frame #0091 until frame #0094 (**c**), when it was completely visible. As shown in frame #0095 (**d**), the vessel was re-tracked steadily.*

Table 3. Results of averaged RMSE and mean absolute deviation (MAE). SMD (onshore) and the PETS 2016 dataset were split into three different motion scenarios for the experiment, and a Kalman filter with a single constant velocity (CV) model was used as the benchmark.

Motion Scenarios	**Algorithm**	**Averaged RMSE (L2 Norm)**		**Averaged MAE (L1 Norm)**	
		X pos.	**Y pos.**	**X pos.**	**Y pos.**
Straight line, constant velocity	Kalman (CV)	14.71	16.06	11.47	14.53
	IMM (CV + CA + CM)	15.10	15.87	12.03	14.28
	MM(Proposed)	**9.10**	**9.58**	**5.75**	**6.02**
Acceleration	Kalman (CV)	27.71	21.29	25.38	20.54
	IMM (CV + CA + CM)	21.17	18.26	18.67	15.88
	MM(Proposed)	**13.93**	**12.21**	**8.04**	**7.04**
Turning	Kalman (CV)	20.99	15.85	19.54	15.03
	IMM (CV + CA + CM)	15.10	13.11	12.93	10.98
	MM (Proposed)	**12.44**	**9.56**	**8.80**	**6.67**

4.3.2. Effect of Detector

As illustrated in Table 4, the detector built on YOLOv3 with Darknet53 as the backbone increased mAP by nearly 40% when compared with SSD, and FPS was five times higher than that of faster R-CNN. It was a good compromise between mAP and FPS, and both of them were equally critical for improving tracking performance. These experiments showed that the quality of the detection algorithm has significant influence on multi target tracking performance.

Table 4. Detector performance in the ablation study. We replaced the detector with SSD and faster regions with CNN features (R-CNN), respectively, for these ablation experiments.

Tracker	Detector	mAP	FPS	Averaged RMSE (L2 Norm)		Averaged MAE (L1 Norm)	
				X pos.	Y pos.	X pos.	Y pos.
M^3C	SSD300 (Mobilenet)	**41.2**	**46**	20.22	17.59	9.54	16.37
	YOLOv3 (Darknet53)	57.9	20	9.10	9.58	**7.75**	6.02
	Faster R-CNN (Resnet50)	**59.1**	4	**8.94**	**6.39**	8.44	**5.83**

4.3.3. Effect of Multicue Data Association

It can be observed from Table 5 that the M^3C pipeline obtained the highest value of MOTA, which is one of the most important indicators of multitarget tracking. These data lead us to the conclusion that combining the motion features with appearance features can effectively improve the tracking accuracy.

Table 5. Results of the ablation study. multimodel (MM), single model (SM), and re-identification (ReID) denote appearance model, multimodel filter, single model filter, Ship-ReID, respectively.

Tracker	Detector	MOTA↑	MOTP↑	MT↑	ML↓	IDS↓	FPS↑
MM	YOLOv3	46.2%	44.5%	12.9%	43.2%	74	28
MM + Attention	YOLOv3	47.1%	43.8%	13.1%	44.7%	73	**32**
SM+ReID	YOLOv3	59.8%	64.5%	21.4%	23.6%	62	20
MM+ReID (M^3C)	YOLOv3	**63.4%**	**74.6%**	**36.2%**	**17.9%**	**24**	**16**

By combining deep feature extraction and a traditional kinematic filtering algorithm, the proposed M^3C approach strikes a balance between tracking accuracy and speed and improves tracking accuracy while guaranteeing real-time performance, as discussed previously.

5. CONCLUSIONS

We proposed a novel tracking by detection approach integrating MM and MC to detect and track surrounding vessels of USVs at sea in this paper. MM was used to solve the problem of unstable tracking of a maneuvering target in the traditional single-model Kalman tracker (such as the CVCV model). MC combines the attitude of the ego vessel and the appearance of the target vessels to solve the problem of frequent IDS that is caused by motion blurring and occlusion. Experiments have demonstrated its efficiency and robustness and it achieved real-time performance. In the future, we plan to focus on the following two directions. First, we will construct a more complete Ship-ReID dataset and optimize the network framework and algorithm. Second, we intend to integrate Ship-ReID with a convolutional neural network of the detector and to co-train them in a unified network, so as to further improve the real-time performance.

REFERENCES

1. Prasad, D.K.; Rajan, D.; Rachmawati, L.; Rajabally, E.; Quek, C. Video Processing from Electro-Optical Sensors for Object Detection and Tracking in a Maritime Environment: A Survey. *IEEE Trans. Intell. Transp. Syst.* **2017**, *18*, 1993–2016.
2. Bloisi, D.; Iocchi, L. ARGOS—A video surveillance system for boat traffic monitoring in venice. *Int. J. Pattern Recognit. Artif. Intell.* **2009**, *23*, 1477–1502.
3. Kang, Y.T.; Chen, W.J.; Zhu, D.Q.; Wang, J.H.; Xie, Q.M. Collision avoidance path planning for ships by particle swarm optimization. *J. Mar. Sci. Technol.* **2018**, *26*, 777–786.
4. Dijk, J.; van der Stap, N.; van den Broek, B.; Pruim, R.; Schutte, K.; den Hollander, R.; van Opbroek, A.; Huizinga, W.; Wilmer, M. Maritime detection framework 2.0: A new approach of maritime target detection in electro-optical sensors. In Proceedings of the Electro-Optical and Infrared Systems: Technology and Applications XV, Berlin, Germany, 9 October 2018; Volume 10795, p. 1079507.
5. Yazdi, M.; Bouwmans, T. New trends on moving object detection in video images captured by a moving camera: A survey. *Comput. Sci. Rev.* **2018**, *28*, 157–177.
6. Patino, L.; Nawaz, T.; Cane, T.; Ferryman, J. Pets 2016: Dataset and Challenge. In Proceedings of the IEEE Computer Society Conference on Computer Vision and Pattern Recognition Workshops (CVPR), Las Vegas, NV, USA, 1–26 June 2016; pp. 1–8.
7. Girshick, R. Fast R-CNN. In Proceedings of the IEEE International Conference on Computer Vision (ICCV), Santiago, Chile, 7–13 December 2015; pp. 1440–1448.
8. Ren, S.; He, K.; Girshick, R.; Sun, J. Faster R-CNN: Towards Real-Time Object Detection with Region Proposal Networks. *IEEE Trans. Pattern Anal. Mach. Intell.* **2017**, *39*, 1137–1149.
9. Dai, J.; Li, Y.; He, K.; Sun, J. R-FCN: Object detection via region-based fully convolutional networks. In Proceedings of the Advances in Neural Information Processing Systems, Barcelona, Spain, 5–10 December 2016; pp. 379–387.
10. Liu, W.; Anguelov, D.; Erhan, D.; Szegedy, C.; Reed, S.; Fu, C.Y.; Berg, A.C. SSD: Single shot multibox detector. In Proceedings of the Computer Vision—ECCV 2016, Amsterdam, The Netherlands, 8–16 October 2016; Lecture Notes in Computer Science. pp. 21–37.
11. Redmon, J.; Farhadi, A. Yolov3: An Incremental Improvement. *arXiv* **2018**, arXiv:1804.02767.
12. Kong, T.; Sun, F.; Liu, H.; Shi, J. FoveaBox: Beyond Anchor-based Object Detector. *arXiv* **2019**, arXiv:1904.03797.
13. Zhu, C.; He, Y.; Savvides, M. Feature Selective Anchor-Free Module for Single-Shot Object Detection. *arXiv* **2019**, arXiv:1903.00621.

14. Tonissen, S.M.; Evans, R.J. Performance of dynamic programming techniques for track-before-detect. *IEEE Trans. Aerosp. Electron. Syst.* **1996**, *32*, 1440–1451.

15. Kieritz, H.; Hübner, W.; Arens, M. Joint detection and online multi-object tracking. In Proceedings of the IEEE Computer Society Conference on Computer Vision and Pattern Recognition Workshops (CVPRW), Salt Lake City, UT, USA, 18–22 June 2018; pp. 1540–15408.

16. Gordon, D.; Farhadi, A.; Fox, D. Re^3: Real-time recurrent regression networks for visual tracking of generic objects. *IEEE Robot. Autom. Lett.* **2018**, *3*, 788–795.

17. Bae, S.H.; Yoon, K.J. Confidence-Based Data Association and Discriminative Deep Appearance Learning for Robust Online Multi-Object Tracking. *IEEE Trans. Pattern Anal. Mach. Intell.* **2018**, *40*, 595–610.

18. Arandjelović, O. Automatic vehicle tracking and recognition from aerial image sequences. In Proceedings of the 12th IEEE International Conference on Advanced Video and Signal Based Surveillance (AVSS), Karlsruhe, Germany, 25–28 August 2015; pp. 1–6.

19. Wojke, N.; Bewley, A. Deep cosine metric learning for person re-identification. In Proceedings of the 2018 IEEE Winter Conference on Applications of Computer Vision (WACV), Lake Tahoe, NV, USA, 12–15 March 2018; pp. 748–756.

20. Yang, X.; Tang, Y.; Wang, N.; Song, B.; Gao, X. An End-to-End Noise-Weakened Person Re-Identification and Tracking with Adaptive Partial Information. *IEEE Access* **2019**, *7*, 20984–20995.

21. Zhao, D.; Fu, H.; Xiao, L.; Wu, T.; Dai, B. Multi-object tracking with correlation filter for autonomous vehicle. *Sensors* **2018**, *18*, 2004.

22. Liu, H.; Tian, Y.; Wang, Y.; Pang, L.; Huang, T. Deep Relative Distance Learning: Tell the Difference between Similar Vehicles. In Proceedings of the IEEE Computer Society Conference on Computer Vision and Pattern Recognition (CVPR), Las Vegas, NV, USA, 27–30 June 2016; pp. 2167–2175.

23. Fefilatyev, S.; Goldgof, D.; Shreve, M.; Lembke, C. Detection and tracking of ships in open sea with rapidly moving buoy-mounted camera system. *Ocean Eng.* **2012**, *54*, 1–12.

24. Jeong, C.Y.; Yang, H.S.; Moon, K.D. Fast horizon detection in maritime images using region-of-interest. *Int. J. Distrib. Sens. Networks* **2018**, *14*.

25. Zhang, Y.; Li, Q.Z.; Zang, F.N. Ship detection for visual maritime surveillance from non-stationary platforms. *Ocean Eng.* **2017**, *141*, 53–63.

26. Sun, Y.; Fu, L. Coarse-fine-stitched: A robust maritime horizon line detection method for unmanned surface vehicle applications. *Sensors* **2018**, *18*, 2825.

27. Bovcon, B.; Perš, J.; Kristan, M. Stereo obstacle detection for unmanned surface vehicles by IMU-assisted semantic segmentation. *Rob. Auton. Syst.* **2018**, *104*, 1–13.

28. Cane, T.; Ferryman, J. Evaluating deep semantic segmentation networks for object detection in maritime surveillance. In Proceedings of the 15th IEEE International Conference on Advanced Video and Signal-Based Surveillance (AVSS), Auckland, New Zealand, 27–30 November 2018; pp. 1–6.

29. Kim, K.; Hong, S.; Choi, B.; Kim, E. Probabilistic Ship Detection and Classification Using Deep Learning. *Appl. Sci.* **2018**, *8*, 936.

30. Marié, V.; Béchar, I.; Bouchara, F. Real-time maritime situation awareness based on deep learning with dynamic anchors. In Proceedings of the 15th IEEE International Conference on Advanced Video and Signal-Based Surveillance (AVSS), Auckland, New Zealand, 27–30 November 2018; pp. 1–6.

31. Cao, X.; Gao, S.; Chen, L.; Wang, Y. Ship recognition method combined with image segmentation and deep learning feature extraction in video surveillance. *Multimed. Tools Appl.* **2019**, 1–16.
32. Leclerc, M.; Tharmarasa, R.; Florea, M.C.; Boury-Brisset, A.C.; Kirubarajan, T.; Duclos-Hindié, N. Ship Classification Using Deep Learning Techniques for Maritime Target Tracking. In Proceedings of the 2018 21st International Conference on Information Fusion (FUSION), Cambridge, UK, 10–13 July 2018; pp. 737–744.
33. Kejriwal, L.; Singh, I. A Hybrid Filtering Approach of Digital Video Stabilization for UAV Using Kalman and Low Pass Filter. *Procedia Comput. Sci.* **2016**, *93*, 359–366.
34. Best, R.A.; Norton, J.P. A new model and efficient tracker for a target with curvilinear motion. *IEEE Trans. Aerosp. Electron. Syst.* **1997**, *33*, 1030–1037.
35. Perera, L.P.; Oliveira, P.; Guedes Soares, C. Maritime Traffic Monitoring Based on Vessel Detection, Tracking, State Estimation, and Trajectory Prediction. *IEEE Trans. Intell. Transp. Syst.* **2012**, *13*, 1188–1200.
36. Li, J.; Dai, B.; Li, X.; Xu, X.; Liu, D. A Dynamic Bayesian Network for Vehicle Maneuver Prediction in Highway Driving Scenarios: Framework and Verification. *Electronics* **2019**, *8*, 40.
37. Hochreiter, S.; Schmidhuber, J. Long Short-Term Memory. *Neural Comput.* **1997**, *9*, 1735–1780.
38. Cho, K.; van Merrienboer, B.; Gulcehre, C.; Bahdanau, D.; Bougares, F.; Schwenk, H.; Bengio, Y. Learning Phrase Representations using RNN Encoder–Decoder for Statistical Machine Translation. In Proceedings of the 2014 Conference on Empirical Methods in Natural Language Processing (EMNLP), Doha, Qatar, 25–29 October 2014; pp. 1724–1734.
39. Bar-Shalom, Y.; Blackman, S.S.; Fitzgerald, R.J. Dimensionless score function for multiple hypothesis tracking. *IEEE Trans. Aerosp. Electron. Syst.* **2007**, *43*, 392–400.
40. Hoiem, D.; Efros, A.A.; Hebert, M. Putting objects in perspective. *Int. J. Comput. Vis.* **2008**, *80*, 3–15.
41. Richardson, E.; Peleg, S.; Werman, M. Scene geometry from moving objects. In Proceedings of the Advanced Video and Signal-based Surveillance (AVSS), Seoul, South Korea, 26–29 August 2014; pp. 13–18.
42. Racine, V.; Hertzog, A.; Jouanneau, J.; Salamero, J.; Kervrann, C.; Sibarita, J. Multiple-Target Tracking of 3D Fluorescent Objects Based on Simulated Annealing. In Proceedings of the 3rd IEEE International Symposium on Biomedical Imaging: Nano to Macro, Arlington, VA, USA, 6–9 April 2006; pp. 1020–1023.
43. Osman, I.H. Heuristics for the generalised assignment problem: Simulated annealing and tabu search approaches. *OR Spektrum* **1995**, *17*, 211–225.
44. Gundogdu, E.; Solmaz, B.; Yücesoy, V.; Koc, A. MARVEL: A large-scale image dataset for maritime vessels. In Proceedings of the Asian Conference on Computer Vision (ACCV), Taipei, Taiwan, 21–23 November 2016; pp. 165–180.
45. Nguyen, D.; Vadaine, R.; Hajduch, G.; Garello, R.; Fablet, R. A multi-task deep learning architecture for maritime surveillance using AIS data streams. In Proceedings of the 2018 IEEE 5th International Conference on Data Science and Advanced Analytics (DSAA), Turin, Italy, 1–3 October 2018; pp. 331–340.
46. Bernardin, K.; Stiefelhagen, R. Evaluating multiple object tracking performance: The clear mot metrics. *Eurasip J. Image Video Proc.* **2008**, *2008*, 246309.
47. Xiang, Y.; Alahi, A.; Savarese, S. Learning to track: Online multi-object tracking by decision making. In Proceedings of the 2015 IEEE International Conference on Computer Vision (ICCV), Santiago, Chile, 7–13 December 2015; pp. 4705–4713.

48. Bewley, A.; Ge, Z.; Ott, L.; Ramos, F.; Upcroft, B. Simple online and realtime tracking. In Proceedings of the 2016 International Conference on Image Processing (ICIP), Phoenix, AZ, USA, 25–28 September 2016; pp. 3464–3468.

49. Henriques, J.F.; Caseiro, R.; Martins, P.; Batista, J. High-speed tracking with kernelized correlation filters. *IEEE Trans. Pattern Anal. Mach. Intell.* **2014**, *37*, 583–596.

50. Yu, F.; Li, W.; Li, Q.; Liu, Y.; Shi, X.; Yan, J. POI: Multiple object tracking with high performance detection and appearance feature. In Proceedings of the European Conference on Computer Vision (ECCV), Amsterdam, The Netherlands, 8–16 October 2016; pp. 36–42.

51. Chen, L.; Ai, H.; Zhuang, Z.; Shang, C. Real-Time Multiple People Tracking with Deeply Learned Candidate Selection and Person Re-Identification. In Proceedings of the 2018 IEEE International Conference on Multimedia and Expo (ICME), San Diego, CA, USA, 23–27 July 2018; pp. 1–6.

52. Labbe, R.R. Kalman and Bayesian Filters in Python. 2018. Available online: **https://github.com/rlabbe/Kalman-and-Bayesian-Filters-in-Python** (accessed on 20 May 2019).

Chapter 7

Financial Analysis of Automated Container Terminal Capacity from the Perspective of Terminal Operating Company

Nam Kyu Park[1] and Yohan An[2]

[1] *Department of International Logistics, Tongmyong University, Busan 48520, Korea*

[2] *Department of Finance & Accounting, Tongmyong University, Busan 48520, Korea*

ABSTRACT

This study examines the financially feasibility of the proper terminal capacity by each berth size of the automatic container terminal (ACT) from the perspective of Terminal Operating Company (TOC). ACT is a highly productive and eco-friendly port facility, but it requires a lot of capital investment. Thus, the investment of ACT should consider the TOC's operating profit preservation to determine the proper terminal capacity. In this study, we attempt to conduct financial analysis using the net present value method and estimate breakeven handling volume of five berth sizes (nine, five, four, three, and two berths). In particular, as the aim of this study is to propose a capacity model of ACT, the model must be able to adapt to a variety of situations reflecting the number of berths and financial discount rate. The case study focused on the new port of Busan, introducing ACT. As a result, the breakeven terminal capacity changes from 560,421 TEU of the 9-berth model to 633,102 TEU of the 2-berth model, applying a 4.5% standard discount ratio. In a sensitivity test considering the change in discount rate and the size of the berth at the same time, the net present value (NPV) has a positive value at the level of at least 550,000 TEU (nine berths and 3.5% discount rate) and up to 650,000 TEU (two berths and 5.5% discount rate). The method of optimizing financial efficiency by analyzing the appropriate loading capacity will be an important support tool in decision-making by providing the analysis results and reasonable information obtained during the analysis process to the TOC, the main stakeholder in the adoption of ACT.

Keywords: *automated container terminal capacity; financial analysis; Terminal Operating Company*

1. INTRODUCTION

The remarkable changes in the global port and logistics market are economies of scale due to the enlargement of ships and cost reduction through minimizing dwell time at ports [1,2]. In response to these changes, major ports in each country are improving the efficiency of port operation by introducing automated container terminal (ACT) systems [3]. ACT systems based on advanced technology are being introduced mainly in countries such as the Netherlands, Germany, USA, and China which have world-class advanced port infrastructure, and Korea is also introducing automated terminals in the new port of Busan [4].

In this context, the ACT is a terminal in which Automated Guided Vehicle (AGV) and Automated Rail Mounted Gantry Crane (ARMGC) are operated. Of the 23 automated container terminals worldwide, 12 are semi-automated and 11 are fully automated. The APMT and RWG in Rotterdam Port that opened in 2015, the LBCT that opened in 2016, the QQCTN in Qingdao Port that opened in 2017, and the 4th Yangsan Port are fully automated, showing recent trends in the types of automated terminals. The Korean government has introduced ACT for stages 2–5 and 2–6 of the new port of Busan by benchmarking the ACT already in operation.

Since 1997, the Korean government has introduced the Terminal Operating Company (TOC) system, which is used by private companies TOC for a certain period of time by leasing port facilities from the state [5]. The TOC system contributes to the improvement of the efficiency and productivity of ports and the service level for users by introducing private enterprise operation methods to port operations [6]. The proper terminal capacity is considering the quality of the service provided by the system and the waiting cost of service users [7]. In addition, the proper terminal capacity, considering enhancing port competitiveness and minimizing total cost, is a different concept from the maximum capacity for profits pursued by TOC [8].

This study aims to estimate the proper terminal capacity by each berth size of ACT from the perspective of TOC's financial feasibility. Compared to existing terminals, ACT pursues complete unmanned operation while dramatically increasing productivity and reducing carbon emissions. However, as the construction of ACT needs much capital investment, not only the government and port authority (PA), but also TOC is responsible for part of the ACT investments. Thus, the investment of ACT should consider the government and PA's investment cost recovery and the TOC's operating profit preservation to determine the proper terminal capacity. For the purpose, this study suggests financial model of ACT for finding the proper terminal capacity considering investments, expenses, and profits incurred in relation to the ACT business through the net present value (NPV) method. This study also deals with a case study in new port of Busan in order to verify the model feasibility. The new port of Busan is attempting to improve productivity by introducing ACT, and advanced terminal operating system and equipment.

This study contributes to the smart port research and practice. First, this study will be the first study to analyze financial feasibility of introducing ACT system. Calculation of terminal capacity due to the introduction of ACT is inconsistent with the interests of the TOC, shipping company and government. TOCs focus on increasing terminal capacity to maximize profits, while shipping companies focus on lowering terminal capacity to reduce shipping costs. On the other hand, the government focuses on the proper terminal capacity to improve productivity by increasing the capacity to handle in the same size berth by introducing more advanced operating methods and equipment with improved performance. Therefore, it is necessary to evaluate the financial feasibility of the proper loading capacity that minimizes conflicts among major stakeholders of ACT. Second, the proper terminal capacity is defined as terminal capacity from a public interest point of view that simultaneously considers enhancing port competitiveness and minimizing total cost. Thus, in terms of enhancing competitive cargo attraction with nearby competing ports, the financial analysis of proper terminal capacity by introducing ACT will provide good implications for policy makers.

2. MATERIALS AND METHODS

2.1. Automated Container Terminal (ACT)

ACT is defined as "a terminal that automates some or all of the core processes of container terminal operation, such as ship terminal, transfer and yard equipment" [7]. The fully unmanned ACT is almost vertically arranged, while the semi-automated terminal is vertically arranged and horizontally arranged. The 23 automation terminals developed so far are dominated by vertical block arrangements. The fully unmanned ACT is a robotic port that performs container terminal and movement by itself, and is a representative logistics facility that realizes the 4th industrial revolution.

As the volume of freight to be processed per ship increases due to large-scale ship size, space for handling and storing many containers simultaneously is required. In addition, as terminal congestion costs such as an increase in the number of vehicles in the terminal are increased, the introduction of ACT is not an option but is accepted as a necessity in order to improve terminal efficiency [9].

Compared to existing terminals, the fully unmanned ACT reduces operating costs such as labor and power costs by more than 37%, improves container throughput and space utilization per berth, and significantly reduces ship's turnaround time and waiting time for external trucks, and also reduces carbon emissions by 50% [10]. In particular, the fully unmanned ACT can increase productivity by nearly 40% compared to the existing ports, and is recognized as a prerequisite for servicing the recently appeared super-large ships of 20,000 TEU or higher. According to Thomas Koch (2003), the ACT investment cost is about 116% higher than that of the existing container terminal, but the operating cost is about 16% lower than that of the existing container terminal, so it is more economical than the existing container terminal from 3 to 4 years after ACT operation [11]. In addition, most of the equipment in existing ports use diesel, but the fully unmanned ACT can minimize the occurrence of pollutants in the port by using battery-type AGV and electric-powered eco-friendly loading and terminal equipment.

The disadvantage of ACT is that the flexibility may be reduced due to the use of the structure, and it will be difficult to respond to failures. In addition, the initial construction cost will be very high. Thus, profitability must be secured to recover huge capital investments.

Cullinane et al. (2004) find that the efficiency of global ports such as Rotterdam, Hamburg, and Amsterdam increases with the introduction of ACT and active investment in port infrastructure [12]. Cullinane and Wong (2006) investigate port efficiency of 69-container terminal in Europe. As a result, the larger the terminal, the better the terminal's efficiency by using advanced equipment such as ACT and highly developed management system [13]. Thus, the introduction of ACT improves port services by eliminating inefficiencies that occur in the process of container handling volumes. The introduction of ACT is based on the environment in which each country is in the type or type of ACT, and mainly focuses on the feasibility of introducing the ACT deployment type, operation method, and operating equipment [14,15]. Therefore, there are many parameters related to the introduction of ACT, but this study analyzes only those related to financial feasibility.

2.2. Determinants of Proper Terminal Capacity

The terminal capacity considering the appropriate service level means the terminal capacity, when the total service cost such as ship/freight waiting cost and construction cost is minimal. The terminal capacity is different from the maximum terminal capacity in terms of TOC's profits meaning that the terminal can actually handle regardless of the service level, and the terminal capacity from a public interest point of view that simultaneously considers enhancing port competitiveness and minimizing total cost as shown in Figure 1 [7].

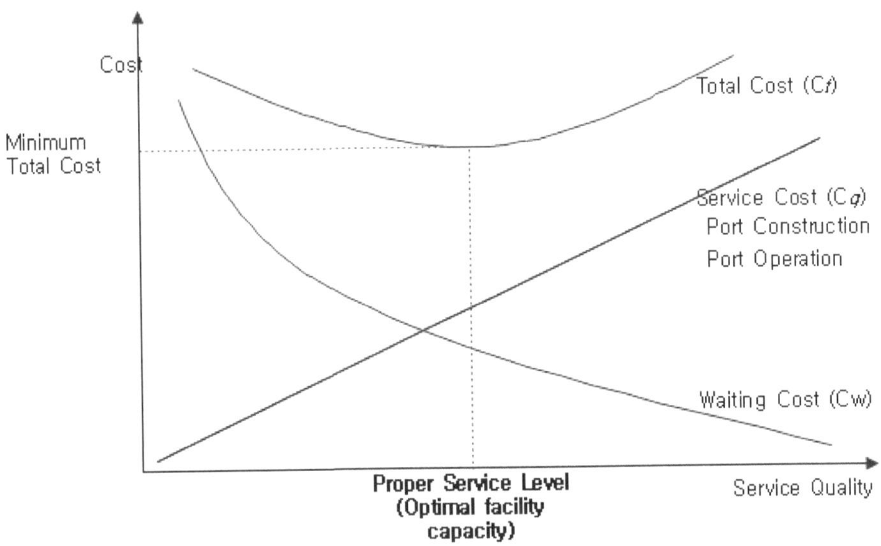

Figure 1. *The relationship between port service quality and cost (Source: Park,S-K; Park,N-K. 2013 [7]).*

As the cost and revenue of ACT are a function of the terminal size, financial analysis must be performed to determine the proper terminal capacity. Financial analysis to determine the proper capacity analyzes total costs and total revenue that vary depending on the size of terminal.

2.3. Financial Analysis Method

Financial analysis is performed for the purpose of comparing the investments and expenses and the revenue recovery [16] incurred in relation to the ACT project from the perspective of TOC. In this study, financial analysis is to analyze the profitability of the business by calculating the financial costs and income of the subject performing the ACT project, and calculating NPV of terminal capacity.

The NPV of investments is the excess of the present value of future cash flow returns over the initial investment. An investment that earns the same rate of return as the desired rate of return has an NPV of zero. The NPV will be greater than zero when an investment earns a rate of return greater than the desired rate of return.

$$\mathrm{NPV} = -I_0 + \sum_{n=1}^{N} \frac{CF_n}{(1+r)^n} \tag{1}$$

In Equation (1), cash flow (CF_n) means revenue (RE_n) minus cost (COn) on year n. Revenue is revenues on year n based on terminal fees during all the operation years after introducing ACT and Cost is costs on year n based on all investments and expenses incurred by ACT related to TOCs. The initial investment cost – I0 is the equipment cost invested in equity capital, liked in TOCs during the year of constructing ACT. Weighted average cost of capital (r) is discount rate.

The equipment investment cost (I_0) of the terminal is calculated as the number of equipment per berth NOEb times equipment price Pe times equity capital ratio Rectimes the number of berths NBt

$$I_0 = NOE_b \times P_e \times R_{ec} \times NB_t \tag{2}$$

The life of equipment is limited to 20 years for Quay Crane (QC), Rail-Mounted Gantry Crane (RMGC), and Automatic Guided Vehicle (AGV), and 10 years for other equipment. After this period, it has to be reinvested because it has zero economic value. RI_e, the reinvestment cost is calculated as the I_0 times compounded value factor of inflation rate rate (Inflation Rate) with 20 years equipment useful life squares (Economic Life). The formula for expressing reinvestment and residual value is as follows.

$$RI_e = I_0 \times (1 + \text{inflation Rate})^{\text{Economic Life}} \tag{3}$$

Terminal revenue REV_t is obtained by multiplying the berth loading and terminal capacity by the container handling fee. The capacity of the berth (LCb) is changed by 100,000 units from 550,000 TEU to 650,000 TEU. Container handling fee (HFc) are calculated by dividing the sales of the five TOCs (PNIT, PNC, HJNC, HPNT, and BNCT) in the new port of Busan for three years by the number of containers processed.

$$REV_t = LC_b \times HF_c \tag{4}$$

Terminal revenue REV_t is obtained by multiplying the berth capacity by the container handling price. The loading and terminal capacity of the berth (LCb) is changed by 100,000 units from 550,000 TEU to 650,000 TEU. Container loading and terminal fees (HFc) are calculated by dividing the sales of the five TOCs in the Busan Port New Port for three years by the number of containers processed.

Here, it is assumed that there are at least 2 to 9 operating units of the berth reflecting the ACT development plan in the new port of Busan. Then, the total revenue of the terminal TRE_t will be the number of berths NBt times the loading capacity per berth LCb times the container handling fee HFc .

$$TRE_t = NB_t \times LC_b \times HF_c \tag{5}$$

The cost of sales COs consists of material cost COm, labor cost COl, rent fee COr, depreciation COd, other expenses CO0, and interest cost COi..

$$CO_s = CO_m + CO_l + CO_r + CO_d + CO_0 + CO_i \tag{6}$$

As material cost COm is calculated as material costs of three terminals in new port of Busan A (COs1), B (COs2), and C (COs3) times a percentage of sales, data for the last 3 years of existing terminals are required.

$$\begin{aligned} CO_m &= Average(CO_{s1} + CO_{s2} + CO_{s3} + \ldots) \\ &\quad \times materialcost\ percentage\ of\ sales\ of\ existing\ terminals \end{aligned} \tag{7}$$

Labor costs (CO_l) consist of sum of labor costs incurred during terminal operation COloand during sales management COlm. The labor cost per berth NPb is calculated by multiplying the number of personnel per berth by the average labor cost ACl.

$$CO_l = CO_{lo} + CO_{lm} = NP_b \times AC_l \tag{8}$$

The rent (CO_r) consists of the quay rental fee CORb and the yard rental fee CORy. The quay rental fee is obtained by multiplying the rental fee per meter ROBm by the berth length LOB. Yard rent is calculated by multiplying the rent per TGS (Twenty Feet Ground Slots) ROYtgs by the total TGS.

$$CO_r = COR_b + COR_v = (LOB \times ROB_m) + \left(total\ TGS \times ROY_{tgs}\right) \tag{9}$$

The depreciation cost ($CO_{e)}$ is calculated using the straight-line method for the useful life of the all equipment: (1) depreciation cost of QC (ROE_{QC}), (2) depreciation cost of RMGC (ROE_{RMGC}), (3) depreciation cost of AGV (ROE_{AGV}) and (4) depreciation cost of other equipments (ROE_{others}).

$$CO_e = ROE_{QC} + ROE_{RMGC} + ROE_{AGV} + ROE_{others} \tag{10}$$

Other costs (CO_o) consist of variable (VC_o puls VC_m) and fixed costs (FC_o plus FC_m), respectively, during terminal operation and sales management.

$$CO_o = (VC_o + FC_o) + (VC_m + FC_m) \tag{11}$$

Equation (11) is decomposed by Equation (12), represented by the sales amount (REV_t) times other variable cost and fixed cost rates for sales and general management (RVC_o plus RFC_o) and other variable cost and fixed cost rates for manufacturing (RVC_m plus RFC_m).

$$\begin{aligned} CO_0 &= (VC_0 + FC_0) + (VC_m + FC_m) \\ &= \{REV_t \times (RVC_0 + RFC_0)\} + \{REV_t \times (RVC_m + RFC_m)\} \end{aligned} \tag{12}$$

The borrowings consist of equipment investment ICEec by equity capital (ICE_{ec}) and cash flows (CF_{cy}) for the current period. Interest (CO_i) is obtained by multiplying the borrowings by weighted average borrowing rate.

$$CO_i = \left(ICE_{ec} + CF_{cy}\right) \times \text{weighted average borrowing rate} \tag{13}$$

3. THE PREMISE OF ANALYSIS FOR CASE STUDY

3.1. ACT Layout for Analysis

As of 2019, Busan Port handles 21.91 million TEU of container traffic, maintaining its status as the world's fifth largest container port. In particular, the new port of Busan has become Korea's largest container port, handling 14.67 million TEU of container traffic in 2018 [17]. Currently, the five TOCs operated in the new port of Busan are partial automated terminals that operate only yard cranes in an unmanned manner. The Korean government is building a smart port to cope with the 4th industrial revolution, and, accordingly, is promoting the establishment of a fully ACT in the new port of Busan [18].

This study selects the new port of Busan to carry out the PCTC (Proper Container Terminal Capacity) of the ACT from perspective of financial analysis, as shown in Figure 2 [19]. In order to deal with the paper's objective, this study applies some assumptions following Korea's government guidelines [20]. The most important factor in establishing an estimate period for costs and benefits is to determine the economic life of critical facilities and equipment. According to the Korean government guidelines, the analysis period for ports is set to 30 years, and all future revenues and expenses are estimated at constant prices at the time of the base year.

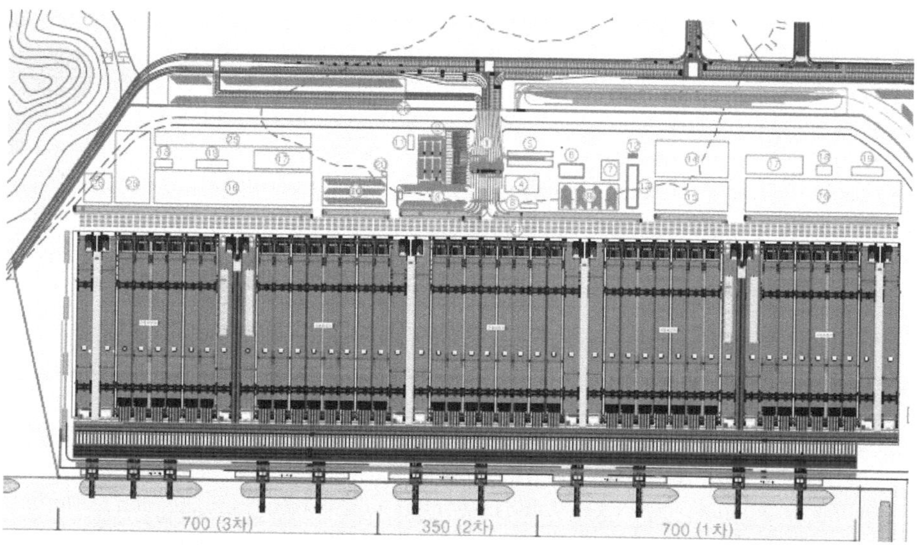

Figure 2. *Layout of ACT in The New Port of Busan (Phase 2–5,2–6) (Source: Ministry of Ocean and Fisheries [**19**]).*

3.2. Basic Assumptions for Financial Analysis

The base year for analysis of this study is set to 1 January 2020, and all construction investments of ACT are assumed to be completed on 30 June 2019. The amount of equipment investment is based on the equipment manufacturer's undisclosed price that considers the type of each yard. Equipment investment is converted to the value at the end of 2020, and it is regarded as the returnable investment. The analysis period applies 30 years from 1 January 2020 to 1 January 2049. The size of the berths to be analyzed is 9, 5, 4, 3, and 2 berths. In the case of TOC, detailed financial information such as labor and maintenance costs is not available because major financial data such as cost statements are not disclosed. In the ratio analysis, the average ratio of the neighborhood terminal during last three years is applied.

The TOC subject to financial ratio analysis are conducted for three major TOCs (PNIT, HJNC, and HPNT) in the new port of Busan. The capital structure is applied at the time of the initial equipment introduction of three major TOCs, and thus the equity ratio is 17.6% and the debt ratio is 82.4%, respectively. It is assumed that the operating cash flows deducting interest expenses is repaid annually, and a sales growth rate of 2.49% is applied according to annual economic growth expectation by the Bank of Korea [21].

Due to the nature of port development and construction projects, costs are intensively incurred at an early stage, whereas benefits occur over a long period of time after completion [22]. Therefore, the expected costs and benefits during the analysis period are evaluated by applying discount rates to present values. Following Korean government guideline, 4.5% discount ratio is applied [20].

Based on the 2019 rental fee of the new port of Busan Port, 40% is based on the quay wall and 60% is based on TGS. Annual inflation rate is applied. The consumer inflation rate during the operation period is the average of 1.39% of the consumer inflation rate for the last 7 years (2011–2017) of the Bank of Korea (excluding the maximum and minimum values) [21]. Moreover, all currency units used in this study are expressed in U.S. dollars applying the average USD-KRW exchange rate of 1132 KRW for the last three years [21].

Corporate tax is applied at 10% below $176,678 (equivalent to 200 million KRW), 20% below $17,667,845, and 22% above $17,667,845 (equivalent to 20 billion KRW), and additionally considered as 10% local tax. There are no tax adjustments, the balance after deducting deficits and interest expenses from operating profit is reflected as a tax base, and an additional 10% corporate tax payment on unrefunded income is not considered.

In Table 1, the weighted average cost of capital (WACC) of 3.48% of the three TOCs in the new port of Busan is applied as the borrowing interest rate based on the audit report at the end of 2018.

Sensitivity analysis is performed on the terminal volume and financial discount rate. Specifically, from the perspective of TOC, the estimated rent level, which is an important cost factor, is calculated as a fixed value by applying the current estimate, and the terminal volume that has the most important effect on the profit factor is calculated as the target of sensitivity analysis. Table 2 shows criteria of sensitivity analysis.

Table 1. Weighted average cost of capital of the three Terminal Operating Companies (TOCs).

	Lender	Balance of Borrowings	Interest Rate (a)	Weights (b)	WACC (=a*b)
Quay 1 (PNIT)	KDB(Korea Development Bank)	$6,218,4503	3.44%	0.43	1.47%
	KDB	$9,717,314	2.92%	0.07	0.20%
Quay 3 (HJNC)	-	-	-	0.00	0.00%
Quay 4 (HPNT)	Hana Bank	$19,891,755	3.68%	0.14	0.50%
	Hana Insurance	$7,629,548	3.68%	0.05	0.19%
	National Credit Union Federation of Korea	$2,1515,323	3.68%	0.15	0.55%
	Municipal Credit Union of Changwon Mirae	$686,659	3.68%	0.00	0.02%
	Municipal Credit Union of Changwon	$686,659	3.68%	0.00	0.02%
	Hana Bank	$22,888,642	3.39%	0.16	0.53%
	Total	$145,200,404	-	1.000	3.48%

Source: *DART* system [23].

Table 2. Sensitivity analysis criteria.

	Review Scope	Remark
Treatment quantity per berth Discount rate	550,000 TEU~650,000 TEU per berth 3.50~5.50%	100,000 TEU per berth 0.5% *p* unit

Source: Author.

4. ESTIMATING COST AND REVENUE FOR CASE STUDY

In this study, the major costs calculated by Equation (6) are classified into five major categories: (1) Equipment investment; (2) Material costs; (3) Labor costs; (4) Rent fee, depreciation, and other costs; and (5) Interest costs. As three kinds of costs excluding equipment investment and interest costs are not directly provided by TOCs, we analyze them using the TOCs' annual audit reports. However, as annual audit reports of TOCs are publicly disclosed on the Data Analysis, Retrieval and Transfer (DART) system and audited by external audit, the estimates of this study are likely to be reliable.

We assume that the base year for analysis of this study is set to 1 January 2020, and all construction investments of ACT are assumed to be completed on 30 June 2019. The amount of equipment investment is based on a self-estimating method that considers the type of each yard. Equipment investment is converted.

4.1. Equipment Investment Cost

Equipment investment is self-estimating by assuming the operation of an automated container terminal in consideration of the yard type. In addition, the reliability of the estimation is secured through consultations from ZPMC (Zhenhua Heavy Industries Co., Shanghai, China) in Shanghai Port and the person in charge of the 4th automated terminal equipment operation in Yangsan Port, China. Based on the five berths, the estimated total equipment investment is $556,501,767 and the converted amount per berth is $111,300,353 based on Equation (2). The useful life of equipment is estimated to be 20 years for all equipment except for other equipment. Panel A of Table 3 shows the estimated equipment investment based on five berths, and Panel B of Table 3 represents total equipment investment per berth using estimation of Panel A.

Table 3. Estimation of equipment investment cost.

Panel A: Equipment Investment Estimate Based on 5 Berths(Unit: U.S. Dollars)						
		Unit Price	**Required Number of Units (Based on 5 Berths)**	**Total Price (Based on 5 Berths)**	**Amount per Berth**	**Economic Useful Life (Depreciation Method)**
Quay unloading equipment	Dual Trolley	$11,925,795	18	$214,664,311	$42,932,862	20 years (Straight-line)
Yard equipment	RAMEN TYPE	$2,826,855	71	$199,293,286	$39,858,657	20 years (Straight-line)
	RTGC for Dangerous Cargo	$2,650,177	4	$10,600,707	$2,120,141	20 years (Straight-line)
	Dual Cantilever type	$3,091,873	24	$72,659,011	$14,531,802	20 years (Straight-line)
Transport equipment	AGV (Electricity NON-LIFTTYPE)	$618,375	89	$55,035,336	$11,007,067	20 years (Straight-line)
Other equipment	Reach Stacker	$441,696	2	$883,392	$176,678	20 years (Straight-line)
	YT (Yard Tractor)	$88,339	6	$530,035	$106,007	10 years (Straight-line)
	Spreader Pulling Car	$26,502	5	$132,509	$26,502	10 years (Straight-line)
	Road Washing Car	$26,502	5	$132,509	$26,502	10 years (Straight-line)
	Patrol Car	$26,502	60	$1,590,106	$318,021	10 years (Straight-line)
	Engineering Car	$26,502	15	$397,527	$79,505	10 years (Straight-line)
	Maintain Car	$26,502	15	$397,527	$79,505	10 years (Straight-line)
	Chassis	$13,251	14	$185,512	$37,102	10 years (Straight-line)
Total				$556,501,767	$111,300,353	

Panel B: Total equipment investment per berth (unit: U.S. Dollars)					
	Based on 9 Berths	**Based on 5 Berths**	**Based on 4 Berths**	**Based on 3 Berths**	**Based on 2 Berths**
Quay unloading equipment	$386,395,760	$214,664,311	$171,731,449	$128,798,587	$85,865,724
Yard equipment	$508,595,406	$282,553,004	$226,042,403	$169,531,802	$113,021,201
Transport equipment	$100,653,710	$55,918,728	$44,734,982	$33,551,237	$22,367,491
Other equipment	$6,058,304	$3,365,724	$2,692,580	$2,019,435	$1,346,290
Total investment per berth	$1,001,703,180	$556,501,767	$445,201,413	$333,901,060	$222,600,707

Source: Author.

4.2. Material Cost

Material cost is estimated by applying the ratio of material cost to sales found in TOCs' audit report for the last three years. In particular, we estimate the material cost through the ratio of the material cost to total sales applied to the terminal capacity, which is the subject of sensitivity analysis. The sales of three major TOCs over the past three years are $961,599,219 and the identifiable material cost is $24,616,940 in applying Equation (7). We find the ratio of material costs is estimated at 2.56%. Total sales and materials costs of TOCs' for the last 3 years and the ratio of material cost are as follows Table 4.

Table 4. Estimated material cost.

Sales for the Last 3 Years	Material Costs for the Last 3 Years	Ratio of Average Material Cost to Sales
$961,599,219	$24,616,940	2.56%

Source: Author.

4.3. Labor Costs

In the same way as the material cost calculation, the labor cost is calculated by applying the salary confirmed in the audit report for the last three years. In addition, the personnel worker cost per berth is calculated from the number of workers included in the 2018 corporate profile data.

The total number of workers in the major TOCs in new port of Busan over the past three years is 831 and the average number of workers per 350 m berth is 84.3. As of 2018, the annual salary per worker is $53,767 in using Equation (8). Due to the nature of ATC, the number of people does not increase in proportion to the number of berths; thus, a 5% diminishing ratio in number of personnel increase per berth increase is applied, and the number of persons and salaries for each berth in 2018 reflecting this are as follows Table 5.

Table 5. Number of workers per berth and salary as of 2018.

Number of Berths	Number of Workers	Total Salary	Manufacturing Salary	Sales and General Management Salary
Based on 9 berths	152	$8,158,996	$6,369,496	$1,789,499
Based on 5 berths	223	$12,011,854	$9,377,313	$2,634,541
Based on 4 berths	291	$15,638,074	$12,208,201	$3,429,874
Based on 3 berths	354	$19,037,655	$14,862,156	$4,175,499
Based on 2 berths	565	$30,369,593	$23,708,678	$6,660,914

Source: Author.

Welfare benefits are calculated by applying the ratio to the salary for each sector, and the retirement benefit is calculated by applying the 1/12 ratio to the salary collectively. It is assumed that the labor cost increases by 1.39% every year, and the labor cost breakdown by berth size during the analysis period taking this into account is as follows Table 6.

4.4. Rent Fee and Other Costs

The rental fee is based on the results of the Korean government guideline [20], and, accordingly, $8,176,486,749 per berth is applied based on Equation (9). For the economic life and depreciation method based on Equation (10), the results of the preceding study [24] are used as a proxy, and are applied by referring to the useful lives of each TOC's audit report refer to Table 3. In addition, reinvestment is made in an amount considering the inflation rate up to the end of the economic life, and the residual value is estimated to the unamortized book value using Equation (3). Table 7 shows the reinvestment amount accrued by year and the reversal of the residual value in the final year.

Table 6. Total labor costs (unit: U.S. Dollars).

Year	9 Berths	5 Berths	4 Berths	3 Berths	2 Berths
2020	$38,909,011	$24,390,459	$20,035,336	$15,389,576	$10,453,180
2021	$39,450,530	$24,730,565	$20,314,488	$15,603,357	$10,598,940
2022	$40,000,000	$25,075,088	$20,597,173	$15,820,671	$10,746,466
2023	$40,557,420	$25,424,028	$20,884,276	$16,041,519	$10,895,760
2024	$41,121,908	$25,778,269	$21,174,912	$16,265,018	$11,047,703
2025	$41,695,230	$26,136,926	$21,469,965	$16,491,166	$11,201,413
2026	$42,275,618	$26,500,883	$21,768,551	$16,720,848	$11,357,774
2027	$42,864,841	$26,870,141	$22,072,438	$16,954,064	$11,515,901
2028	$43,461,131	$27,244,700	$22,379,859	$17,189,929	$11,675,795
2029	$44,067,138	$27,623,675	$22,690,813	$17,429,329	$11,839,223
2030	$44,680,212	$28,008,834	$23,007,067	$17,672,261	$12,003,534
2031	$45,303,004	$28,398,410	$23,327,739	$17,917,845	$12,170,495
2032	$45,933,746	$28,794,170	$23,652,827	$18,167,845	$12,340,106
2033	$46,573,322	$29,195,230	$23,981,449	$18,420,495	$12,512,367
2034	$47,221,731	$29,601,590	$24,316,254	$18,677,562	$12,686,396
2035	$47,879,859	$30,014,134	$24,654,594	$18,937,279	$12,863,074
2036	$48,546,820	$30,431,979	$24,998,233	$19,201,413	$13,042,403
2037	$49,222,615	$30,856,007	$25,346,290	$19,469,081	$13,224,382
2038	$49,908,127	$31,285,336	$25,698,763	$19,739,399	$13,408,127
2039	$50,603,357	$31,721,731	$26,056,537	$20,015,018	$13,594,523
2040	$51,308,304	$32,163,428	$26,419,611	$20,293,286	$13,784,452
2041	$52,022,968	$32,611,307	$26,787,986	$20,575,972	$13,976,148
2042	$52,747,350	$33,065,371	$27,160,777	$20,863,074	$14,170,495
2043	$53,481,449	$33,525,618	$27,538,869	$21,152,827	$14,368,375
2044	$54,227,032	$33,992,933	$27,922,261	$21,447,880	$14,568,021
2045	$54,981,449	$34,466,431	$28,311,837	$21,746,466	$14,771,201
2046	$55,747,350	$34,946,113	$28,705,830	$22,049,470	$14,977,032
2047	$56,523,852	$35,432,862	$29,106,007	$22,356,890	$15,185,512
2048	$57,310,954	$35,926,678	$29,510,601	$22,667,845	$15,396,643
2049	$58,109,541	$36,426,678	$29,922,261	$22,983,216	$15,611,307
Total salary	$1,436,735,866	$900,639,576	$739,813,604	$568,260,601	$385,986,749

Source: Author.

Table 7. Reinvestment and residual value (unit: U.S. Dollars).

Year	Based on 9 Berths	Based on 5 Berths	Based on 4 Berths	Based on 3 Berths	Based on 2 Berths
2020~2028	-	-	-	-	-
2029	$2,318,905	$1,545,936	$3,864,841	$3,091,873	$6,956,714
2030~2038	-	-	-	-	-
2039~2038	$440,303,887	$293,536,219	$733,840,106	$587,071,555	$1,320,911,661
2049	($218,820,671)	($145,880,742)	($364,700,530)	($291,760,601)	($656,461,131)
Total reinvestment	$223,802,120	$149,201,413	$373,004,417	$298,402,827	$671,407,244

Source: Author

For fixed costs, the ratio of material cost to sales as identified in the audit report for the last three years is applied. Other manufacturing costs of TOCs for the last three years are calculated by Equation (11) assuming a variable cost ratio of 60%.

Based on Equation (12), the variable cost portion of the fixed cost is estimated through the ratio of other expenses (variable expenses) to sales applied to the stevedoring volume subject to sensitivity analysis. Considering the economies of scale, a 3% decrease is applied to the increase in variable cost compared to the increase in the number of berths.

In the case of fixed costs, a 3% reduction rate per increase in the number of berths is applied in consideration of the economies of scale, and the details of the fixed costs during the analysis period (manufacturing costs, sales, and general management costs added) are as follows Table 8.

Table 8. Other cost (unit: U.S. Dollars).

Year	9 Berths	5 Berths	4 Berths	3 Berths	2 Berths
2020	$28,877,208	$18,680,212	$15,471,731	$11,999,117	$8,263,251
2021	$29,280,035	$18,940,813	$15,687,279	$12,166,078	$8,378,092
2022	$29,687,279	$19,204,064	$15,905,477	$12,335,689	$8,494,700
2023	$30,100,707	$19,471,731	$16,127,208	$12,507,951	$8,613,074
2024	$30,520,318	$19,742,933	$16,351,590	$12,681,979	$8,733,216
2025	$30,945,230	$20,017,668	$16,579,505	$12,858,657	$8,855,124
2026	$31,376,325	$20,296,820	$16,810,071	$13,037,986	$8,977,915
2027	$31,812,721	$20,579,505	$17,044,170	$13,219,081	$9,103,357
2028	$32,256,184	$20,865,724	$17,281,802	$13,403,710	$9,229,682
2029	$32,705,830	$21,156,360	$17,522,968	$13,590,106	$9,358,657
2030	$33,160,777	$21,451,413	$17,766,784	$13,779,152	$9,489,399
2031	$33,622,792	$21,750,000	$18,014,134	$13,970,848	$9,621,025
2032	$34,090,989	$22,053,004	$18,265,018	$14,166,078	$9,755,300
2033	$34,566,254	$22,359,541	$18,519,435	$14,363,074	$9,891,343
2034	$35,046,820	$22,671,378	$18,777,385	$14,562,721	$10,029,152
2035	$35,535,336	$22,986,749	$19,038,869	$14,765,901	$10,168,728
2036	$36,030,035	$23,307,420	$19,303,887	$14,971,731	$10,310,071
2037	$36,531,802	$23,631,625	$19,572,438	$15,180,212	$10,453,180
2038	$37,040,636	$23,961,131	$19,845,406	$15,391,343	$10,598,940
2039	$37,556,537	$24,295,053	$20,121,908	$15,606,007	$10,746,466
2040	$38,079,505	$24,633,392	$20,401,943	$15,823,322	$10,896,643
2041	$38,610,424	$24,976,148	$20,686,396	$16,043,286	$11,048,587
2042	$39,147,527	$25,324,205	$20,974,382	$16,266,784	$11,202,297
2043	$39,693,463	$25,676,678	$21,265,901	$16,493,816	$11,357,774
2044	$40,245,583	$26,034,452	$21,562,721	$16,723,498	$11,515,901
2045	$40,806,537	$26,396,643	$21,863,074	$16,955,830	$11,676,678
2046	$41,374,558	$26,764,134	$22,166,961	$17,192,580	$11,839,223
2047	$41,950,530	$27,136,926	$22,476,148	$17,431,979	$12,004,417
2048	$42,535,336	$27,515,018	$22,788,869	$17,674,028	$12,171,378
2049	$43,127,208	$27,898,410	$23,106,007	$17,920,495	$12,340,989
Total of other costs	$1,066,314,488	$689,779,152	$571,299,470	$443,083,039	$305,124,558

Source: Author.

4.5. Interest Cost

It is assumed that 82.4% of ACT's total investment is financed through borrowing based on corporate profile data, and the interest cost applied with an interest rate of 3.48% of the total financed amount incur. The balance of the current operating cash flow less interest for the current period assumes repaid from borrowings at the end of the previous year. Using Equation (13), Table 9 shows the amount of interest expense accrued by year, assuming that operating cash flows occur on a 650,000 TEU basis per berth and that operating cash fluctuates complexly when volume fluctuates.

Table 9. Total interest costs per berths (unit: U.S. Dollars).

Year	Based on 9 Berths	Based on 5 Berths	Based on 4 Berths	Based on 3 Berths	Based on 2 Berths
2020	$28,886,926	$16,308,304	$13,098,057	$9,860,424	$6,597,173
2021	$29,187,279	$16,761,484	$13,515,901	$10,215,548	$6,858,657
2022	$28,591,873	$16,743,816	$13,563,604	$10,295,936	$6,940,813
2023	$27,024,735	$16,216,431	$13,209,364	$10,079,505	$6,826,855
2024	$24,408,127	$15,136,042	$12,417,845	$9,539,753	$6,501,767
2025	$21,554,770	$13,943,463	$11,540,636	$8,938,163	$6,136,042
2026	$18,452,297	$12,631,625	$10,572,438	$8,271,201	$5,728,799
2027	$15,086,572	$11,195,230	$9,507,951	$7,535,336	$5,277,385
2028	$11,445,230	$9,626,325	$8,342,756	$6,727,032	$4,779,152
2029	$7,512,367	$7,918,728	$7,069,788	$5,840,989	$4,232,332
2030	$3,273,852	$6,063,604	$5,684,629	$4,874,558	$3,633,392
2031	-	$4,054,770	$4,180,212	$3,821,555	$2,978,799
2032	-	$1,883,392	$2,551,237	$2,679,329	$2,266,784
2033	-	-	$790,636	$1,441,696	$1,494,700
2034	-	-		$103,357	$657,244
2035~2049	-	-	-	-	-
Total	$243,860	$168,083	$142,683	$113,454	$80,270

Source: Author.

4.6. Estimating Revenue

The main revenue of TOC is the import of unloading fees, and the unloading unit price is calculated from the unloading volume compared to the sales for the last three years (2016–2018) using Equation (4). The unit price of sales is calculated including the private pier to secure the reliability of the population. In addition, the unloading revenue based on Equation (5) is calculated by applying the average unloading unit price to the unloading volume subject to sensitivity analysis. However, the average unloading unit price is based on the average annual economic growth rate of 2.49% [21].

The total sales of the five TOCs in new port of Busan over the past three years is $1,881,070,227, the total amount of unloading volume is 40,804,521 TEU, and average unloading unit price is $46 per TEU which shown in Panels A–C of Table 9, respectively. In addition, 5 years (2020–2024) from the first introduction of ACT is considered as the ramp up period. It is assumed that the product will reach 100% in 2024 by adding 10% from 60% of the quantity calculated from 2020 presented in Panel D of Table 10.

Table 10. Total sales, unloading volume, and average unloading unit price of TOCs.

Panel A: Total Sales by TOC in the new port of Busan for the last 3 years (unit: U.S. Dollars).				
	2016	2017	2018	Total
Quay 1 (PNIT)	$100,850,646	$96,850,481	$101,356,938	$299,058,065
Quay 2 (PNC)	$195,555,672	$223,822,824	$225,736,504	$645,115,001
Quay 3 (HJNC)	$105,346,780	$98,266,256	$111,611,095	$315,224,132
Quay 4 (HPNT)	$120,640,091	$112,821,083	$114,033,422	$347,494,596
Quay 5 (BNCT)	$80,956,617	$92,840,186	$100,381,630	$274,178,433
Total	$603,349,806	$624,600,831	$653,119,590	$1,881,070,227
Panel B: Unloading volume by TOC in the new port of Busan for the last 3 years (Units: TEU)				
	2016	2017	2018	Total
Quay 1 (PNIT)	2,418,702	2,689,256	2,477,785	7,585,742
Quay 2 (PNC)	4,626,435	4,531,302	4,938,446	14,096,183
Quay 3 (HJNC)	1,925,545	2,219,402	2,770,535	6,915,482
Quay 4 (HPNT)	2,322,166	2,069,198	2,061,131	6,452,495
Quay 5 (BNCT)	1,541,859	1,940,079	2,269,682	5,751,620
Total	12,834,708	13,449,235	14,517,579	40,801,521

Panel C: Calculating average unloading unit price (unit: U.S. Dollars)		
Unloading Income for the Last 3 Years	Handling of Cargo in the Last 3 Years	Average Unloading Unit Price
$1,881,070,227	40,801,521 TEU	$46/TEU

Panel D: Ramp up period					
Year	2020	2021	2022	2023	2024
Unloading throughput achievement rate	60%	70%	80%	90%	100%

Source: Author.

5. DISCUSSION

In terms of the operating balance of TOCs in the new port of Busan for ACT investment, the calculation results of NPV calculated by Equation (1) and breakeven volume considering unloading volume changes per berth and discount rate are as follows. In Table 11, the value of the quantity per berth is based on $8,176,486,749 as the standard rent fee per berth, and the financial discount rate is calculated by applying 4.5% based on basic assumption of financial analysis.

Table 11. NPV per berth considering changes of processing quantity and discount ratio.

Panel A: 9 Berth Model (Unit: U.S. Dollars)						
		Discount Ratio				
		3.50%	4.00%	4.50%	5.00%	5.50%
Handling quantity per berth (TEU)	550,000	$29,176,678	($19,780,919)	($61,402,827)	($96,759,717)	($126,768,551)
	560,000	$99,241,166	$44,426,678	($2,496,466)	($42,656,360)	($77,020,318)
	570,000	$168,961,131	$108,387,809	$56,248,233	$11,358,657	($27,298,587)
	580,000	$238,392,226	$172,202,297	$114,969,081	$65,454,064	$22,594,523
	590,000	$307,630,742	$235,909,011	$173,655,477	$119,579,505	$72,571,555
	600,000	$376,639,576	$299,424,912	$232,185,512	$173,577,739	$122,446,996
	610,000	$445,646,643	$363,071,555	$290,960,247	$227,919,611	$172,750,000
	620,000	$514,432,862	$426,489,399	$349,499,117	$282,019,435	$222,806,537
	630,000	$583,203,180	$490,032,686	$408,287,986	$336,477,032	$273,314,488
	640,000	$651,682,862	$553,280,035	$466,773,852	$390,628,975	$323,513,251
	650,000	$720,409,011	$616,847,173	$525,645,760	$445,220,848	$374,199,647

(Cont.)

Table 11. (*Cont.*)

Panel B: 5 berth model (unit: U.S. Dollars)						
		Discount Ratio				
		3.50%	4.00%	4.50%	5.00%	5.50%
Handling quantity per berth (TEU)	550,000	($144,772,968)	($158,209,364)	($168,899,293)	($177,302,120)	($183,803,887)
	560,000	($105,433,746)	($122,452,297)	($136,366,608)	($147,674,028)	($156,795,053)
	570,000	($66,458,481)	($86,943,463)	($103,984,099)	($118,111,307)	($129,779,152)
	580,000	($27,814,488)	($51,669,611)	($71,751,767)	($88,628,975)	($102,781,802)
	590,000	$10,462,898	($16,673,145)	($39,719,965)	($59,278,269)	($75,859,541)
	600,000	$48,514,134	$18,168,728	($7,783,569)	($29,971,731)	($48,935,512)
	610,000	$86,356,890	$52,887,809	$24,107,774	($645,760)	($21,938,163)
	620,000	$124,129,859	$87,590,106	$56,024,735	$28,743,816	$5,155,477
	630,000	$161,732,332	$122,151,060	$87,827,739	$58,043,286	$32,178,445
	640,000	$199,267,668	$156,717,314	$119,697,880	$87,461,131	$59,365,724
	650,000	$236,671,378	$191,175,795	$151,480,565	$116,811,837	$86,501,767

Panel C: 4 berth model (unit: U.S. Dollars)						
		Discount Ratio				
		3.50%	4.00%	4.50%	5.00%	5.50%
Handling quantity per berth (TEU)	550,000	($148,431,095)	($156,196,113)	($162,064,488)	($166,371,908)	($169,401,060)
	560,000	($116,733,216)	($127,469,965)	($136,008,834)	($142,717,314)	($147,903,710)
	570,000	($85,338,339)	($98,941,696)	($110,060,071)	($119,090,989)	($126,371,908)
	580,000	($54,271,201)	($70,643,993)	($84,258,834)	($95,542,403)	($104,856,007)
	590,000	($23,505,300)	($42,566,254)	($58,607,774)	($72,083,039)	($83,379,859)
	600,000	$6,980,565	($14,698,763)	($33,106,890)	($48,721,731)	($61,954,947)
	610,000	$37,314,488	$13,068,021	($7,661,661)	($25,378,975)	($40,516,784)
	620,000	$67,505,300	$40,771,201	$17,787,102	($1,974,382)	($18,968,198)
	630,000	$97,535,336	$68,353,357	$43,150,177	$21,373,675	$2,550,353
	640,000	$127,457,597	$95,848,057	$68,443,463	$44,669,611	$24,030,919
	650,000	$157,360,424	$123,389,576	$93,840,989	$68,116,608	$45,703,180

Panel D: 3 berth model (unit: U.S. Dollars)						
		Discount Ratio				
		3.50%	4.00%	4.50%	5.00%	5.50%
Handling quantity per berth (TEU)	550,000	($137,865,724)	($141,069,788)	($143,135,159)	($144,280,035)	($144,689,929)
	560,000	($111,645,760)	($117,488,516)	($121,906,360)	($125,150,177)	($127,431,979)
	570,000	($87,948,763)	($96,019,435)	($102,437,279)	($107,478,799)	($111,377,208)
	580,000	($64,485,866)	($74,703,180)	($83,053,004)	($89,833,922)	($95,299,470)
	590,000	($41,235,866)	($53,530,919)	($63,753,534)	($72,223,498)	($79,213,781)
	600,000	($18,181,095)	($32,495,583)	($44,541,519)	($54,657,244)	($63,136,042)
	610,000	$4,658,127	($11,622,792)	($25,443,463)	($37,166,078)	($47,098,057)
	620,000	$27,325,088	$9,128,092	($6,426,678)	($19,718,198)	($31,073,322)
	630,000	$49,912,544	$29,849,823	$12,604,240	($2,220,848)	($14,966,431)
	640,000	$72,424,028	$50,522,085	$31,608,657	$15,270,318	$1,150,177
	650,000	$94,796,820	$71,083,039	$50,526,502	$32,696,996	$17,220,848

(*Cont.*)

Table 11. (*Cont.*)

Panel E: 2 berth model (unit: U.S. Dollars)						
		Discount Ratio				
		3.50%	4.00%	4.50%	5.00%	5.50%
Handling quantity per berth (TEU)	550,000	($110,424,028)	($110,590,989)	($110,226,148)	($109,448,763)	($108,358,657)
	560,000	($91,124,558)	($93,373,675)	($94,849,823)	($95,701,413)	($96,052,120)
	570,000	($73,885,159)	($77,863,958)	($80,881,625)	($83,108,657)	($84,688,163)
	580,000	($58,081,272)	($63,549,470)	($67,904,594)	($71,333,039)	($73,991,166)
	590,000	($42,457,597)	($49,357,774)	($55,001,767)	($59,590,106)	($63,294,170)
	600,000	($26,978,799)	($35,264,134)	($42,157,244)	($47,871,908)	($52,591,873)
	610,000	($11,634,276)	($21,265,901)	($29,373,675)	($36,186,396)	($41,897,527)
	620,000	$3,601,590	($7,344,523)	($16,638,693)	($24,523,852)	($31,206,714)
	630,000	$18,764,134	$6,535,336	($3,918,728)	($12,855,124)	($20,489,399)
	640,000	$33,861,307	$20,384,276	$8,796,820	($1,165,194)	($9,730,565)
	650,000	$48,882,509	$34,174,912	$21,473,498	$10,499,117	$1,015,018

Source: Author.

When applying a financial discount of 4.5%, NPV is positive (+) at 570,000 TEU for nine berths, 610,000 TEU for five berths, 620,000 TEU for four berths, 630,000 TEU for three berths, and 640,000 TEU for two berths shown in Figure 3. It means that TOCs must secure a minimum of 570,000 TEU and a maximum of 640,000 TEU of unloading capacity to invest ACT.

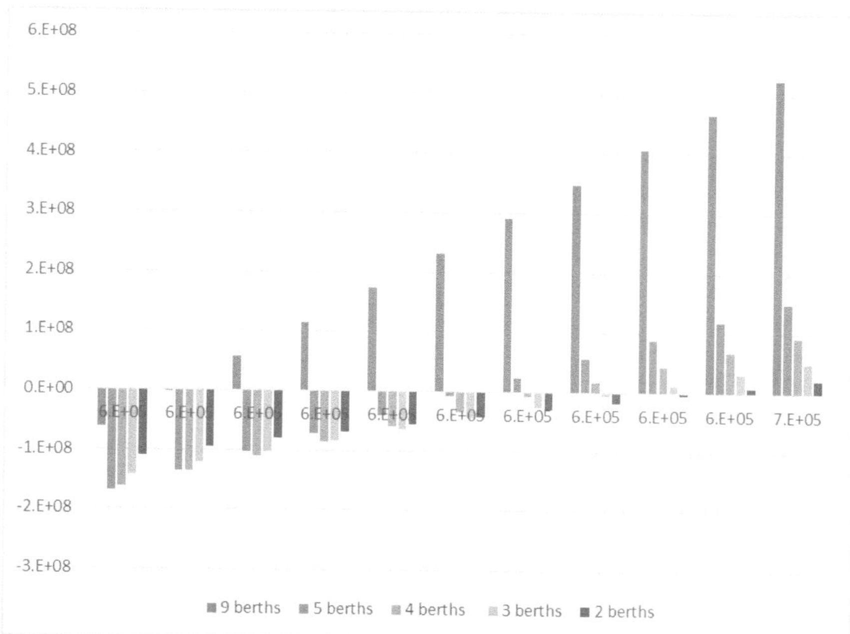

Figure 3. *The net present value (NPV) changes per berth size in applying 4.5% discount ratio (Source: Author).*

In the sensitivity analysis that considers the changes of discount rate and the processing quantity per berth, NPV has a positive value at the level of at least 550,000 TEU (nine berths and 3.5% discount rate) and up to 650,000 TEU (two berths and 5.5% discount rate).

The breakeven point (BEP) is the point at which revenues equal total cost and profit is zero and it is based on the revenue-output and cost-output function. Many of the planning activities that take place within a company are based on anticipated levels of output. Thus, determining the BEP is a safe and practical way of gauging the viability of TOC. It is a pointer that helps TOC decides whether to invest the ACT project or discontinue it.

In terms of the operating balance of the terminal operator, the volume of breakeven unloading that matches the investment cost of each berth and the operating and financial cash flows is shown in Table 12. When applying the standard discount rate of 4.5%, the amount of unloading per berth at breakeven is a minimum of 560,421 TEU (nine berths) and a maximum of 633,102 TEU (two berths).

Table 12. The breakeven handling volume per berth.

Berth Type	Initial Investment and Reinvestment Amount (Unit: U.S. Dollars)	Handling Containers per Berth at Breakeven
9 berth models	$1,673,110,424	560,421 TEU
5 berth models	$929,506,184	602,446 TEU
4 berth models	$748,904,594	613,014 TEU
3 berth models	$557,703,180	623,364 TEU
2 berth models	$371,802,120	633,102 TEU

Source: Author.

6. CONCLUSIONS

Efficient port investment and operation are important factors in enhancing national competitiveness by increasing port productivity. Due to the Fourth Industrial Revolution, the port environment has changed a lot, and it is expected that there will be more rapid changes in the future. Advanced ports such as European Container Terminal in Rotterdam, Netherlands and Container Terminal Altenwerder in Hamburg, Germany are increasing the use of ACT as an important part of smart ports. As ACT is the key element of smart ports due to the modernization of cargo equipment and large-scale ship size, estimating proper terminal capacity model by berth size is critical because it provides management with a standard for cargo throughput. Introduction of ACT is essential, but the financial feasibility analysis for proper terminal capacity to ensure the investment return must be implemented.

This study focused on the new port of Busan, Korea's largest container port. In this area, ACT is being introduced as a smart port prepared by the Korean government for the Fourth Industrial Revolution. Therefore, this study is a relevant to present the financial feasibility of the proper terminal capacity of TOC due to the introduction of ACT.

In this study, the base year of the analysis is 1 January 2020, and the analysis target period is set as a total of 30 years from 1 January 2020 to 1 January 2049. The financial discount rate is based on the government guideline of 4.5%, and the proper terminal capacity is calculated by applying a total of five berth sizes (nine, five, four, three, and two berths). This study finds that the proper terminal capacity depends on the berth size and discount ratio. As a result, the breakeven terminal capacity is from 560,421 TEU of the 9-berth model to 633,102 TEU of the 2-berth model, applying a 4.5% standard discount ratio. In a sensitivity test considering the change in discount rate and the size of the berth at the same time, NPV has a positive value at the level of at least 550,000 TEU (nine berths and 3.5% discount rate) and up to 650,000 TEU (two berths and 5.5% discount rate).

The purpose of financial analysis for the introduction of ACT and determination of proper terminal capacity is to analyze the feasibility of development for various scales from the financial perspective to confirm the risk of development and the securing of TOC profitability in advance. In the process of determining the proper terminal capacity by financial analysis, the breakeven handling containers volume of each berth can be confirmed by analyzing the change pattern and sensitivity of NPV by berth size.

The method of optimizing financial efficiency by analyzing the appropriate loading capacity will be an important support tool in decision making by providing the results of analysis and reasonable information obtained in the analysis process to the TOC, the main stakeholder of ACT introduction.

In the future, not only for Busan Port, but also other ports considering the introduction of ACT, applying the financial feasibility estimation method suggested in this study is expected to benefit analyzing the proper terminal capacity. However, this study focuses only on financial analysis and does not consider technical analysis or other parameters such as security assessment other than financial variables. Therefore, subsequent study is expected to conduct a more comprehensive perspective suggesting the completeness of the financial analysis model can be improved.

This study has a limitation regarding the estimation. Because TOCs do not disclose some financial data such as cost statements, detailed financial information (i.e., labor and maintenance costs) cannot be obtained. This study has no choice but to use estimates of the financial data presented in the audit report of each TOCs. Therefore, as there is a possibility that an error may occur in the estimate, it is necessary to be supplemented or improved.

REFERENCES

1. Molavi, A.; Lim, G.J.; Race, B. A framework for building a smart port and smart port index. *Int. J. Sust. Trans.* **2020**, *14*, 686.
2. Pessôa, M.V.P.; Becker, J.M.J. Smart design engineering: A literature review of the impact of the 4th industrial revolution on product design and development. *Res. Eng. Des.* **2020**, *31*, 175.
3. You, J.W.; Kim, Y.S. A Study on Decision-Making Model for Port Selection: Container Terminal's Perspectives. In Proceedings of the Korean Institute of Navigation and Port Research Conference, Busan, Korea, 20 November 2019; pp. 138–139.
4. Jun, W.K.; Lee, M.K.; Choi, J.Y. Impact of the smart port industry on the Korean national economy using input-output analysis. *Transp. Res. Part A Policy Pract.* **2018**, *118*, 480.
5. Cullinane, K.; Song, D.W. Container terminals in South Korea: Problems and panaceas. *Marit. Policy Manag.* **1998**, *25*, 63.
6. Kil, K.-S.; Kim, E.-S. Improvement Scheme of Lease Rate Assessment System for Terminal Operating Company's Wharf. *J. Korea Port Econ. Assoc.* **2016**, *32*, 127.
7. Park, S.-K.; Park, N.-K. A Study on the Estimation Model of the Proper Cargo Handling Capacity based on Simulation in Port—Port Cargo Exclusive Pier Example. *J. Korea Inst. Inf. Commun. Eng.* **2013**, *17*, 2454.

8. Jun, C.; Lee, J.-P. Research on the estimation of the required cargo containers to offset initial costs, including realistic profits, at container ports. *J. Shipp. Log. Study* **2006**, *49*, 67.
9. Jeon, Y.-H.; Choi, H.-R.; Ryu, D.-H.; Yoon, S.-J. A Study on Core Factor Analysis for the Development of Automated Container Terminal in Korea. *J. Korea Port Econ. Assoc.* **2008**, *24*, 97.
10. Chen, J.; Huang, T.; Xie, X.; Lee, P.T.-W.; Hua, C. Constructing Governance Framework of Green and Smart Port. *J. Mar. Sci. Eng.* **2019**, *7*, 83.
11. Koch, T. *Automated Container Terminals: Benefits for Operators and their Clients*; International Symposium Busan: Busan, Korea, 2003.
12. Cullinane, K.; Song, D.W.; Ji, P.; Wang, T.F. An application of DEA windows analysis to container port production efficiency. *Rev. Netw. Econ.* **2004**, *3*, 184.
13. Cullinane, K.; Wang, T.F. Port governance in China. *Res. Transp. Econ.* **2006**, *17*, 331.
14. Park, N.-K.; Choi, H.-R.; Lee, C.-S. The Operating Objective of Automated Container Terminal by the AHP Techniques. *J. Shipp. Log. Study* **2004**, *40*, 113.
15. Cho, S.; Won, S.; Choi, S. A Comparative Study on Automated Container Terminal according to Stevedoring System. *J. Korea Port Econ. Assoc.* **2014**, *30*, 121.
16. Castro-Santos, L.; Bento, A.R.; Silva, D.; Salvação, N.; Guedes Soares, C. Economic Feasibility of Floating Offshore Wind Farms in the North of Spain. *J. Mar. Sci. Eng.* **2020**, *8*, 58.
17. Kim, D.-H.; Lee, K.-B. Forecasting the Container Volumes of Busan Port using LSTM. *J. Korea Port Econ. Assoc.* **2020**, *36*, 53.
18. Kang, H.-M.; Hong, S.-T.; Moon, S.-Y. A Study on the Efficiency Analysis of Port Automated Terminals. *Korea. Log. Rev.* **2019**, *29*, 11.
19. Ministry of Oceans and Fisheries. *A Study on Recalculation of Proper Port Loading Capacity*; MOF: Sejong, Korea, 2020.
20. Korea Development Institute. *Standard Guidelines of Preliminary Feasibility Studies for Projects in the Road and Railroad Sector*, 8th ed.; KDI: Seoul, Korea, 2008; pp. 392–397.
21. Economic Statistics System. Bank of Korea. Available online: **https://ecos.bok.or.kr/jsp/vis/keystat/#/detail** (accessed on 11 October 2020).
22. Lee, K.-H.; Hwang, D.-K.; Kim, M.-H. A Study on the Evaluation of Economic Benefit for Port Hinterland's Investment in Busan New Port. *J. Korea Port Econ. Assoc.* **2008**, *24*, 153.
23. Data Analysis, Retrieval and Transfer (DART) System, The Korea Financial Supervisory Service. Available online: **http://dart.fss.or.kr/** (accessed on 30 June 2020).
24. Kil, K.-S. Standardization of the Lease Fee Assessment System of Busan Port Container Terminals. *J. Korea Port Econ. Assoc.* **2011**, *27*, 65.

Chapter 8

Exploring the Impact of Different Port Governances on Smart Port Development Strategy in Taiwan and Spain

Sheng-Che Lin[1], Hsien-Kuo Chang[1] and Ying-Feng Chung[2]

[1] *Department of Civil Engineering, National Yang Ming Chiao Tung University, Hsinchu City 30010, Taiwan*

[2] *Taiwan International Ports Corporation, Ltd., Kaohsiung City 80443, Taiwan*

ABSTRACT

With the rapid development of new technologies, such as automation, big data, and artificial intelligence, a port that utilizes these technologies can significantly improve its operational efficiency and make it economically more competitive. Despite its increasing importance, there is no standard definition of a smart port, resulting in each country having its own smart port development strategy. In this study, we explore how varied port governance of port authorities may impact the smart port development strategy. First, we established a set of evaluation criteria, using the Borda count and fuzzy analytic hierarchy process, to obtain the key indicators for the development of smart ports in Taiwan. Then, we compared the preferred indicators of Taiwan's smart port development strategy with those of Spanish ports. An obvious difference in this ranking comparison is found. This indicates that Taiwan's strategy focuses on optimizing maritime safety, while Spain's strategy focuses on improving operational economics. Finally, we analyzed the differences between the smart port development strategies of these two countries using the latest port governance classification table. The results of the above analysis are also verified by comparing the differences in port legislation between them. It turns out that the private sector is deeply involved in the operation of ports in Taiwan compared with Spain. Our main contribution lies in providing a clear pathway for developing a smart port in Taiwan in the future and investigating the reasons for different smart port development strategies in different countries.

Keywords: *smart port; evaluation indicators; port governance; FAHP; port development strategy; Taiwanese ports; Spanish ports*

1. INTRODUCTION

Sea transportation is a port-to-port mode of transportation, which is not only an important means of transportation for global trade, but also an important node in the supply chain. According to the United Nations [1], sea transportation trade accounts for 90% of global trade by weight, which means that the efficiency of port operations will directly affect the efficiency of world trade. In recent years, owing to the global economic demand for maritime transport; the changing international, political, and economic situation; the impact of pandemics; and international competition, ports need to optimize their functionality in terms of economy, environment, energy, and security to enhance their competitiveness. In this context, a new concept of port governance (PG) has been proposed: the smart port (SP), which aims to realize a highly efficient and cost-effective port operation model through Industry 4.0 technologies [2]. To improve the competitiveness of ports, smart port development (SPD) has been currently regarded as the top priority for every international benchmark port. Singapore, Germany, and the Netherlands have all recently included SPD in their national plans [3,4,5]. In addition, Taiwan and Spain, two maritime-based countries, also need to change their port development strategies because their shipping accounts for a high proportion of the country's total trade [6,7]. Taiwan is an island off the southeastern China coast, where its shipping accounts for more than 99% of the total trade volume [8]. Spain, on the other hand, is in southwestern Europe. It is not only the gateway to the Atlantic Ocean and the Mediterranean Sea, but also the main route between Europe and Asia because of the Suez Canal. Spanish shipping accounts for more than 80% of the total trade volume, and Spanish ports are important distribution centers for international logistics [9].

Although there are some studies on SPD [6,7,10,11,12], the definition of SP varies from country to country, and the existing studies on SPD may not meet the needs for Taiwan's ports [7,12]. Thus, this study aims to first develop an evaluation method of SPD for Taiwan's ports using the Borda count (BC) and fuzzy analytic hierarchy process (FAHP). Secondly, we aim to compare the main criteria for SPD obtained from those of the Spanish ports presented by Alberto et al. [6]. This is not only because of the different port management models of the Taiwan Port Authority (TPA) and the Spanish Port Authority (SPA)—both of which are government-owned, with the management and control of one port being delegated to a local government agency, while the other is managed by a corporatized entity [13]—but also because such a comparison can expand more studies and discussions, such as the differences in port management between European and Asian countries [14], or the differences in PG in a country between a single PA and multiple PAs [15]. In addition, both countries have a long history of ports and works on port development have been widely discussed internationally [11,14,15,16,17,18], so information on port development in both countries is relatively easy to obtain and examine. In recent years, both countries have made a strong push to automate and digitize their ports as an SP [8,19]. Thus, the objective of this paper is to fill the above-mentioned literature gaps with the following two points:

- Using the BC and FAHP method to establish the evaluation indicators of Taiwan's SP and reveal its priority.
- Comparing the preferred indicators for SPD strategies between Taiwan and Spain and analyzing the differences from the perspective of the PG model through comparative analysis.

The remainder of the paper was structured as follows. Section 2 summarizes the theoretical background. Here, we present a conceptual model that informs our research hypotheses and survey design. Section 3 describes the details of our methodological approach. Section 4 reports our survey results and the empirical analysis of survey data. Finally, Section 5 concludes our research by discussing its practical and policy implications, acknowledging its limitations, and suggesting directions for future research.

2. LITERATURE REVIEW

2.1. The Role and Development History of Ports

With the advancement of technology and social development, the role and functions of ports are constantly evolving. In 1923, the Convention and Statute on the International Régime of Maritime Ports [20] defined a port as a berth for ocean-going vessels engaged in international trade; in 1989, the Dictionnaire Juridique des ports maritimes [21] defined a port as a natural or artificial berth. It was not until 2001, when the Brussels—2nd European White Paper on Transport [22] defined ports as nodes in the logistics supply chain of the transformation of transport modes, that ports were gradually transformed from mere berths or service berths to the important hubs of logistics and trade they are today. In the Brussels 2011—3rd European Transport White Paper [22], the concept of the network core is more explicitly included, considering the port as an integrated, network core logistics platform. As port functions evolve, more specific areas of port evolution are revealed. The United Nations Conference on Trade and Development (UNCTAD) defined third-generation ports (3GP) in a report published in 1992 [23]. UNCTAD mentioned that third-generation ports are considered as distribution centers for goods, offering more diversified services such as inventory, packaging, and other types of activities, generating additional employment opportunities and regional economic development. The fourth generation of ports was later mentioned in Port Express, No. 19, 1999 [1]. 4GPs were defined as the information and knowledge center of the supply chain to enable real-time information transfer and reduce logistics time beyond the past. Finally, fifth-generation ports (5GPs) incorporate the service functions of all of the first fourth-generation ports. The new concept is introduced by Train Motorways of Sea (TRAINMOS II) in Smart Cities and Smart Ports [24]. In 5GPs, communication technology (ICT) is used in multifaceted port management, focusing on issues other than logistics efficiency improvement such as environmental protection and security in ports. In addition, Lee et al. [25] adds in their study that the concept of customer-centric community ports is one of the important features of 5GPs, showing a trend towards more diversified and comprehensive port development.

2.2. Summary of Smart Ports

Before discussing SPs, we need to distinguish the difference between green ports (GPs) and SPs. Literally, a GP should focus on the maintenance of the port environment, such as protecting the flora and fauna in the port area, as well as efficient use of energy and reduction of marine pollution, while a SP should focus on using newly developed technologies to optimize port management and reduce various costs of the port. These two types of ports seem to be separate. However, the authors of [11,26,27] believe that GPs and SPs are inseparable. GPs and SPs will be integrated into operation and development because the trend of port development now is to take GP as the main goal and SP development as the main technical means. The green development of the port is based on low energy consumption, low emission, and low pollution as the main goal. The development of SPs is based on high-tech intelligent technology to improve the comprehensive efficiency and competitiveness of port operation. The green development of the port needs the support of port intelligence and technological innovation, and green development is also an important concept for the development of SPs. Without the concept of green development, it is difficult to realize sustainable development and the intelligent mode of the port. In general, green development is an important concept of SPs. Therefore, in this study, our evaluation index of SPs in Taiwan includes not only the use of intelligent technology, but also whether the port uses some technical means to reduce the environmental pollution of the port.

The term "smart port" was first introduced in 1994, when the concept of the intelligent transport system was introduced in the ports of maritime transportation hubs, where the most advanced information, communication, electronic control, and computer processing technologies were integrated into the traffic and transportation management system. However, to understand the nature and main structure of an SP, it is necessary to first explain the definition of smartness. In the field of technology, Spangler et al. [28] argue that smartness means that something behaves in accordance with the principles of automatic computing, such as self-protection, self-repair, and self-optimization. In urban planning, "smart growth" is a management approach to promote economic development without destroying the environment [29], while smart cities represent a management model for infrastructure integration and resource optimization [30]. In summary, this paper uses a quote from the Paris Innovation Review [31] to unify the concept of smartness: "Becoming smart means being more attractive and competitive, and doing more with less." In other words, smartness stands for the concept of optimizing resources and finding the best solution to a problem in decision making.

To determine the definition of an SP, we refer to five recent studies on the subject. Bao [32] believed that an SP is the use of technology derived from Industry 4.0 to facilitate the exchange of information across the port. Wang et al. [10] argued that an SP usually uses high-level technologies such as the Internet of Things (IoT), ICT, and big data to improve the productivity and efficiency of ports. Anahita et al. [2] asserted that an SP is not only technology-oriented, but also defined as ports where the four fields of digital communication, new technology, environmental education, and public service intersect. Moreover, Lin [7] mentioned that though there is no unified framework for an SP; the overall planning in the international development trend can still be observed to be oriented towards the similar three major goals:

1. Improving the safety, efficiency, and sustainability of the port;
2. Innovation and integration of port functions;
3. Enhancing port service capabilities.

To date, although there is still no unified global standard to define an SP, the above literature review shows that the development of an SP is not only limited to optimizing the application of information and ICT, but also includes multiple services for social education, coastal environmental protection, and reducing energy consumption, aiming to create an SP to be "efficient", "safe", and "sustainable".

2.3. Overview of International SPD Strategies and Evaluation Indicators' Selection

To understand the extent of SPD, we first need to have a basic idea of the SPD strategy for each international benchmark port to ensure that our evaluation indicators are aligned with international trends. The information can be seen in Table 1. The Singapore government implemented the Smart Nation 2025 ten-year plan back in 2014 [4], which aims to establish a nationwide sensor network to enhance data collection, transmission, and sharing. In addition, the Maritime and Port Authority of Singapore (MPA) has partnered with IBM to develop Sense-making Analytics for Maritime Event Recognition (SAFER), a port and maritime event recognition system that will reduce the occurrence of maritime events through the use of seven smart technology systems. The systems are intended to reduce the occurrence of maritime disasters and have already completed three of seven trials in 2018. The Dutch port of Rotterdam and the German port of Hamburg are pioneers in the development of smart ports in Europe. In 2018, the Port of Rotterdam Authority (PRA) installed many sensors within 42 km of the port for the "Port Shipping Connect 2025" project [3]. Through these sensors, a large amount of weather data is collected, transmitted back, and analyzed, enabling the port management to make fast and accurate operational decisions, such as reducing ship waiting times, helping to determine the best ships to reduce ship waiting times, and helping to determine the best time to call at the port. The Port of Hamburg is the second largest container port in Europe and the improvement in logistics efficiency is considered by the Hamburg Port Authority (HPA) to be the most important factor for the port. In 2015, the HPA presented a smart port management program for achieving economic and ecological co-prosperity, emphasizing the logistics supply chain and energy development of the port [5]. In addition, since 2017, the Spanish port industry has also shown great interest in SPD [11]. It holds annual conferences to discuss the most relevant and innovative SP topics in the world. The Port Authority of Valencia (PAV), the greatest port in Spanish, presents the SP 2030 strategy in London, with the strategic objective of connectivity, efficiency, and responsibility. Its aim is to use the port community system (PCS) to reach cooperation and information transfer between the port and the community to improve the competitiveness and sustainability of its region of influence [33].

Table 1. Overview of international SPD strategies.

Port & Country	SPD Strategy
Singapore port (Singapore)	The Maritime and Port Authority of Singapore (MPA) and IBM jointly developed the SAFER system in 2017. Aiming to reduce navigation accidents through intelligent traffic monitoring.
Rotterdam port (Dutch)	The goal is to achieve connected shipping by 2025, which will enable ships to navigate the waterways of ports automatically.
Hamburg port (Germany)	Smart Energy: To reduce port-related energy consumption and pollution, and to encourage the use of renewable energy. Smart Logistics: To improve the economic efficiency of ports and apply IT technology to optimize port efficiency.
Valencia port (Spain)	Strengthen cooperation between ports and cities through PCS. Reduce pollution and increase efficiency through automated processes, aiming to make ports into sustainable, efficient, carbon-neutral smart ports.
Qingdao port (China)	The goal is to achieve a fully automated terminal that can increase operating capacity by 30% and save 70% of labor compared with a traditional terminal.
Xiamen port (China)	It is the first fully automated container terminal in China, featuring full intelligence and zero emissions. It is electrically powered and can save more than 25% of energy compared with traditional terminals.
Kaohsiung port (Taiwan)	With a port design that targets automated terminals, future terminal operators can introduce automated equipment such as automated guided vehicles (AGVs) and remote-control loaders.

In Asia, seven of the top ten ports in the world in terms of cargo and container throughput are in China, which has also promoted digitization and automation in port construction and policy [34]. The advantages of zero-touch, low-risk, high-efficiency technology have not only increased traditional terminal operations by 30% and saved 70% of labor costs, but also avoided the risk of epidemic transmission with relatively low manpower requirements. Production and management can still be carried out through automated operations at the port. Meanwhile, Taiwan established an automated port gate system and RFID container e-seal, which saves one-quarter of the time compared with traditional paper-based manual customs clearance and tracks logistics status in real time, in 2012. In 2018, the Taiwan Port Authority (TPA) proposed the Transform Sustainable Modern and Advanced ports with Revolutionary Technology (Trans-SMART) plan, which is divided into two action plans for sea and land, including land testing to optimize logistics, security, and efficiency functions and the use of big data analysis and intelligent technology to enhance the international competitiveness of Taiwan's ports [8].

Then, the selection and establishment of evaluation indicators is the most essential item in research. We reviewed three studies on SP and compared the method, main indicators, number of sub-indicators, weighting method, and study subjects, as shown in Table 2. Anahita et al. [2] developed four main indicators including 68 sub-indicators of SPD through an extensive literature analysis and collected official statistics from ports of various countries to arrive at the SP index for the top 14 international ports after normalization. Chen et al. [11] also constructed 6 major indicators and 20 sub-indicators affecting green ports and SPs and used the interpretative structural model method to establish the relationship between each indicator. Alberto et al. [6] invited a group of experts and scholars to agree on evaluation indicators for SP through the Delphi method and established weights for each indicator. Indicators such as digitization, ICT, and automation were considered relatively important in the SPD.

Table 2. Literature review of the establishment of SP evaluation indicators.

Item	Anahita et al. [2]	Alberto et al. [6]	Chen et al. [11]
Method	Literature Analysis	Delphi Method	Literature Analysis
Main indicators	Operation, Environment, Energy, Safety, and Security	Operational economic, Social, Political, Environment	Greenness, Agility, Personalization, Cooperation, Intelligence, Liberalization
Sub indicators	68 pcs	32 pcs	20 pcs
Weighting method	equal distribution	expert consensus	-
Study subjects	14 international ports	10 Spanish ports	14 international ports

Based on the above review, we conclude that the main indicators, number of sub-indicators, and weighting of each indicator are different among the three studies. Thus, we used the intersection of the assessment indicators of these three studies to find common key indicators in this study. At the same time, we used a theoretical method to determine the reasonable weight of each indicator with different importance to improve the disadvantage that each indicator has the same weight.

2.4. A Brief Review of Port Governance

The world's perception of port governance (PG) has changed over the past four decades. The devolution of publicly operated transport operations to the private sector in many countries starting in 1980 is a phenomenon that illustrates the impact of governance on the port industry. For example, the British government privatized ports [35], tendered management facilities with preferential policies, or delegated the management of ports and terminals to other government departments [36]. To define governance, the World Bank (2001) [37] considered governance structures as consisting of the systems, structures, and processes developed by a group to achieve common goals. The legislation imposed by the government on a business or nonprofit organization also shapes the governance structure in which the organization operates. Port governance, on the other hand, refers to the means or methods by which government or those in authority in the port industry own, manage, and control port operations [38].

Typically, PG can be divided into two cases according to its subject, namely port governance and port authority governance [18,39]. In the first case, port governance covers the way services and terminals are managed and their relationship with the port [37]; in the second case, it refers to the internal governance of the PA and its relationship with the government and external agencies [40]. In the latter, the governance process is defined as the interaction between policy makers and the PA: the greater the autonomy of the PA, the greater its own responsibility in terms of port management performance [41]. In exploring the relationship between PG and PA, Verhoeven [42] concludes that PG patterns are influenced by the size of the corresponding PA. The role of the PA has also attracted considerable academic interest in the literature on port governance, management efficiency, and logistics performance. Thus, owing to the strong link between PG and PA, in our study, PG is defined as the second case mentioned above, i.e., governed by the PA.

2.5. Impact of PG Model on Port Development Strategy and How to Analyze

Under the influence of globalization, Sanchez and Pinto [43] adapted the decision-making process of PG [44] and found that the technological innovation may change government strategies for port development, leading to the emergence of new governance models, which in turn led to port new development policies. Thus, the emergence of smart ports seems to be causally related to PG to some extent. This also aroused our interest to explore the relationship between PG and SPD.

The concept of PG models has evolved over time and space because of the convergence of several elements [42], including decentralization, corporate governance, operational status, functional autonomy, functional initiative, investment responsibility, and financial autonomy. This evolution has been reinforced since the 1990s with the development of decentralization programs around the world because it brings a lot of success in the port industry. For instance, Goss [45] pointed out that privatization of ports helps PA to maintain competition, and thus improve operational efficiency. Baird [35] also showed that the key factor for the increasing competitiveness of U.K. ports is the removal of control over the port labor market. Thus, it can be argued that changes in PG models have driven different port development strategies in ports and created different port performance. Currently, 80% of the world's ports are landlord ports [39]. Unlike the tradition of full government responsibility, in this model, PA is like the landlord of the port, holding the ownership and leasing out the port infrastructure to private operators [37]. The PA is only responsible for economic development; long-term development of the land; and maintenance of the port infrastructure such as channels, berths, and terminals, while the private operator is responsible for the operation of the terminals.

However, even though the main challenges and objectives of ports around the world are very similar today, the reform of port governance has taken different paths. The same conclusion can be drawn in the case of the EU. The diversity of rules makes it impossible to identify a fixed European governance model [42]. Ng and Pallis [46] also find that the role of ports in PG practices varies within a group of countries, depending on the political traditions of each country. Therefore, it can be concluded that PG models evolve according to the socio-economic framework and the performance of ports [41]. That is, the evolution of PG is unique in each country, which leads to different port development strategies. For example, Bennathan and Walters [47] divided PG into two doctrines. The Anglo-Saxon doctrine and the European doctrine, which are distinguished by whether port profitability is the primary goal of PA. Lee and Flynn [14] investigated the characteristics of major container port developments in Asia and Europe and compared their port pricing systems, finding that there is a third doctrine driving port development policy, the Asian PG doctrine, in addition to the Anglo-Saxon and European doctrines. Likewise, Laxe et al. [18], by comparing the timeline of port reforms (legislation) and external causes of PG changes, found that PG in Latin America is currently still conceptually in the old definition and, therefore, illustrates that its port development strategy is less adaptive to port changes than its European counterparts.

Therefore, to analyze a country's port development strategy, knowing its PG model is crucial; however, this matter is not simple. Although it would be easier to distinguish between different types of PG models using only classification tables [35,37,48], these frameworks are too simplistic to reflect the infinite diversity of today's highly competitive port environment [44]. The division of PG models into neatly labeled packages is becoming increasingly unimportant as it does not adequately reflect the regional nature of a port system [39]. To sum up, based on our literature review, an extensive literature review with observational and comparative analysis would be a better choice for PG analysis rather than a standardized classification table.

3. RESEARCH METHODOLOGY

We used a series of methods in three stages to obtain the criteria for SP development in Taiwan ports. The three stages of research methodology in this study are developing, determination, and reflection, in which the key methods used in each stage are the Borda count (BC), FAHP, and PG analysis. Figure 1 illustrates the flow sequence of our research methodology.

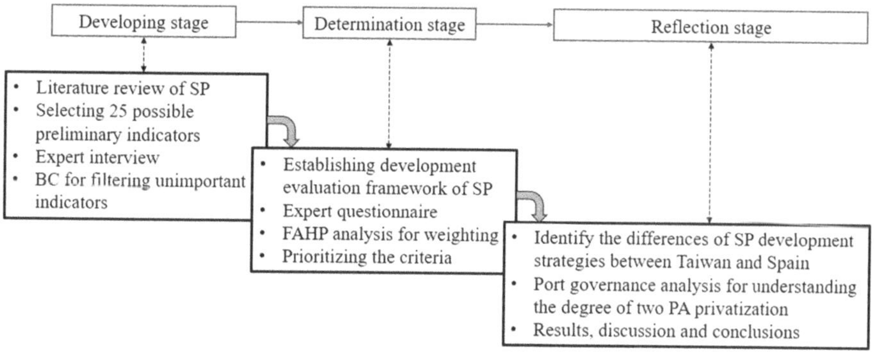

Figure 1. *Flow sequence of the research methodology.*

3.1. Developing Stage

To make the questionnaire more valid for FAHP analysis, the number of final indicators should not be too large. Saaty [48] mentioned that the maximum number of factors in each layer of FAHP should not exceed seven items. Miller [49] also pointed out that human psychological responses are limited to 7 ± 2 and suggested that the design of the assessment framework should be 3~5 elements per layer. Based on the above principles, we first collected all of the different indicators of SP from some previous works and finally decided on 25 pcs as possible preliminary indicators. Although those indicators have been approved in some European and American ports using a qualitative method, their applicability to Asian ports still needs to be discussed through a rigorous approach.

Therefore, at this stage, some indicators that are considered extremely unimportant are removed in the preliminary indicators through Borda count (BC). BC has a higher degree of fairness to elect a more consensual candidate compared with a plurality voting system [50]. It indicates that the BC is more likely to integrate the opinions of experts in group-decision making problems. The rule of BC is to give the candidates points according to their places on each ballot. Its calculation is expressed as Equation (1). Each indicator is assigned a rank from each decision-maker equal to the score of the indicator he or she prefers, so that, with n indicators, each one receives n − 1 point for a first preference, n − 2 for a second, and so on. After quantifying all rankings, the final score for each indicator was obtained by summing.

$$f_B(x) = \sum_{Y \in A} (i : xP_i y) \tag{1}$$

where fB(x) where is the final score of indicator x; *y* is the total number of preference xPi cases in the set *A*; and xPi is the point of indicator x in the *i*-th *y*.

The interview of BC was scored by 10 experts, including senior engineers and university scholars in harbor engineering (as shown in Table 3). The job tenure and their diverse expertise indicate that the respondents have sufficient knowledge and expertise, with the ability to offer reliable observations and comprehensive information about the selection of smart port evaluation indicators in Taiwan. Most respondents have more than 20 years of work experience and are familiar with port affairs. Interviewees were asked to rank these preliminary indicators of each dimension considering whether they are suitable for local conditions or follow the international trend of SP development. The interview with one respondent lasted one hour, and this time was used to explain to the respondent how to answer the questionnaire and to answer their concerns about the questions. A total of 10 questionnaires were distributed and 10 usable questionnaires were received. This results in a response rate of 100%, which is satisfactory for this type of survey-based study [51]. After BC calculation, the scores of the whole indicators were obtained. Then, we use a z-score to determine the importance of the indicators. The detailed filtering process will be described in Section 4.1 and the z-score is defined as follows:

$$z = \frac{x - \overline{x}}{S} \tag{2}$$

where x is the raw score of the sample, $\overline{x}$ is the mean of the total samples, and s is the standard deviation of the sample.

Table 3. Respondents' profile.

Job Title	Number	Affiliation	Job Tenure	Expertise
Vice director	1	Taiwan port authority	≥20 years	Port development planning
Senior manager	3		11–20 years	Port operation and energy management
Senior engineer	3	Construction consultant	≥20 years	Port engineering and design
Academia	3	University professor	≥20 years	Logistics management and environmental assessment

3.2. Determination Stage

After the first stage, we then established the SPD evaluation framework and developed an expert questionnaire, which is designed to form the pairwise comparison matrices for grouped sub-criteria and dimensions. The respondents are the same as those in the developing stage.

Multi-criteria decision making (MCDM) is widely used in many fields such as textbook selection, engineering risk assessment, transportation system design, market research, and energy policy planning. The most common applications are the selection of factors or indicators and the evaluation of factor weights. Analytic hierarchy process (AHP), fuzzy analytic hierarchy process (FAHP), and analytic network process (ANP) are three extensively used methods for MCDM and have been refined since they were proposed [48,52,53]. AHP was first introduced to numerically select decision options from unquantifiable factors by Saaty [48]. FAHP is an extended application of AHP, which features fuzzy numbers to align the analysis results more with natural human semantics. ANP is often used to solve problems that cannot be expressed in terms of hierarchical structure. The relationship of interdependence between the criteria should be seriously considered in ANP, but is not needed for AHP and FAHP. Thus, ANP requires more complex calculations than AHP and FAHP. Although these three methods are still widely used in the field of maritime policy and management [54,55], FAHP, which can reflect the real semantic meaning of human beings by simple calculation, is more suitable for the needs of this study rather than AHP and ANP, and is thus chosen for further analysis in this study. Thus, FAHP is the key method to determine the weight and the priority of all selected criteria in our study. The FAHP procedure was adopted from Yu and Wei [56] and a brief introduction to its theory and calculation is provided as follows.

3.2.1. Establishing a Fuzzy Number

Zadeh [57] defined a fuzzy set as a class of objects with a continuum of grades of membership ranging between zero and one. According to Liu et al. [58], 99 out of 109 papers surveyed used triangular fuzzy number (TFN), and the high percentage of 91% indicated that triangular fuzzy number is an accepted method in academics. A TFN is represented by three points and denoted simply as (l, m, u). The parameters l, m, and u denote the smallest possible value, the most promising value, and the largest possible value, respectively. A triangular membership function is defined as Equation (3):

$$UA(x) = \begin{cases} \frac{x-l}{m-l}, & l \le x \le m \\ 1, & x = m \\ \frac{u-x}{u-m}, & m \le x \le u \\ 0, & l \ge x \ge u \end{cases} \quad (3)$$

3.2.2. Determining the Linguistic Number

The concept of linguistic variables is very practical in handling ill-defined or complex situations reasonably described in conventional quantitative expressions. We followed the equivalent fuzzy numbers and linguistic comparison terms of [59] shown in Table 4.

Table 4. Fuzzy comparison measures.

TFN	Linguistic	TFN	Linguistic
$\tilde{1} = (1,1,1)$	Equal	$\tilde{6} = (5,6,7)$	Fairly good
$\tilde{2} = (1,2,3)$	Weak advantage	$\tilde{7} = (6,7,8)$	Very good
$\tilde{3} = (2,3,4)$	Not bad	$\tilde{8} = (7,8,9)$	Absolute
$\tilde{4} = (3,4,5)$	Preferable	$\tilde{9} = (8,9,10)$	Perfect
$\tilde{5} = (4,5,6)$	Good	-	-

3.2.3. Building the Fuzzy Positive Reciprocal Matrix

The fuzzy positive reciprocal matrix may be expressed in the following form:

$$\widetilde{A} = [\widetilde{a}_{ij}] \tag{4}$$

where n is the number of sub-factors of each dimension, and a˜ij represent the pairwise comparison between criterion i and j using fuzzy numbers and where i, j=1,2,…,n

$$\widetilde{a}_{ij} = \begin{cases} 1, \ \forall i = j, \\ 1/\widetilde{a}_{ji}, \ \forall i \neq j \end{cases} \tag{5}$$

3.2.4. Calculating the Fuzzy Weights

The fuzzy weights of each criterion in the fuzzy positive reciprocal comparison matrix are obtained using the geometric mean method, where the geometric mean of the fuzzy comparison value of criterion i to each other criterion is derived as follows:

$$Z_i = (a_{i1} \times a_{i2} \times \ldots \times a_{in})^{1/n} \ \forall i = 1, 2, \ldots, n \tag{6}$$

The fuzzy weight of the ith criterion, as represented by a TFN, is then computed as follows:

$$W_i = Z_i \times (Z_1 \times Z_2 \times \ldots \times Z_n)^{-1} \ \forall i = 1, 2, \ldots, n \tag{7}$$

3.2.5. Defuzzification

Because the criteria weights are still in the form of fuzzy triangular values, there is a need to translate them into non-fuzzy values through a process of defuzzification. A common method is the maximum mean method, centre of area (CoA) method, and alpha-cut method. CoA proved to be better for defuzzification than the other [58], and is thus used in this study. The defuzzification weight of the ith evaluation criterion of CoA is defined as follows:

$$DF_i = \frac{U_i + 2 \times M_i + L_i}{4} \ \forall i = 1, 2, \ldots, n \tag{8}$$

3.2.6. Normalization and Synthetic Analysis

Although the previous step yields a non-fuzzy number, normalization is required to represent the relative importance between factors. The normalization weight of the ith criteria, NWi, is expressed by the following equation.

$$NW_i = \frac{DF_i}{\sum_{i=1}^{n} DF_i} \ \forall i = 1, 2, \ldots, n \tag{9}$$

3.3. Reflection Stage

Although in the literature review, we mentioned that the existing PG classification tables do not satisfy all PG models around the world, it is possible to find some variables that can distinguish trends in the application of classical models in different regions of the world [15]. Similarly, while Brooks [60] acknowledges that there are many differences and regional variations in PG, they still insist on proposing and applying discrete categories of PG. However, further analysis of PG practices, styles, and patterns requires a more continuous and fluid approach to the subject, such as exploring its variation over time and space [39]. Therefore, at this stage, a set of analytical methods is going to be conducted to fill this gap. First, we will examine the PG characteristics of Spain and Taiwan through a latest and simple PG classification table. Second, to compensate for the discrete nature as well as the incompleteness of the PG classification framework, we will use a comparative analysis to examine the extent to which PAs are allowed to intervene in port services in these two countries from the aspect of port legislation or regulation. This is because national laws and specific port legislation have a direct impact on the decision-making process of PG model mechanisms [61,62]. Likewise, national policies have an impact on the decision-making of ports and port governance models [63].

Finally, using the results of the above analysis, we will explore whether the different PGs in these two countries have contributed to the differences in SPD. Given the relative novelty of the study and the analytical approach adapted from the main literature within the last 20 years [44,64,65], we use the PG classification table of Caldeirinha et al. [15] in our study. The variables we used include PA type, PA power, PA focus, PA relationship, PA geo-organization, PA function, PA competition pattern, PA managers' selection, PA management/legal structure, PA nature, and PA financing, as shown in Table 5.

Table 5. Classification of the main models of PG.

Main Port Governance Models Characteristics	Private	Liberal	Controlled	Centralist
PA type	Full private	Landlord	Landlord	Tool/Service port
PA power	Private power	Devolution/Autonomy	Controlled devolution	Centralist
PA focus	Core	Regional	Regional	Core
PA's relations	Competition	Competition	Cooperation	Integration
PA geo-organization	Each port PA	Each port PA	PA regional fusion	One PA
PA functions	Land manager	Development	Facilitator	Operator
PA competition model	Liberalization	Liberalization	Limitation	Monopoly
PA managers' selection	Technical	Mix	Political	Political
PA management/Legal structure control type	Minimal	Internal control	Finance and invest	Total control
PA nature	Full private	Government company	Government company	Government department
PA financing	Pay to government	Balanced accounts	Government aid	Public money

Using these 11 variables from the literature, four major governance models can be identified, categorized as private, liberal, control, and centralized, which can give us a little insight into the degree of port privatization in a country. Then, the subsequent comparative analysis of PG is collated from an extensive literature review, while the main criteria of SPD in Taiwan, obtained by FAHP at the last stage, were compared with those in Spain [6] to derive the SPD differences between the two countries.

4. RESULTS AND FINDINGS

4.1. Determination of Final Indicators

In this section, we determined the final indicators based on the BC method. Table 6 indicates the item code, cumulative score, calculated z-score of each indicator from BC, and the determination of pass or fail. We used the first letter of the name of these dimensions as the item code. The local z-score is obtained from the in-group comparison (IGC) of sub-indicators in each dimension. The global z-score is considered by the whole-group comparison (WGC) of 25 indicators. Z-score is the value after standardization of the data, which indicates the distance between that raw score of the indicator and the sample mean in units of the standard deviation. Tam [66] showed that an indicator with a lower negative z-score indicates an insignificant one in a normal distribution. For a sample size of 25 indicators, minus one is valid as a threshold value of z-score to remove some insignificant indicators. Therefore, the threshold of the z-score was set to −1 for both IGC and WGC in the study. If both z-scores of an indicator are lower than −1, it cannot pass the threshold. Failed indicators being quite unimportant through local and global z-scores were removed from 25 indicators.

Table 6. The local and global z-scores for each indicator and the decision to pass.

Dimension	Preliminary Criteria	Score	Z-Score		Pass/Fail
			IGC	WGC	
Operational economics	O-1. Logistics process digitalization	17	1.10	0.65	P
	O-2. Degree of mechanical systems automation	16	0.73	0.73	P
	O-3. Docking line efficiency	14	0.00	0.00	P
	O-4. Use of storage capacity	13	0.37	−0.37	P
	O-5. Number of reception capacity of large ship	10	−1.46	−0.79	P
Environmental protection	E-1. Reduction of greenhouse gas emissions	18	0.95	0.86	P
	E-2. Number of automations of environmental quality monitor	16	0.48	0.45	P
	E-3. Amount of recycled wasted	13	0.24	−0.17	P
	E-4. Certification of environment management system	12	−0.48	−0.38	P
	E-5. Degree of ecological protection [1]	7	−1.66	−1.41	F
Energy Sustainability	S-1. Annual electricity and fuel conservation rate	21	1.43	1.48	P
	S-2. Number of energy conservation	17	0.61	0.65	P
	S-3. Usage of shore power system	12	−0.41	−0.38	P
	S-4. Certification of energy management	11	−0.61	−0.59	P
	S-5. Renewable energy production [2]	9	−1.02	−1.00	F
Maritime safety	M-1. Working environment safety	23	1.49	1.89	P
	M-2. Navigation aid systems	14	0.00	0.00	P
	M-3. Real-time monitor security system	14	0.00	0.00	P
	M-4. Hazardous cargo management system	13	−0.17	−0.17	P
	M-5. Compliance with ISPS training requirements [3]	6	−1.32	−1.62	F
Community service	C-1. Work environment equality and friendly in the workforce	23	1.22	1.89	P
	C-2. Data and information transparency and accessibility	18	0.54	0.86	P
	C-3. Connectivity with neighbor community	16	0.27	0.27	P
	C-4. Social media engagement	8	−0.81	−1.21	P
	C-5. Free WIFI service [4]	5	−1.22	−1.82	F

[1–4] These four indicators marked as the grayscale bottom failed and were removed from 25 preliminary indicators through the BC method. Thus, 21 indicators remained for the FAHP analysis.

Table 6 showed that all the indicators in the operational economics dimension passed both thresholds and were accepted for further analysis. Only four indicators failed and were removed from 25 preliminary indicators. These are "Degree of ecological protection", "Renewable energy production", "Compliance with ISPS training requirements", and "Free WIFI service", which are specially marked as the grayscale bottom for easy identification. Thus, 21 indicators remained for the FAHP analysis.

4.2. Weight and Prioritizing of the Criteria

FAHP commonly includes three layers: goal, dimension, and criterion, as shown in Appendix A. The goal is set to be the "Taiwan Smart Port Development Evaluation Model". The main dimension includes five aspects: operational economics, environmental protection, energy sustainability, maritime safety, and community service. Each dimension contains four to six sub-criteria, which are distinguished by numbers. The evaluation variables of each sub-criterion are shown in Table A1. The number indicates the cited literature whose sources are listed below this table. After the 21 indicators were selected, their weights were obtained by FAHP analysis and are presented in Table 7. The consistency ratio of the pairwise comparison matrix is examined to be a value of 0.1 which is commonly used in FAHP. The obtained consistency ratio of 0.051 indicates that all weights obtained are consistent and acceptable.

Table 7. Weights and ranking of 21 criteria.

Second Layer		Third Layer		
Dimension	**FAHP**	**Criteria**	**Weight**	**Ranking**
Operational economics	0.3511	O-1. Logistics process digitalization	0.0768	4
		O-2. Degree of mechanical systems' automation	0.0684	6
		O-3. Docking line efficiency	0.1213	2
		O-4. Use of storage capacity	0.0486	10
		O-5. Number of reception capacity of large ship	0.0360	11
Environmental protection	0.1458	E-1. Reduction of greenhouse gas emissions	0.0694	5
		E-2. Number of automations of environmental quality monitor	0.0251	13
		E-3. Amount of recycled wasted	0.0296	12
		E-4. Certification of environment management system	0.0217	14
Energy Sustainability	0.1025	S-1. Annual electricity and fuel conservation rate	0.0486	9
		S-2. Number of energy conservation	0.0220	15
		S-3. Usage of shore power system	0.0172	17
		S-4. Certification of energy management	0.0147	18
Maritime safety	0.3457	M-1. Working environment safety	0.1000	3
		M-2. Navigation aid systems	0.1270	1
		M-3. Real-time monitor security system	0.0548	8
		M-4. Hazardous cargo management system	0.0639	7
Community service	0.0550	C-1. Work environment equality and friendly in the work force	0.0238	16
		C-2. Data and information transparency and accessibility	0.0123	20
		C-3. Connectivity with neighbor community	0.0128	19
		C-4. Social media engagement	0.0061	21

For the second layer, the weight of each dimension is presented in the second column of Table 7. The largest weight is 0.3511. The result shows that the dimension of operational economics is the most important among the five dimensions. The dimension of maritime safety has a weight of 0.3457, which is slightly lower than 0.3511. The next three sorted dimensions are environmental protection, energy sustainability, and community services, which have weights of 0.1458, 0.1025, and 0.0550, respectively. The result shows that the dimensions of operation economics and maritime safety have similar importance and the dimension of community service is the least important.

For the third layer, the weight of each criterion is listed in the fourth column of Table 7, where the corresponding weight of each dimension is considered. Thus, the sum of all these weights is one. Prioritizing of all criteria is represented by numbers from 1 to 21, according to the largest to smallest weights. The result listed in the last column of Table 7 reveals that the top three most important criteria are navigation aid systems, docking line efficiency, and working environment safety, of which the corresponding weights are 0.1270, 0.1213, and 0.1000, respectively.

4.3. Comparative Analysis of the SPD Strategies in Taiwan and Spain

4.3.1. Comparison of Main Criteria

To understand the differences in SPD strategies between Taiwan and Spain, we compare the criteria priorities of SPD in Taiwan obtained by FAHP with those of Spanish ports. The number of evaluation criteria for the SPD in the two studies is 21 and 32. Higher priority criteria represent important guidelines for SPD. Table 8 shows the comparison of the top 3 and 16th priority criteria for SPD between the two countries.

Table 8. Main indicators between two countries in smart port development.

Rank	Taiwan	Spain
1st	M-2. Navigation aid systems	O-1. Logistics process digitalization
2nd	M-1. Working environment safety	O-2. Degree in mechanical systems automation
3rd	O-3. Docking line efficiency	O-3. Docking line efficiency
⋮	⋮	⋮
16th	C-1. Work environment equality and friendly in the workforce	M-2. Navigation aid systems

For the Taiwanese ports, navigation aid systems and working environment safety in the maritime safety dimension are evaluated as the two most important criteria for SPD. Then, the third important criterion is docking line efficiency in the operational economics dimension. Lin [7] pointed out that the infrastructure costs of building an SP are usually high, and the benefits are difficult to estimate. However, according to Anahita et al. [2], the degree of SPD is positively correlated with national R&D funding expenditures, indicating that, the more a country is open and committed to the innovation of new systems, the more it can promote the SPD. Thus, based on Taiwan's existing software strength, the development of navigation aid systems is relatively simple to implement and has real benefits for Taiwan. Nevertheless, the top 3 criteria for the Spanish ports are all in the operational economics dimension. These are logistics process digitalization, degree of mechanical systems automation, and docking line efficiency. The navigation aid system in the maritime safety dimension was ranked 16th out of 32 priority criteria. The ranking is significantly different from the top ranking for the Taiwanese ports. The ranking difference can be interpreted as SPA emphasizing much more port operation efficiency compared with TPA in SPD. The ranking difference arouses our great interest in the study.

4.3.2. Comparison of PG

In this subsection, we use the PG model proposed by Caldeirinha et al. [15] to explain this difference. In classifying PA, four main models of governance emerge, namely, the private model, the liberal model, the control model, and the centralized model. The private model is based on the Anglo-Saxon model and consists of an autonomous private PA and a private operator with minimal government control over the contract. The liberal model is based on the Hanseatic model and consists of a public PA and a private concessionaire or leaser with considerable autonomy and a corporate oversight body controlling internal finances. The controlled model, on the other hand, is based on the Latin model and relies on a public capital PA and an entirely private public service concession terminal operator with controlled decentralization and a financial and investment strategy controlled by the government. The centralized model relies on the PA as the government agency that controls all or part of the nation's ports, with tight control over finances, economics, operations, and investment or strategy [15,44].

Table 9 shows the comparison of variables between TPA and SPA regarding PA. The data sources for this table are from [67,68] and the official port legislation website. Although the labels provided in the Table 5 do not perfectly match the governance models of these two countries, this comparison still reveals some characteristics of their PG. The words under grayscale represent that they are different from the original labels and show the governance characteristics of the two PAs. Overall, based on the results of the table's classification, it can infer that TPA governance seems to be more privatized than SPA because TPA has more labels on the liberal model and SPA has more on the controlled model. Furthermore, Taiwan has only one PA because the country is too small, not for political reasons. Thus, it would be easier for port management if the old PA were merged.

Table 9. Classification of the main models of PG in TPA and SPA.

Main Port Governance Models Characteristics	Taiwanese Port Authority (More Like Liberal)	Spanish Port Authority (More Like Controlled)
PA type	Landlord, but partial hypocrisy [5]	Landlord based on Latin tradition [6]
PA power	Devolution	Devolution
PA focus	Core [7] (Only one PA responsible for a whole country)	Regional [8] (46 public interest ports, managed by 28 PAs)
PA's relations	Integration [9]	Cooperation
PA geo-organization	One PA [10]	PA regional fusion
PA functions	Development	Facilitator
PA competition model	Liberalization	Limitation
PA managers selection	Mix	Mix
PA management	Internal control	Internal control
PA nature	Government company	Government company
PA financing	Balanced accounts	Own receipts (Set on yearly 2.5% profitability of PAs)

[5–10] The variables marked as the grayscale bottom show the difference between TPA and SPA.

In terms of the grayscale labels, we can obtain a better understanding of the differences between the two PAs. TPA is currently in a hypocritical state of PG, as it established the goal of moving to a private governance model in the 2012 port reform, but the PA has publicly owned bureaus that retain most of the governance structure [67]. This led to difficulties in policy implementation, and although the PA was consolidated into one department for efficiency, the newly created branches under the PA resulted in increased time and cost to adopt a decision. Meanwhile, the governance model of the SPA is considered by scholars as a more centralized governance framework (Latin model), in which the PA is more influenced by central government decisions and planning [68]. The Spanish national port system consists of 46 ports of public interest, managed by 28 PAs. Their coordination and control of efficiency correspond to the national port public authorities, responsible for implementing the government's port policies [69]. Thus, TPA's PG seems to fall between the liberal and controlled model, while SPA's PG falls between the controlled and centralist model, which also suggests that TPA is more privatized than SPA.

4.3.3. Verification and Discussion

To verify the above results, we would like to discuss them more precisely in the context of port legislation. Given that the two PAs are essentially a landlord PG, the PAs are only responsible for the maintenance of port facilities, while port operations are the responsibility of private companies or terminal operators. This decentralization is usually through concession agreements [68], or contracts whereby the government transfers operating rights to a private company, which then engages in activities under the terms of the contract with government approval [70]. This means that, in the landlord port model, the PA is actually the concessionaire. Consequently, in this liberalized environment, concession contracts are one of the most important tools through which PAs under the landlord type can directly or indirectly influence port activities [68]. Therefore, we set out to find evidence and examples of indirect involvement of both countries in port activities using concessions, starting from information or laws on port concessions, and found that Spanish ports have more power to intervene in port operations.

According to Article 109, Section 3 of Royal Legislative Decree 2011 [71], the Spanish government would take the necessary measures to ensure the provision of adequate port services when port services from private companies are insufficient. This article gives the SPA the right to participate in the operation of port logistics and the responsibility to provide port services directly. For example, the holder of a license issued by a PA borrows pilotage services in a private legal system. Given that anyone can challenge the granting of a license, the port authority must assume, directly or indirectly, the obligation to provide the service [72]. Thus, in the Spanish port system, the relationship between the provider, the public administration, and the recipient of the service may have different legal implications in terms of legacy liability. However, there is no similar law in Taiwan's port regulations that enables the TPA to participate indirectly in port operations. Most cargo handling and warehousing operations originally in charge of the public sector were transferred to private companies [73]. Cargo management and shipping has already been already controlled by private shipping companies and cargo contractors since TPA transformed from a public institution to a private company in 2012. Although both Taiwanese and Spanish ports are classified as landlord ports, SPA has less privatization than TPA and still can involve several port services as an operator based on their port legislation. Through the PG analysis on TPA and SPA, we found two reasons for different strategies of SPD.

- TPA has little power to improve cargo stevedoring efficiency, so TPA can only focus on maritime safety to prioritize SPD.
- SPA must be directly involved in the port logistics market, thus most of the proposed policies need to improve the operational economics of the ports.

Based on the above discussion, we finally inferred that a different degree of port privatization is the key factor indirectly affecting the SPD strategies.

5. CONCLUSIONS

5.1. Practical and Policy Implications

In this paper, the key indicators of SPD in Taiwanese ports were studied and compared with those of Spanish ports. The FAHP method was applied to determine the ranking criteria of the SPD strategy for TPA. The result showed that advancing the navigation aid systems for SPD in Taiwan is at the top of the list. Two main reasons were determined for the SPD strategy for the ports in Taiwan. First, the primary operator in Taiwan's ports has been transformed from a public sector to a private sector since 2012 to enhance its competitiveness. As a result, TPA mainly focuses on the optimization of maritime safety and is not interested in improving operational economics. Second, its cost is relatively low in Taiwan compared with other countries. According to the president of TPA, Taiwan has been building its shipping database since 2011, so it has the advantage of focusing on upgrading, developing, and integrating the original system compared with other countries that spend more money on infrastructure.

The other aim of this paper is to compare the main priority criteria of SPD in Taiwan and Spain and explore the reasons for their differences. The difference in the top priority criteria of SPD is that SPA is most concerned with operational economics, but TPA is more concerned with maritime safety. Through the PG analysis, it is found that the different degrees of port privatization in the two countries result in different PG types, which leads to differences in SPD strategies.

5.2. Limitations and Future Research

The obtained criteria for SPD in this study were based on the survey from port experts, the empirical analysis of survey data, and collected data from literature. The result may include a potential bias of self-selection in this survey method. Differences in experience and understanding of SP among respondents of the questionnaire can also lead to skewed results. Despite these limitations, we provided an applicable evaluation model for SPD. The results can be used to assist the government and port authorities to formulate, adapt, and develop effective SPD strategies based on their own country's port governance model. In addition, as this study was conducted only in Taiwan, similar studies should be conducted in other ports in different geographical regions using the methodology of this study in the future to make the theoretical and practical contributions of this study more meaningful. Another future research direction could be to study the differences in SPD strategies between various countries, such as Asian ports and U.S. ports, based on the PG analysis method in current research.

Our study only discusses the impact of PG of PA on port development, but in fact, ports are a complex system, and thus involve many stakeholders such as terminal operators, customers, and supply chain participants. It is also mentioned in the literature [39] that, under the current most common mode of operation, the landlord type, if the PA does not perform its due duties, these private companies will also take on more responsibilities of the port because of their interests, which may also cause innovations in PG and port development, so future studies can select one of the stakeholders for analysis.

APPENDIX A

Table A1. Selected indicators and evaluation variables in the developing stage.

Goal: Taiwan Smart Port Development Evaluation Model		
Dimension	**Criteria**	**Evaluation Variables**
Operational economics	Logistics process digitalization	• Automated clearance lane ratio • Use of RFID container seal rate • Use of single customs window (SCW) or not
	Degree of mechanical systems' automation	• Container terminal automation ratio
	Docking line efficiency	• Berth occupancy rate • Annual TEUs/meter of container quay • Annual cargo/meter of quay
	Use of storage capacity	• Annual TEUs/total storage or yard area • Annual cargo/total storage or yard area
	Number of reception capacity of large ship	Two conditions must be met at the same time: • Pier draft $\geq$14.5 m • The pier has $\geq$250 m of parking space
Environmental protection	Reduction in greenhouse gas emissions	• (This year, 2008)/2008 emissions
	Number of automations of environmental quality monitor	• Number of water quality, air pollution, and noise sensors
	Amount of recycled wasted	• Recycle amount/total waste yield
	Certification of environment management system	• ISO14001 certification or not
Goal: Taiwan Smart Port Development Evaluation Model		
Dimension	**Criteria**	**Evaluation Variables**
Energy Sustainability	Annual electricity and fuel conservation rate	For office and work area usage: • Total electricity and oil consumption should not exceed the previous year
	Number of energy conservation	• LED streetlight number/port lighting equipment • Inductive streetlight/port lighting equipment
	Usage of shore power system	• Proportion of shore power used by ships in port • Number of onshore power stations (OPSs)
	Certification of energy management	• ISO50001 certified or not
Maritime safety	Working environment safety	• Average training hours of workers • Frequency of workplace accidents • Number of AEDs
	Navigation aid systems	• Number of meteorological stations • Number of undersea robots • Whether or not VTMIS is introduced to the vessel maneuvering system
	Real-time monitor security system	• The number of CCTV surveillance devices in AI
	Hazardous cargo management system	• Time consuming query for storage location • Number of regular checks and audits
Community service	Work environment equality and friendly in the work force	• Annual employee turnover rate • Number of female employees/totals • Number of employees over 50 years old/totals • Number of employees with physical and mental disabilities $\geq$ 3% of the total number of employees
	Data and information transparency and accessibility	• Official website for open-source data • Provide historical port statistics
	Connectivity with neighbor community	• Number of public and educational lectures • Waterfront park area
	Social media engagement	• Social media management • Regularly publish and update related news

REFERENCES

1. UNCTAD. *Ports Newsletter N°19*; UNCTAD: Geneva, Switzerland, 1999; Volume 19, pp. 9–10.
2. Molavi, A.; Lim, G.J.; Race, B. A Framework for Building a Smart Port and Smart Port Index. *Int. J. Sustain. Transp.* **2019**, *14*, 686–700.
3. Port of Rotterdam. Port of Rotterdam Authority Launches New Pronto Application. Available online: **https://www.portofrotterdam.com/en/news-and-press-releases/port-of-rotterdam-authority-launches-new-pronto-application** (accessed on 8 August 2019).
4. Port of Singapore. Port Development of Singapore. Available online: **http://www.mpforum.org/uploads/xcpdf/2018-06-19/5b29085fa4083.pdf** (accessed on 13 June 2018).
5. Port of Hamburg. Paves the Way with Europe's First smartROAD. Available online: **https://newsroom.cisco.com/c/r/newsroom/en/us/a/y2015/m06/port-of-hamburg-paves-the-way-with-europe-s-first-smartroad.html** (accessed on 3 June 2015).
6. González, A.R.; González-Cancelas, N.; Serrano, B.M.; Orive, A.C. Preparation of a Smart Port Indicator and Calculation of a Ranking for the Spanish Port System. *Logistics* **2020**, *4*, 9.
7. Lin, J.Y. *Application of Internet of Things Technology to Smart Port and Terminal Operations*; Ministry of Transportation and Communications Institute: Taipei, Taiwan, 2019.
8. Taiwan International Ports Corporation, Ltd. *Taiwan International Business Port Future Development and Construction Plan Master Plan Report*; Taiwan International Ports Corporation, Ltd.: Kaohsiung City, Taiwan, 2020; pp. 1–67.
9. Del Estado, P. *Informe de Gestio'n del Sistema Portuario de Titularidad Estatal*; Ministerio de Fomento: Madrid, Spain, 2008.
10. Jun, W.K.; Lee, M.K.; Choi, J.Y. Impact of the smart port industry on the Korean national economy using input-output analysis. *Trans. Res. Part. A Policy Prac.* **2018**, *108*, 80–493.
11. Chen, J.; Huang, T.; Xie, X.; Lee, P.T.-W.; Hua, C. Constructing Governance Framework of a Green and Smart Port. *Mar. Sci. Eng.* **2019**, *7*, 83.
12. Chang, C.S. Smart Sustainable Harbor Development. *J. Chin. Inst. Eng.* **2021**, *94*, 1. Available online: **http://www.cie.org.tw/cms/JournalFiles/11003_all.pdf** (accessed on 21 June 2022).
13. Brooks, M.R.; Cullinane, K. (Eds.) *Devolution, Port Governance and Port Performance*; Elsevier: Amsterdam, The Netherlands, 2006; Volume 17.
14. Lee, P.T.W.; Flynn, M. Charting a New Paradigm of Container Hub Port Development Policy: The Asian Doctrine. *Transp. Rev.* **2011**, *31*, 791–806.
15. Caldeirinha, V.R.; Felício, J.A.; Cunha, S.F.; Luz, L.M. The nexus between port governance and performance. *Marit. Policy Manag.* **2018**, *45*, 877–892.
16. Taiwan International Ports Corporation, Ltd. Port of Kaohsiung Environmental Report. Available online: **Chrome-extension://efaidnbmnnnibpcajpcglclefindmkaj/https://www.twport.com.tw/Upload/A/RelFile/CustomPage/2920/f367460f-55e8-416c-a4ef-9c3597715001.pdf** (accessed on 20 January 2020).
17. Mendo, J.B.; Orive, A.C.; Nicoletta, G.C.; Serrano, B.M. Update of the strategic framework for the Spanish port system using a SWOT analysis. *Cuadernos de Administración* **2020**, *36*, 96–111.
18. Laxe, F.G.; Sánchez, R.J.; Alonso, L.G. The adaptation process in port governance: The case of the Latin countries in South America and Europe. *J. Shipp. Trade* **2016**, *1*, 177.
19. Grupo XXI de Communication Empresarial. Transporte XXI. Available online: **https://www.transportexxi.com/puertos-2020/** (accessed on 1 April 2020).

20. Convention and Statute on the International Regime of Maritime Ports. Available online: **https://www.ssatp.org/sites/ssatp/files/publications/HTML/legal_review/Annexes/Annexes%20II/Annex%20II-13.pdf** (accessed on 9 December 2020).
21. Bécet, J.-M.; Rézenthel, R. Dictionnaire Juridique des Ports Maritimes et de l'Environnement Littoral. Available online: **https://www.pur-editions.fr/product/1135/dictionnaire-juridique-des-ports-maritimes-et-de-l-environnement-littoral** (accessed on 9 September 2004).
22. White Paper Roadmap to a Single European Transport Area—Towards a Competitive and Resource Efficient Transport System. Available online: **https://eur-lex.europa.eu/legal-content/EN/ALL/?uri=CELEX%3A52011DC0144** (accessed on 28 March 2011).
23. UNCTAD. *Port Marketing and the Third Generation Port*; UNCTAD: Geneva, Switzerland, 1992; Volume 2, p. 23.
24. TrainMoS II. *Smart Cities and Intelligent Ports*; UNCTAD: Geneva, Switzerland, 2013; Volume 3, p. 4:12.
25. Lee, P.T.W.; Lam, J.S.L.; Lin, C.W.; Hu, K.C.; Cheong, I. Developing the fifth-generation port concept model: An empirical test. *Int. J. Logist. Manag.* **2018**, *29*, 1098–1120.
26. Girard, L.F. Toward a smart sustainable development of port cities/areas: The role of the "Historic Urban Landscape" approach. *Sustainability* **2013**, *5*, 4329–4348.
27. Kia, M.; Shayan, E.; Ghotb, F. The importance of information technology in port terminal operations. *Int. J. Phys. Distrib. Logist. Manag.* **2000**, *30*, 331–344.
28. Spangler, W.S.; Kreulen, J.T.; Chen, Y.; Proctor, L.; Alba, A.; Lelescu, A.; Behal, A. A smarter process for sensing the information space. *IBM J. Res. Dev.* **2010**, *54*, 1–13.
29. Taewoo, N.; Theresa, A.P. Conceptualizing Smart City with Dimensions of Technology, People, and Institutions. In Proceedings of the 12th Annual International Digital Government Research Conference: Digital Government Innovation in Challenging Times, College Park, MD, USA, 12–15 June 2011; pp. 282–291.
30. Hall, R.E.; Bowerman, B.; Braveman, J.; Taylor, J.; Todosow, H.; Wimmersperg, U.V. The Vision of a Smart City. In Proceedings of the 2nd International Life Extension Technology Workshop, Paris, France, 28 September 2000.
31. Yann, A. What Is a Smart Port? *Paris Innov. Rev.* **2017**. Available online: **http://www.parisinnovationreview.com/articles-en/what-is-a-smart-port** (accessed on 21 June 2022).
32. Bao, S.K. Concept of Intelligent Port and Its Systematic Structure. *Navig. China* **2013**, *36*, 120–123. Available online: **http://www.cqvip.com/qk/97883x/201302/46487027.html** (accessed on 21 June 2022).
33. PAV. Valencia Port Presents Its Smart Ports 2030 Strategy in London. Available online: **https://www.valenciaport.com/en/valenciaport-presents-its-smart-ports-2030-strategy-in-london/** (accessed on 21 November 2021).
34. Taiwan International Ports Corporation, Ltd. *Study Report on Automated Container Terminal Operations in Xiamen, Shanghai and Qingdao Ports*; Taiwan International Ports Corporation, Ltd.: Kaohsiung City, Taiwan, 2017. Available online: **https://report.nat.gov.tw/ReportFront/ReportDetail/detail?sysId=C10601932** (accessed on 21 June 2022).
35. Baird, A.J. Port Privatisation: Objectives, Extent, Process and the U.K. Experience. *Int. J. Marit. Econ.* **2000**, *2*, 177–194.
36. Kent, P.E.; Hochstein, A. Port reform and privatization in conditions of limited competition: The experience in Colombia, Costa Rica and Nicaragua. *Marit. Policy Manag.* **1998**, *25*, 313–333.

37. World Bank. *Port Reform Toolkit*; World Bank: Washington, DC, USA, 2001.
38. Talley, W.K. *Port Economic*; Routledge: London, UK, 2009; p. 232.
39. Notteboom, T.E.; Haralambides, H.E. Port management and governance in a post-COVID-19 era: Quo vadis? *Marit. Econ. Logist.* **2020**, *22*, 329–352.
40. Verhoeven, P. Chapter 6—Port Privatisation in the United Kingdom and Continental Europe: An Evaluation of Past Experience and New Drivers. In *Port Infrastructure Finance*; CRC Press: Boca Raton, FL, USA, 2014; Volume 20, p. 111.
41. Brooks, M.R.; Pallis, A. Assessing port governance models: Process and performance components. *Marit. Policy Manag.* **2008**, *35*, 411–432.
42. Verhoeven, P. European Port Governance. Brussel. 2011. Available online: **Chrome-extension://efaidnbmnnnibpcajpcglclefindmkaj/https://www.espo.be/media/espopublications/espofactfindingreport2010.pdf** (accessed on 21 June 2022).
43. Sanchez, R.J.; Pinto, F. El gran Desafío para los Puertos: La Hora de Pensar una Nueva Gobernanza Portuaria ha Llegado. In *FAL Bulletin: Facilitación del Transporte y el Comercio en América Latina y el Caribe*; CEPAL, Naciones Unidas: Santiago, Chile, 2015; Volume 37, pp. 1–9. Available online: **https://www.cepal.org/es/publicaciones/37847-gran-desafio-puertos-la-hora-pensar-nueva-gobernanza-portuaria-ha-llegado** (accessed on 21 June 2022).
44. Brooks, M.R.; Cullinane, K. Governance Models Defined. In *Research in Transportation Economics*; Brooks, M.R., Cullinane, K., Eds.; Elsevier: Amsterdam, The Netherlands, 2007; Volume 17, pp. 405–435.
45. Goss, R.O. Economic policies and seaports: Are port authorities necessary? *Marit. Policy Manag.* **1990**, *17*, 257–271.
46. Ng, A.K.Y.; Pallis, A.A. Port Governance Reforms in Diversified Institutional Frameworks: Generic Solutions, Implementation Asymmetries. *Environ. Plan. Econ. Space* **2010**, *42*, 2147–2167.
47. Bennathan, E.; Walters, A.A. Port Pricing and Investment Policy for Developing Countries. *Econ. J.* **1980**, *90*, 944–946.
48. Saaty, R.W. The analytic hierarchy process—what it is and how it is used. *Math. Model.* **1987**, *9*, 161–176.
49. Miller, G.A. The magical number seven, plus or minus two: Some limits on our capacity for processing information. *Psychol. Rev.* **1956**, *63*, 81–97.
50. Emerson, P. The original Borda count and partial voting. *Soc. Choice Welf.* **2011**, *40*, 353–358.
51. Deng, Z.Y.; Zeng, H.K. The inner characteristics and application of the hierarchical analysis method (AHP). *Chin. J. Statis.* **1989**, *27*, 5–22.
52. Buckley, J.J. Fuzzy hierarchical analysis. *Fuzzy Sets Syst.* **1985**, *17*, 233–247.
53. Saaty, T.L. *The Analytic Process*; RWS Publication: Pittsburgh, PA, USA, 1996.
54. Zhang, X.; Lam, J.S.L. A fuzzy Delphi-AHP-TOPSIS framework to identify barriers in big data analytics adoption: Case of maritime organizations. *Marit. Policy Manag.* **2019**, *46*, 781–801.
55. Gavalas, D.; Syriopoulos, T.; Tsatsaronis, M. Assessing key performance indicators in the shipbuilding industry: An MCDM approach. *Marit. Policy Manag.* **2021**, *49*, 463–491.
56. Yu, F.J.; Wei, Y.R. Combining fuzzy analytical hierarchy process and similarity ideal solution to the sequential preference method for the sequential ordering of the remaining Chinese car options. *J. Sci. Eng. Tech.* **2017**, *13*, 1–11. Available online: **chrome-extension://efaidnbmnnnibpcajpcglclefindmkaj/http://libwri.nhu.edu.tw:8081/Ejournal/AL03130101.pdf** (accessed on 21 June 2022).

57. Zadeh, L.A. The Concept of a Linguistic Variable and Its Application to Approximate Reasoning. *Inf. Sci.* **1975**, *8*, 199–249.
58. Liu, Y.; Eckert, C.M.; Earl, C. A Review of fuzzy AHP methods for decision-making with subjective judgments. *Expert. Syst. Appl.* **2020**, *161*, 113738.
59. Gumus, A.T. Evaluation of Hazardous Waste Transportation Firms by Using a Two Step Fuzzy-AHP and TOPSIS Methodology. *Expert. Syst. Appl.* **2009**, *36*, 4067–4074.
60. Brooks, M.R. The governance structure of ports. *Rev. Netw. Econ.* **2004**, *3*, 168–183.
61. Caldeirinha, V.; Felício, J.A.; da Cunha, S.F. Government Policies and Portuguese Port Governance in the Period from 2005 to 2015. *Res. Transp. Bus. Manag.* **2017**, *22*, 11–20.
62. Rodrigue, J.-P. The Governance of Intermediacy: The Insertion of Panama in the Global Liner Shipping Network. *Res. Transp. Bus. Manag.* **2017**, *22*, 21–26.
63. Pallis, A.A.; Vaggelas, G.K. A Greek Prototype of Port Governance. *Res. Transp. Bus. Manag.* **2017**, *22*, 49–57.
64. Baltazar, R.; Brooks, M.R. Chapter 17—Port Governance, Devolution and the Matching Framework: A Configuration Theory Approach. *Res. Transp. Econ.* **2006**, *17*, 379–403.
65. Bandara, Y.M.; Nguyen, H.-O.; Chen, S.-L. Determinants of Port Infrastructure Pricing. *Asian J. Shipp. Logist.* **2013**, *29*, 187–206.
66. Tam, H.P. Introduction of extreme value judgment method. *NTTU Edu. Res. J.* **2008**, *19*, 131–150.
67. Tseng, P.-H.; Pilcher, N. Port governance in Taiwan: How hypocrisy helps meet aspirations of change. *Res. Transp. Bus. Manag.* **2017**, *22*, 38–48.
68. Ferrari, C.; Parola, F.; Tei, A. Governance models and port concessions in Europe: Commonalities, critical issues and policy perspectives. *Transp. Policy.* **2015**, *41*, 60–67.
69. Molina, B.; Segura, A.; Serrano, J.P.; Alonso, F.J.; Molina, L.; Pérez-Borrego, Y.A.; Ugarte, M.I.; Oliviero, A. Cognitive performance of people with traumatic spinal cord injury: A cross-sectional study comparing people with subacute and chronic injuries. *Spinal Cord.* **2018**, *56*, 796–805.
70. Notteboom, T. Chapter 19—Concession agreements as port governance tools. *Res. Transp. Econ.* **2006**, *17*, 437–455.
71. Ministerio de Economía y Hacienda. Real Decreto Legislativo 3/2011, de 14 de Noviembre, Por el Que se Aprueba el Texto Refundido de la Ley de Contratos del Sector Público. Available online: **https://www.boe.es/buscar/act.php?id=BOE-A-2011-17887** (accessed on 14 November 2011).
72. Ángel, E.; Ernesto, M.; Jesus, M.M. Juridical regime of pilotage's service in Spain. *J. Marit. Res.* **2013**, *10*, 85–92.
73. Yu, T.F. Research Report on the Competitiveness of Taiwan's International Business Port. Supervisory. Available online: **https://cybsbox.cy.gov.tw/CYBSBoxSSL/edoc/download/49592** (accessed on 15 April 2013).

Chapter 9
Towards a Reference Architecture for Cargo Ports

Virginia M. Romero and Eduardo B. Fernandez

Department of Electrical Engineering and Computer Science, Florida Atlantic University, Boca Raton, FL 33431, USA

ABSTRACT

Cyber-Physical Systems (CPS) are physical systems whose operations are coordinated, monitored, and controlled by computing and communication functions. These systems are typically heterogeneous, including Internet of Things and information technology subsystems, and can present a myriad of implementation details, making them very complex systems. An important type of CPS is a maritime container terminal (cargo port), which is a facility where cargo containers are transported between ships and land vehicles for onward transportation and vice versa. A cargo port performs four basic functions: receiving, storing, staging, and loading for both import and export containers. We present here process patterns that describe the functional aspects of cargo ports and a pattern that describes their structural properties (patterns are encapsulated solutions to recurrent problems). These patterns describe semantic aspects found in any cargo port and can be adapted to describe other CPSs. We decompose these process patterns into use cases that describe their interactions with the system. We then integrate the process patterns with structural patterns to assemble a partial reference architecture (RA) that shows the interactions of all the patterns while also indicating the typical stakeholders found in all ports. We validate the proposed reference architecture, highlighting its theoretical and practical value. Software and system designers of cargo ports need to start from a conceptual and abstract view that is subsequently refined to add more details. The use of reference architectures and patterns is an effective way to organize and describe the functional and non-functional aspects of a system, as well as to unify the design of all its aspects. This is, until now, the only published RA for cargo ports, and it can be a useful guideline for the designers of any type of cargo port.

Keywords: *maritime transportation; cargo port; cyber-physical systems; IoT; software patterns; process patterns; design science*

1. INTRODUCTION

Cyber-physical systems (CPSs) are systems that integrate computational resources and communication capabilities with the control of activities in the physical world. Another definition is [1]: "CPSs are integrations of computation with physical processes. Embedded computers and networks monitor and control the physical processes, usually with feedback loops where physical processes affect computations and vice versa". The components of a CPS can be centralized or distributed and typically include embedded devices, sensors, actuators, and wireless links (see Figure 1). CPSs are often safety-critical because their failure could endanger lives or cause significant economic losses. CPS may include as components Internet of Things (IoT) and Information Technology (IT) subsystems and can be studied as a system of systems, although we do not do that here. CPSs can be structured into at least five architectural levels [1]: a process layer involving field instrumentation, a basic control layer, a supervisory control layer, a process management control layer involving application servers, and a corporate network layer. CPSs require a high level of adaptability because of continuously evolving conditions and need situation awareness and modifiability [2]. They may include legacy systems, and they increasingly include humans in the loop. Typically, they use combinations of commercial-off-the-shelf (COTS) components and real-time operating systems, as well as products from different vendors using different protocols. Many CPS also need to follow government or industrial regulations [3]. The design of CPS systems requires consideration of several disciplines, such as embedded systems, computers, communications, and others, and the software is embedded in devices whose principal function is not only computation. However, we are concerned here only with software aspects that are pervasive in modern systems.

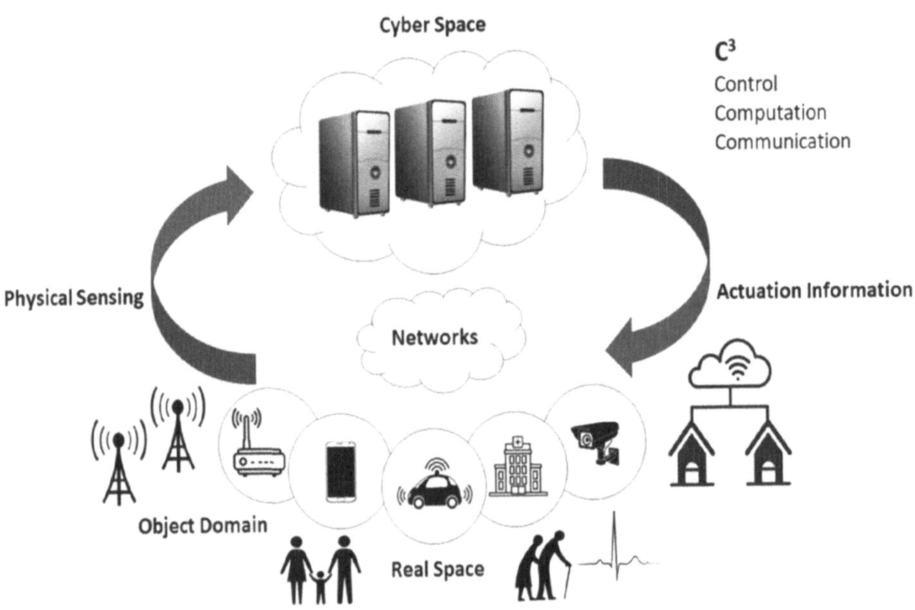

Figure 1. *A cyber-physical system.*

CPSs can be very complex systems and may comprise a multitude of systems, components, and implementations; they can be classified as massively interconnected, complicated systems [4]. To be able to understand their interactions and perform their design, we need to consider their global architectures and apply abstraction, focusing on their fundamental properties. A reference architecture (RA) is a generic software architecture, based on one or more domains, that does not include implementation details [5,6]. An RA is reusable, extensible, and configurable; it also provides guidelines for interoperability and portability. RAs can be seen as patterns for complete architectures and can be instantiated into concrete software architectures by adding platform aspects [5]. A software pattern is an encapsulated solution to a recurrent software problem and defines a vocabulary that concisely expresses requirements and solutions, as well as provides a communication vocabulary for the involved stakeholders [7,8]. A reference architecture built using patterns is a useful way for designers to describe the system and analyze security and other non-functional aspects. Specifically, for CPSs, these abstraction artifacts have become not only very useful to understand and building these complex systems but also have the potential to unify the design of the computational, communication, and control aspects of CPSs, especially in the presence of the many implementation details of their component units.

There are many types of CPSs, such as those in aerospace, automotive, chemical processes, healthcare, manufacturing, and transportation. Our objective is not to define an RA for any type of CPS, but we will consider a specific type of system, a transportation system, specifically a maritime container cargo port. As we discuss in Section 7, an analysis of the relevant literature found that there are not many reference architectures for cargo ports ([9] and Section 7). We show here a cargo port RA built of patterns described using UML models. We consider this effort the first attempt to define a precise and semiformal architecture for these systems. We believe that an abstract, semiformal approach is the most practical design approach given the complexity of the systems we are considering. That is not to say that parts of the architectures cannot be formally modelled; UML models can be complemented with formal descriptions such as OCL [10], and with them, we can make this architecture more formal if needed [11]. Furthermore, patterns and RAs can be converted into ontologies, which are even more precise descriptions and allow for proving some formal properties [12]. Note that for CPSs, it is also possible to define reference architectures that include several levels [4,13,14]. The lower levels of those architectures have many common aspects, but we are concerned only with the semantic aspects of port operations, so we will focus only on the higher levels.

We first assemble a set of process patterns (a CPS process describes a physical process, e.g., the purification of water). A software process is more general; it describes a sequence of steps to accomplish some objective, e.g., teaching a course. Here, process patterns (which describe software processes) describe the functions performed by cargo ports. These patterns include *ship arrivals and departures*; *port drayage* (the transport of containerized cargo by specialized trucking/railway companies between a maritime port terminal and an inland distribution point or a rail terminal [15]); *container terminal physical structure* (the physical units used in the port); and *port loading/unloading* (the activities of loading and unloading containers) [16]. Each process pattern includes one or more use cases. A use case represents a complete interaction between the users and the system; each use case is associated with a set of actors that correspond to some of the stakeholders of the system. The complete patterns have been published in conferences; we show here only their important sections; some have been modified. These patterns result in a partial RA for a cargo port. We used this architecture as the basis for a security reference architecture, where defenses against identified threats have been added [17], so this article is a complement to that article.

We make it clear that we are not presenting a model for a specific port or type of port. Patterns and RAs are generic models intended for system and software designers. Our models are abstractions of fundamental aspects found by studying the structure and operations of several cargo ports around the world. Because of this, regulations, standards, and trade models (e.g., single-window strategy) are not relevant. A complete model is also beyond the possibilities of two people, so this is a partial model, and we do not claim this model to be complete.

Our contributions include:

- A set of process/structure patterns that describe the functional aspects of a reference architecture for cargo ports. These patterns describe semantic aspects found in any cargo port and can be adapted or extended to describe other CPSs.
- A decomposition of the cargo port process patterns into use cases that describe its actor interactions with the system. The use cases are dynamic models that complement the structural models of the port.
- A partial reference architecture that shows how to integrate these process patterns with structural patterns, indicating the typical stakeholders found in all ports. We know of no other reference architecture for cargo ports (see **Section 7**).
- A validation of the proposed RA that highlights its theoretical and practical value.

Although we describe only a partial RA, we show the method to build it, which can be used by others to complete it. From an academic point of view, this partial RA is sufficient, but in an industrial environment, it should be completed. Our main audience includes software architects working with CPS and port experts building cargo port systems. The rest of the paper is structured as follows: Section 2 presents some background information on patterns and reference architectures as well as a description of the basic aspects of container cargo ports and their operations. In Section 3, we introduce the process patterns and list their main actors and use cases (UCs). Section 4 describes in detail each process pattern, while Section 5 shows a partial RA for a cargo port using these patterns. Section 6 validates the RA, while Section 7 inspects related work and puts our work in perspective. We end with conclusions and future work in Section 8.

2. BACKGROUND

2.1. Patterns and Reference Architectures

We present a summary of the concepts of software patterns and reference architectures, which are fundamental to understanding our work. A pattern is a solution to a recurrent problem in a given context; it embodies the knowledge and experience of software experts and can be used for new applications and to guide inexperienced designers. Patterns are described using structured templates that contain sections that define specific information about the problem being solved and its solution, along with recommendations about its use. A frequent way to describe a pattern solution comes in the form of a UML class diagram, complemented with some sequence diagrams that describe dynamic aspects of the use cases and possibly activity or state diagrams. A problem section defines a recurrent problem and the set of forces that constrain its solution. Once the solution is presented, a section on the consequences indicates how well the forces were satisfied by the solution as well as the possible negative aspects of it, such as the extra overhead. An implementation section provides hints on how to use the pattern in an application, indicating what steps are needed and possible realizations. A section on related patterns indicates other patterns that complement the pattern or that provide alternative solutions [18]. This level of detail and precision allows system designers and developers to use them as guidelines to build systems, and for users to understand the effect of the mechanisms they represent. Patterns are also useful for communication between designers and to evaluate and reengineer existing systems [7]. Patterns are applied to designs by instantiating them, which requires tailoring them to the rest of the design, e.g., removing some classes, changing names, or deleting attributes; that is, a pattern behaves like a type in a programming language. The use of patterns for system design is grounded in the principles of design science [19], and several pattern conferences each year introduce new patterns as well as applications of this design style.

There are several types of patterns, such as design and architecture patterns, that can be used to build flexible and extensible systems [7,8]. *Security patterns* are used to build secure systems by describing ways to control their threats, fix their vulnerabilities, and provide security attributes [20]. *Misuse patterns* describe how attacks are performed from the point of view of an attacker [21,22]. They define the environment where the attack is performed, what security mechanisms are needed as countermeasures to stop it, and how to find forensic information to trace the attack once it happens. *Process patterns* describe the workflows of the activities required to perform some business objective, such as order parts, schedule activities, or similar.

A reference architecture (RA) is a standardized, generic software architecture based on a particular domain that does not contain implementation details [5,6,23]. It can be seen as a type of pattern for whole architectures that can be instantiated into concrete architectures by adding platform aspects [24]. There is no general agreement about what an RA should contain [23]. Avgeriou [5] presents an example and describes what should be included in one. Some groups define other types of reference architectures, including hardware and other aspects, e.g., the Smart Grid Architecture Model (SGAM) [4,25] (see the section on related work), but we are not concerned with those types here; we think of this RA as a metamodel for concrete architectures defined as a PIM (Platform Independent Model) in an MDE (Model Driven Engineering) methodology.

Security patterns can be added to the RA to handle their identified threats, resulting in a security reference architecture (SRA). An SRA is an abstract architecture describing a conceptual model of security for systems and providing a way to specify security requirements for a range of concrete architectures. The SRA is an extended RA where security solutions have been added in appropriate places to provide some degree of security [17,26]. As indicated earlier in another paper [17], we have defined an SRA for cargo ports that complements this RA.

2.2. Cargo Ports

We present an overview of cargo ports as a general background on this type of port. As indicated, our models are abstractions of real ports, and we have not intended to describe every detail of ports.

2.2.1. Overview

For millennia, mankind has shipped goods across the oceans, from one land to another. The loading and unloading of a ship have always been very labor-intensive. A ship could easily spend more time in port than at sea while dock workers handled cargo in and out of tight spaces below decks. The introduction of containerization has greatly simplified this process, and nowadays goods can be moved seamlessly between ships, trucks, and trains [27]. The U.S. Bureau of Transportation reports that more than seventy-seven percent of freight tonnage entering the U.S. came by water, compared to 22 percent by land and only 0.3 percent by air [28]. By any measure, marine transportation is the primary means of moving goods and raw materials to and from the U.S. While the U.S. represents only 4.5 percent of the world's population [29], it accounts for 9 percent of the worldwide container traffic, with one container out of every eleven engaged in global trade either bound for or originating in the U.S. [30]. The largest amount of cargo traffic in the world is done by China [31]. The largest container ships can currently carry over 20,000 TEUs (a TEU is the volume of a 20 ft. container).

Port automation has been playing an increasingly important role with the introduction of robots, artificial intelligence, and other digital tools that keep the goods flowing into and out of major ports. This technology is widely seen as the most efficient way for seaports to cope with rising global shipping traffic and massive new ships that haul more and more containers. By digitizing and automating activities once handled by human crane operators and cargo haulers, seaports can reduce the amount of time ships sit in port and thus boost port productivity by up to 30%, according to some estimates [32]. IoT advances have had a strong impact on the automation of port activities, allowing the remote control of cranes and vehicles.

2.2.2. Container Terminal Overview

Container terminals provide many services, i.e., container loading/unloading to and from vessels and feeder ships for import or export purposes, internal container movement from ships to stacking areas and vice versa, stacking containers in dedicated areas distributed in the terminal area, container inspection for customs requirements, refrigerated containers handling and storage, etc. All the above processes require several shared and reusable resources and equipment to fulfill the tasks involved in handling and transporting containers, such as dock cranes, yard cranes, transport vehicles, yard stacking deposits, automatic stacking cranes or automatic storage/retrieval systems, railway tracks, and human operators. All processes and operations are usually planned, scheduled, monitored, and controlled by a central supervisor and make use of information technologies and IoT to allow fast shipping operations, optimize the usage of facilities, and reduce lag times [33].

A typical modern container terminal installation is represented in Figure 2 using UML package notation. It includes [34]:

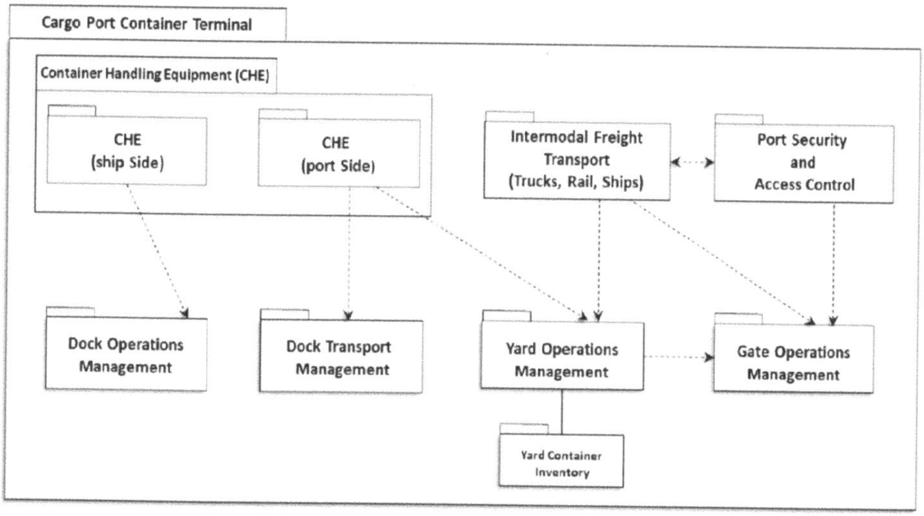

Figure 2. *Cargo port system package diagram.*

An area for port physical security and access control; this incorporates the physical entrance to the terminal, video surveillance, gates, and the infrastructure necessary to check the credentials of the maritime transportation workers requiring unescorted access to the secure areas of the port. Credentials are typically biometric identification cards, also known as TWICs (Transport Worker Identification Cards);

The cargo handling equipment at shipside and landside, which includes the container terminal cranes, transport vehicles, and similar conveyances;

An area for intermodal transportation of the containers, such as commercial long-haul trucks, railways, or even other ships;

A terminal operating center, incorporates the financial, communications, customs, IT security, and other back office functions.

2.2.3. Container Terminal Operations

Figure 3 presents a modern container terminal [34]. Container terminals work day and night, and their workload depends on the number of containers in transit. Containers arrive at the terminal by trains, ships, or trucks and are stored in the terminal yard. Then, they leave the terminal by the same means to reach their final destination [35]. A ship-to-shore crane (STS) is a crane that services a container ship by shifting on a rail to reach the assigned stowage within the same ship and also to move from one ship to the next once the first one has been completed. STS cranes provide the single most important operation (called a move) associated with the loading and unloading of a ship and are the only means of moving containers to or from a ship [36]. To unload a ship, one or several STS cranes pick up containers from the ship and put them on trailers or shuttle trucks that move them to the assigned yard positions within the terminal storage area. Automated guided vehicles (AGV) are also used for the transport of containers from the dock to the yard. To load a ship, the STS crane unloads a container from the trailer and puts it in the ship. In automated or semi-automated container terminals, this operation is performed by crane operators controlling ship-to-shore cranes (STS) from a control room. Optical character recognition (OCR) technology is used to automatically read and record the container's ISO code identification number as it is handled by an STS crane. By automating the handoff of containers from cranes to vehicles, the process becomes safer for people as there is no need for checkers or tally workers under the cranes who are responsible for recording container numbers. More advanced features include damage images, door direction, hazardous labels, and reading and reporting the drayage truck numbers onto which the container is transferred [37].

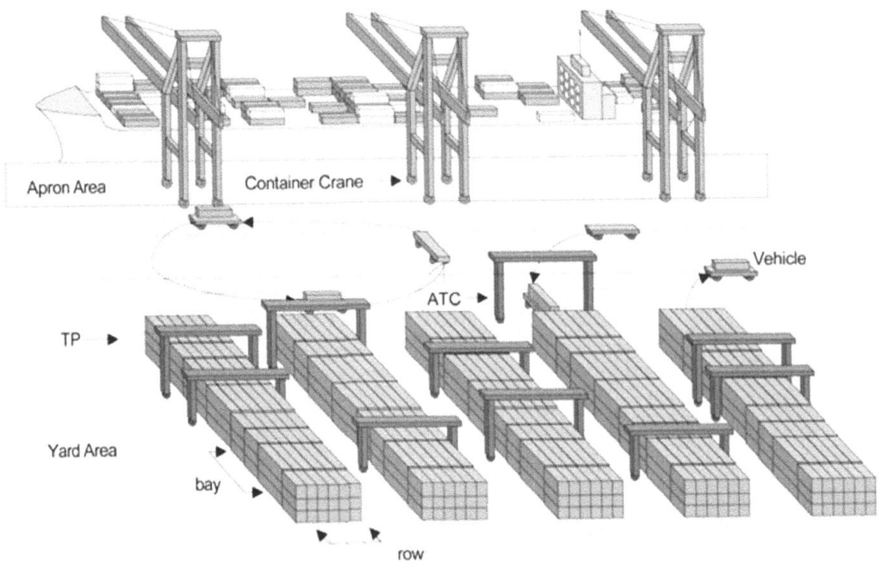

Figure 3. *Modern container terminal.*

Containers stacked in a ship are secured in several ways to prevent them from being damaged at sea. Locking corner castings are placed between stacked containers in non-cellularized ships to align the containers and provide a place to brace them. The cross braces are then secured to the floor of the ship, and, finally, the hatch covers are put back in place. Cellularized ships do not require corner castings or cross braces since permanent guides and locks (which allow containers to be stored more densely than in non-cellularized cargo vessels) are already on board [35,36].

Operations in the storage yard are more flexible than STS crane operations. In a storage yard, gantry cranes, top-pick loaders, or straddle carriers are used to stack containers. Automated stacking cranes (ASC) are also used. Other transport vehicles include automated lifting vehicles and unmanned shuttle carriers able to locate the containers and pick them up from the ground [37]. Automated vehicles may share a common software control module at the equipment level, often referred to as the equipment control system (ECS), for handling, for example, safety features and intra-vehicle coordination. An ECS is defined as the software that monitors and controls all events and processes at the equipment level, either for a single piece of container handling equipment (CHE) such as an AGV or a group of CHE. ECS coordinates interactions between different types of automated equipment [33,37,38].

The container storage area is usually separated into different stacks (or blocks) that are differentiated into rows, bays, and tiers. Some stack areas are reserved for special containers like reefers (refrigerated containers), which need electrical connections, dangerous goods, or overheight/overwidth containers, which do not allow for normal stacking. Often, stacks are separated into areas for import, export, and empty containers. Even with these many subdivisions, the efficiency of storage yard equipment is greatly increased by being able to store only one portion of the yard at a time. To prevent multiple restows or misplaced containers, the efficient assignment of the location or address of the container is of primary importance. Without efficient ways to assign container addresses, multiple restows are likely [39]. Containers are assigned specific addresses before entering the storage yard. In automated or semi-automated container terminals, intelligent stacking cranes operate using the optimum path, overlapping horizontal and vertical motion. Multiple ASC cranes in the yard are coordinated to avoid waiting times, so stacking cranes are able to deliver variable volumes of containers in a short time. Automated equipment is operated without unnecessary accelerations and deaccelerations, and huge energy savings can be made. With intelligent automation, all the operations of the ACT can be comfortably handled from a central control room, which makes terminals safer and the working environment more ergonomic. Remote operation and on-board cameras provide better views than what is possible from a crane cabin in situations like landing containers on a ship or vehicle or handling hatch covers close to the ground [37,38].

Another important element of a container terminal is the movement of containers between the STS cranes and the storage yard. STS cranes unload a ship and place the containers in trailers or shuttle trucks. These shuttle trucks move the containers to storage locations in the yard. This operation is a closed loop, as their only function is to shuttle the containers from the ship to the storage yard. These trucks are local, and they usually do not leave the terminal. A collection of shuttle trucks is called a gang, and they may be automated guided vehicles (AGV) in larger terminals. The road network for these unmanned robotic transport vehicles is defined by electric wires or transponders in the ground, which enable the accurate positioning of these vehicles [40]. The number of shuttle trucks also needs to be considered. Too many trucks in the system cause long lines at the crane and long waiting times for service; conversely, fewer trucks in the system will result in idle stacking equipment. Containers, which are stored in the storage yard, leave the terminal on input/output trucks to reach their final destinations. Because of the high cost of keeping a ship in port, it is important to keep the dock crane operating without delay to turn the ship around as quickly as possible [38].

A great variety of container terminals exist, mainly depending on what type of handling equipment is used in the system. The decision on which equipment to use depends on several factors. Space restrictions, economic reasons, and even historical reasons play an important role. A basic factor is the size of the space that can be used for a terminal. If space is restricted, gantry cranes to store the containers are preferred. A decision for AGVs and automated gantry cranes can be made in cases of high labor costs and new terminal construction. Historical and cultural reasons must be considered if container terminals are enhanced or modernized. Because space is becoming a scarce resource, a tendency toward higher storage is to be expected [41].

3. BUILDING A REFERENCE ARCHITECTURE (RA) FOR A CARGO PORT

In this section, we present several process patterns of cargo ports used to build the architecture and their relationships; we define their use cases and identify their primary actors. A process pattern may contain several use cases.

3.1. Process Patterns

Figure 4 presents a pattern diagram relating some of the patterns we would find in a reference architecture for cargo ports (a pattern diagram shows relationships between patterns). This diagram includes three process patterns: *Ship arrivals or departures* describe the port traffic; *port drayage* delivers (from local sources) or receives (from ships) the containers to be stored in the yard, while *port loading/unloading* performs the ship container handling activities. The *container terminal's physical structure* describes the physical zones and buildings found in a container terminal, and it is an architecture pattern, not a process pattern.

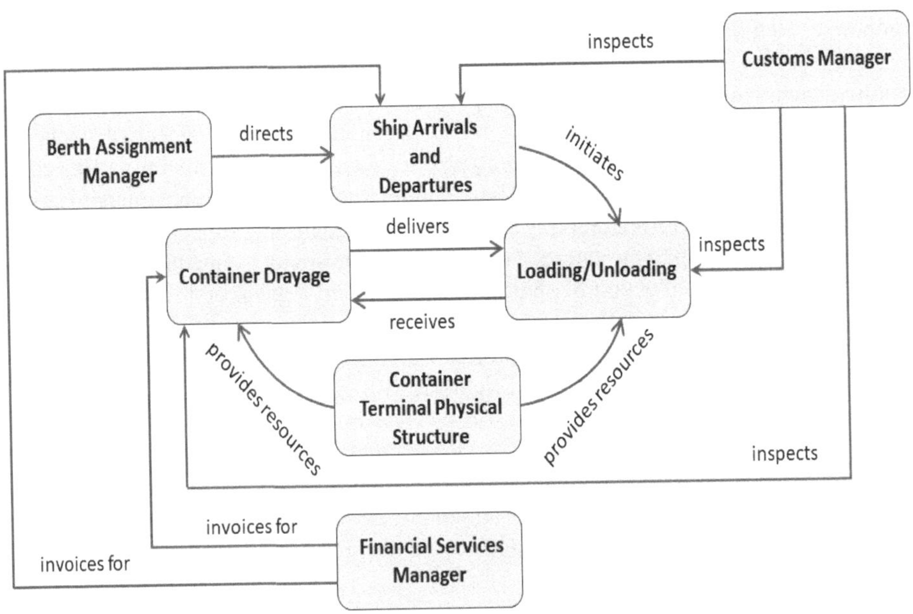

Figure 4. *Relationship of cargo port patterns.*

Customs Manager, Financial Services Manager, and Berth Assignment Manager patterns are not shown for conciseness but are presented in this figure to give a broader perspective. The Customs Manager pattern may include use cases such as "inspect ship" and "inspect container"; the Financial Services Manager pattern may include use cases such as "bill for service to ship" and "bill for transportation services"; the Berth Assignment Manager pattern may include use cases such as "receive cargo manifesto" and "assign berth".

3.2. Use Case Model

A use-case model shows how users interact with the system to solve a problem [42]. A use case represents a complete interaction between a user and the system and describes how the system should respond under various conditions to requests from the user [43]. Each main function in a cargo port is represented as a use case, and each use case is associated with an actor or a set of actors. An actor represents a user or automated system that may interact with the system [42,44]. Generally, an actor is a role rather than a specific person; an actor can be distinguished through his tasks, and a single actor could be associated with one or more use cases and vice versa [44,45]. In these models, actors correspond to some of the stakeholders (see Section 3.3). Figure 5 presents a general overview of the system functions using a UML use case diagram, where the use cases are related to the process patterns (in color). In automated container terminals, use cases *UC5* (*Assign Trailer Driver*) and *UC10* (*Assign Crane Operators*) (shaded in purple) in Figure 5 may not apply, as these activities are substituted by automated guided vehicles (AGVs) and unmanned container terminal cranes. Use cases not related to the patterns in this figure are part of other cargo port patterns not covered here.

3.3. Stakeholders

Following is a description of the stakeholders (we show only the stakeholders relevant to the use cases described here and according to the level of abstraction of this paper). A more complete port description would have more of them in a cargo port, some of which are also actors in the use cases we will show later:

1. ***Port Authority:*** responsible for the overall administration of the property, terminals, and other facilities on the port complex.
2. ***Port Operations Manager:*** responsible for the efficient use of port facilities and resources with specific responsibilities for health, safety, and security. The individual works closely with regulatory authorities, port operations, and personnel of shipping lines. Manages resource allocation within the terminals and monitors expenditures.
3. ***Terminal Operator (leasing company):*** legal entities that lease from port authorities their terminal facilities. Either the port authority or the terminal operator will supply the cranes and other cargo-handling equipment. It depends on the lease agreement between the port authority and each terminal operator.
4. ***Terminal Operations Manager:*** responsible for the planning and administration of the operations at a port terminal in order to optimize resource use, minimize costs, and maintain quality standards. Activities associated with the transportation of cargo include the operation of vessels, cargo handling equipment, locomotives, trucks, vehicles, and storage and warehousing facilities related to the transportation of cargo. Assigned by the terminal leasing company.

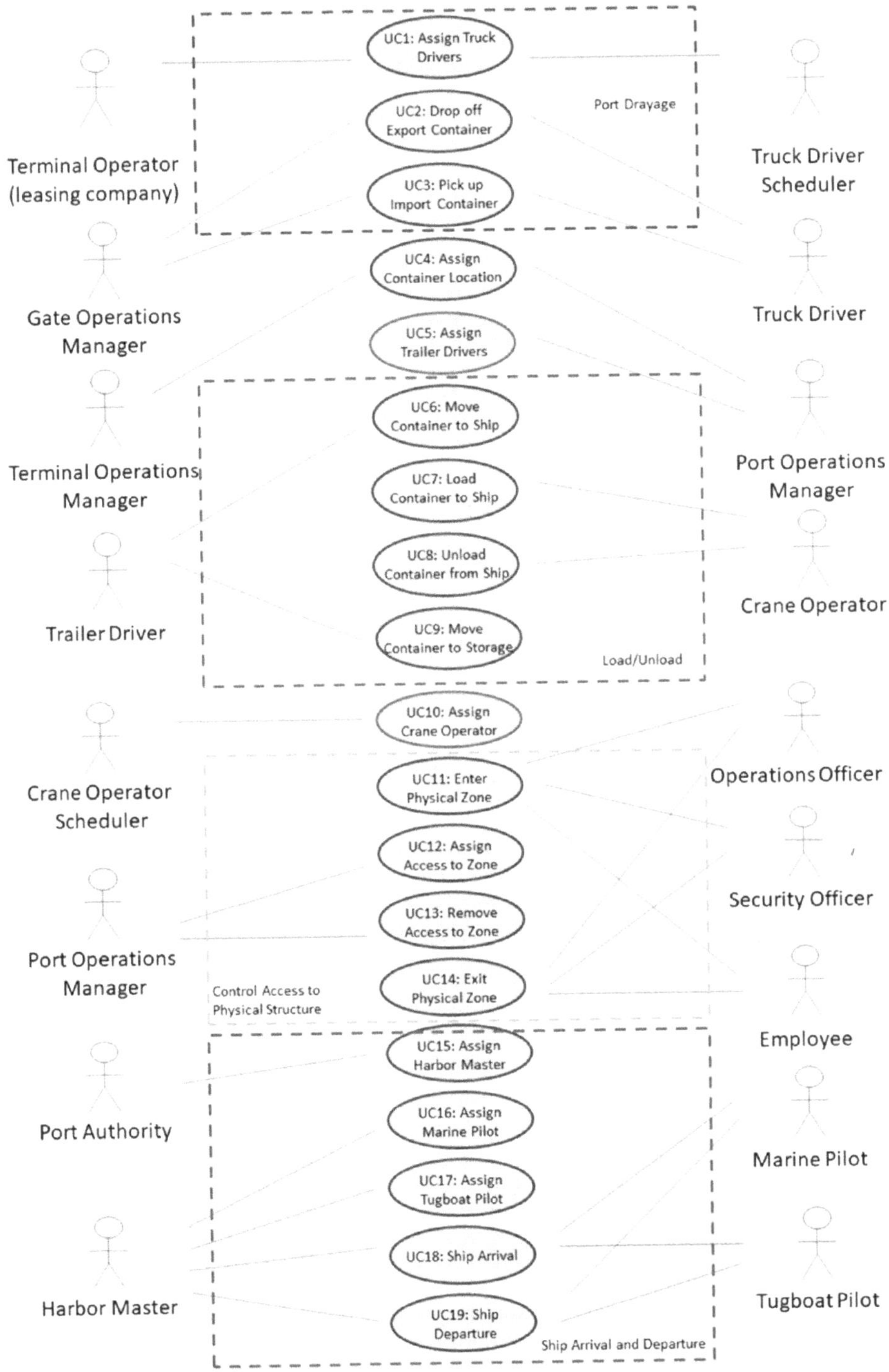

Figure 5. *Use cases for a cargo port.*

5. ***Gate Operations Manager:*** manage all gate, inter-, and intra-terminal transfer activities. Liaison with customers, port authorities, road container haulers, and truck drivers regarding terminal gate container transactions and transfer activities. Assigned by the terminal leasing company.
6. ***Drayage Company:*** legal entities in charge of intermodal transportation, i.e., taking containers in and out of warehouses, rail terminals, ocean ports, and harbors. Assigned by shipping lines or terminal operators.
7. ***Drayage Truck Driver Scheduler:*** an individual responsible for scheduling truck drivers for the drayage transaction (either drop or pick up a container). Assigned by the drayage company.
8. ***Drayage Truck Driver:*** an individual responsible for dropping off or picking up a container.
9. ***Terminal Trailer Driver:*** an individual responsible for moving the container to/from the ship to the storage area inside the terminal.
10. ***Crane Operator Scheduler:*** an individual responsible for assigning crane operators to load/unload containers to/from the ship. Depending on the contract that the terminal operator has with the port authority, the crane operator scheduler can be assigned by the terminal operations manager or the port operations manager.
11. ***Crane Operator:*** an individual that controls the operation of the STS crane when loading and unloading containers to/from a ship from either the crane's cabin or a remote control room.
12. ***Operations Officer:*** a member of a marine operations team involved in the safe transit and handling of vessels into and out of the port. The individual may also be involved in the berthing of vessels, port control and marine services, operation of harbors and marinas, conservancy, and environmental protection.
13. ***Security Officer:*** an individual responsible for the protection of people, properties, and information at the port.
14. ***Gate Control Employee:*** a member of the port authority organization.
15. ***Harbor Master:*** an individual or system establishing a berthing schedule consisting of berthing times and berthing positions of containerships in port container terminals. The berth schedule must be constructed in a way to satisfy requests from carriers on berthing times and minimize handling efforts during ship operation.
16. ***Marine Pilot:*** an individual employed by a port or harbor to ensure the safe navigation of ships in their waters. Pilots board ships entering or exiting the port and navigate them safely, avoiding hazards. Pilots may be faced with high-risk cargo, poor maneuverability, and communication difficulties.
17. ***Tugboat Pilot:*** the pilot of a small vessel that helps larger crafts steer in tight spaces where their engines cannot safely reach full power.

3.4. Use Cases (UCs)

A description of the use cases in the system follows. The relationship of these use cases to the process patterns in the previous section is shown in Figure 5, where patterns are delineated by dashed colored lines.

UC1—Assign Drayage Truck Drivers: Actions taken by the drayage company truck driver scheduler to assign truck drivers to drop off or pick up containers from the cargo port. The drayage company may be assigned by the terminal operator (leasing company) or the shipping line. Actors: terminal operator (leasing company), drayage company truck driver, scheduler.

UC2—Drop off Export Container: Actions taken by the drayage truck driver and the gate operations manager to allow entrance to the port and drop off an export container. Actors: gate operations manager, drayage truck driver.

UC3—Pickup Import Container: Actions taken by the drayage truck driver and the gate operations manager to allow entrance to the port and pick up an import container. Actors: gate operations manager, drayage truck driver.

UC4—Assign Container Location: Actions taken by the terminal operations manager or port operations manager controlling the logistics of the container terminal to assign container locations. Actors: terminal operations manager; port operations manager.

UC5—Assign Trailer Drivers: Actions taken by the port operations manager to assign trailer/shuttle drivers to take the containers from the STS cranes to the drop-off point in the storage yard. Actors: port operations manager.

UC6—Move Container to Ship: Actions taken by the trailer/shuttle driver to take the container from the storage yard to the ship. Actors: trailer driver.

UC7—Load Container to Ship: Actions taken by crane operators to load outbound containers onto ships utilizing STS cranes. Actors: crane operator.

UC8—Unload Container from Ship: Actions taken by crane operators to offload inbound containers from ships utilizing STS cranes. Actors: crane operator.

UC9—Move Container to Storage: Actions taken by the trailer/shuttle driver to take the container from the ship to the storage yard. Actors: trailer driver.

UC10—Assign Crane Operator: Actions taken by the crane operator scheduler to assign a crane operator to a specific STS crane. One crane operator may be assigned more than one STS crane. Actors: crane operator scheduler.

UC11—Enter Physical Zone: Actions taken by an operations officer, a security officer, or an employee to have access to a specific area in the cargo port. Actors: operations officer, security officer, employee.

UC12—Assign Access to Zone: Actions taken by the port operations manager to allow access by an individual to a specific zone. Actors: port operations manager.

UC13—Remove Access to Zone: Actions taken by the port operations manager to remove access for an individual to a specific zone. Actors: port operations manager.

UC14—Exit Physical Zone: Actions taken by an operations officer, a security officer, or an employee to exit a specific area in the cargo port. Actors: operations officer, security officer, employee.

UC15—Assign Harbor Master: Actions taken by port operations to assign a harbor master to schedule vessel arrivals, departures, and berthing assignments. Actors: port authority.

UC16—Assign Marine Pilot: Actions taken by the harbor master to assign a marine pilot to direct the ship to the port area. Actors: harbor master.

UC17—Assign Tugboat Pilot: Actions taken by the harbor master to assign a tugboat pilot in case the ship requires the assistance of a tugboat for performing maneuvers in the port area and completing entrance and mooring operations. Actors: harbor master.

UC18—Ship Arrival: Actions taken by the harbor master, marine pilot, and tugboat pilot to bring the ship to port. Actors: harbor master, marine pilot, tugboat pilot.

UC19—Ship Departure: Actions taken by the harbor master, marine pilot, and tugboat pilot when the ship is departing the port. Actors: harbor master, marine pilot, tugboat pilot.

4. PATTERNS IN A CARGO PORT

Now we show the patterns that we used as building blocks for our cargo port reference architecture. Patterns may contain one or more use cases. We present partial and updated descriptions of the patterns using a POSA template [7]; for full descriptions, see the corresponding references.

4.1. Cargo Port Drayage

The cargo port drayage pattern includes several use cases: *UC1: assign truck drivers*; *UC2: drop off export containers;* and *UC3: pickup import containers*. In this subsection, we discuss *UC2: drop off export container* in detail. This pattern is an update of [15].

4.1.1. Intent

Provide all the needed functions for the delivery and pickup of containers at a maritime cargo port.

4.1.2. Context

Port drayage refers to the movement of containers between a port terminal and an inland distribution point or rail terminal. A typical drayage assignment involves either delivering an export container to a marine terminal or picking up an import container that has arrived by ship [46]. According to the Intermodal Association of North America (IANA), there are more than 60 million drayage movements each year in North America.

4.1.3. Problem

Drayage companies take containers in and out of warehouses, rail terminals, ocean ports, and harbors. Drivers arriving at the entrance gate are anticipating one of the following transactions: picking up a container or dropping off a container. There may be several variations in this process, such as dual transactions, trouble window transactions, and equipment issues. It requires extra time to inspect and accept a container, to receive instructions on where to take the container within the terminal, and to leave the load and either start another transaction or leave. Depending on the type of transaction, gate processing time and associated queueing vary.

- The solution to this problem is guided by the following forces:
- Flexibility: several internal and external roles may be involved, i.e., truck operators, storage area workers and supervisors, crane operators, etc. Resources and devices

used, as well as their corresponding operations, must be flexible to accommodate this variety of roles [**47**];

- Usability: the software used to identify the individuals entering the port, their trucks, and their container contents should be easy to use by roles that do not have technical backgrounds;
- Alerting: any attempt to deviate from the normal operations of the terminal must produce an alert and should be logged;
- Logging: any activity should be recorded and logged for future auditing. In general, every visit should be logged to keep track of any access to the facility. All containers must be registered and logged;
- Location tracking—we need to be able to find every container. A container in the wrong place can delay operations.

4.1.4. Solution

Every maritime port container terminal should include port security and access control functions. Drivers arriving at a maritime terminal entrance gate intend to either drop off a loaded export or an empty export container and/or pick up a loaded or empty import container. The variety of external users and the fact that the contents of the containers are usually not visible present many threats. Continuous checks are required, not only of the individuals entering the terminal but also of the contents of their trucks. We must authenticate drivers and their loads before they enter the terminal. We must log every container move. All activities need to be recorded for future auditing in case of a security violation. All truck and container moves need to be recorded using videos to improve security and operations at the port. The containers need to be tagged, tracked, and their locations recorded for identification and to prevent the illegal entry of weapons, chemicals, or biological agents, as well as human trafficking. In general, every visit should be logged. Other transactions may be possible (such as moving trucks without containers or chassis), but these movements of empty containers or bare chassis are the result of loaded movements.

4.1.5. Structure

Figure 6 presents the class diagram of a cargo port drayage operation. *Quay* represents the platform for loading and unloading ships. *GateAccess* represents the entrance to the terminal. Each quay has only one GateAccess since each port terminal has only one gate for entrance. *DrayageTransaction* represents a single drayage transaction, where a transaction implies the delivery or pickup of a *container* to/from the quay. DrayageTransaction consists of entities *truck*, *truck driver*, and *container*. For each drayage transaction, there is one truck driver in charge of one truck and one to two containers, depending on the size of the containers and whether they drop off and pick up a different one. GateAccess must verify that the drayage operation is a legitimate one. *ImportArea* and *ExportArea* represent the surfaces on the terminal assigned to stack the containers. Each container has a unique location in the storage yard, which is represented by *ImportLocation* or *ExportLocation*. These unique locations are kept and updated in the log.

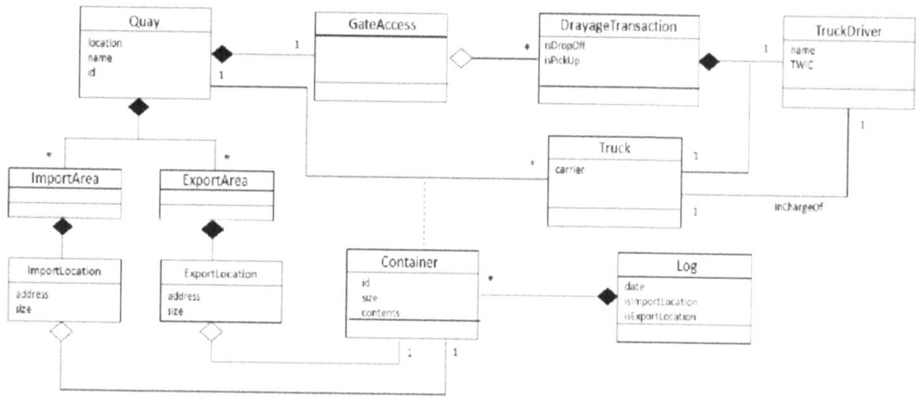

Figure 6. *Class Diagram of Cargo Port Drayage.*

4.1.6. Dynamics

Figure 7 describes the dynamic aspects of the Secure Cargo Port Drayage pattern using an activity diagram for the following use case.

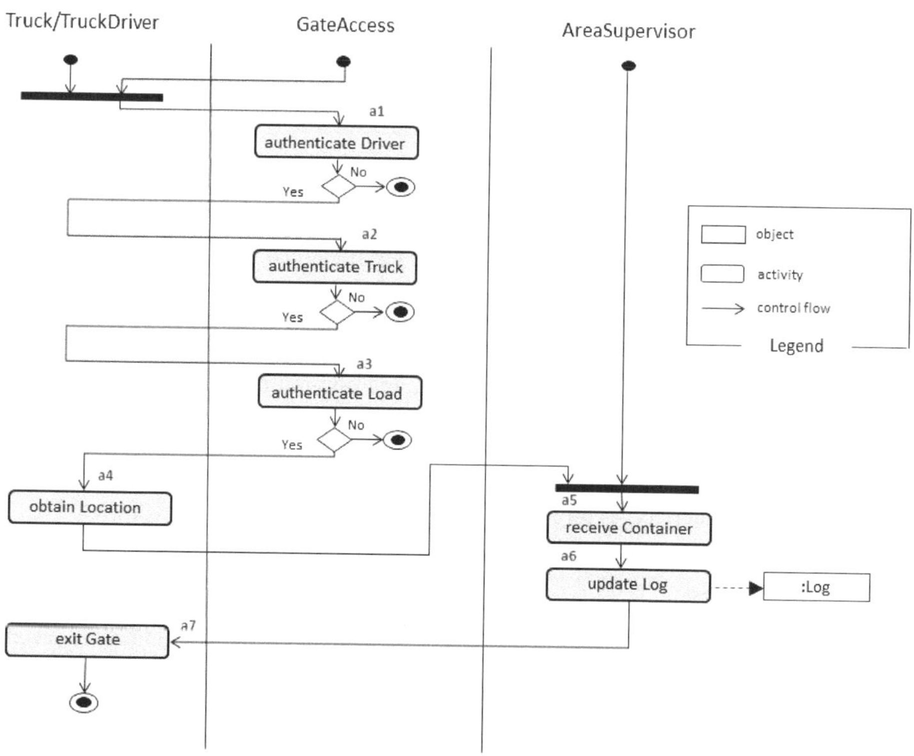

Figure 7. *Activity Diagram for use case drop-off export container.*

Use Case: Drop-off Export Container:

Summary: a truck requests entrance to the maritime cargo port. GateAccess verifies the identity of the driver, the truck, and its load. GateAccess also verifies that the driver's transaction is a legitimate one. The driver obtains the location and proceeds to drop the container off in the storage yard.

Actors: TruckDriver, GateAccess.

Precondition: The drayage operation information has been previously entered into the system.

Description:

1. TruckDriver arrives at the terminal entrance.
2. GateAccess authenticates the driver.
3. GateAccess authenticates the truck.
4. GateAccess authenticates the container load.
5. TruckDriver obtains the location for the drop and proceeds to the storage yard.
6. TruckDriver drops the container in the appropriate location in the storage yard, exits the terminal, or picks up another container. The log is written, indicating the container ID and its location.

Postcondition: A container has been deposited in the port, and the transaction is recorded.

Other typical use cases include picking up import containers, assigning locations to the containers, and assigning drivers to the trucks. Typical roles include gate attendant, storage yard supervisor, storage yard worker, truck driver, and gate access worker.

4.1.7. Implementation

This pattern can be implemented at all gate entrance operations for all ports. Roles and access rights need to be standard across port terminal locations. A code for each port drayage operation can be implemented.

To prevent the forgery and image manipulation of documents, we can use image processing applications that detect irregularities in font and design of inserted words, spaces between letters, discrepancies in size, crowding, and non-uniformities in the background [48].

Access to the storage yard should be limited only to the area indicated by the authenticating system [49]. The different areas may have physical separations and gates. Any deviation should be alerted, reported, and logged.

4.1.8. Known Uses

The ports of Los Angeles and Long Beach, California, USA, are using a similar process for their port drayage operations. In fact, we used their port drayage process as the basis for this pattern [50,51].

The ports of Miami and Fort Lauderdale (Port Everglades), Florida, USA, also apply a similar process for their port drayage operations.

4.1.9. Consequences

The advantages of this pattern include:

- Flexibility: role-based access control (RBAC) [**50**] allows us to accommodate several roles that participate in the system; it allows for all types of users. People performing the same tasks are given the same rights;
- Usability: easy to usc by individuals that do not have technical backgrounds;
- Alerting: we can use the alarm monitoring, security logger, and auditor patterns to record all activities that are security relevant. Any deviation from normal activities will be logged, and if necessary, an alert will be displayed;
- Logging: the places where the containers have been placed are registered. All transactions are logged and later audited to assure compliance with security regulations;
- Location Tracking: logging will keep track of locating the containers.
- The pattern has some liabilities:
- All the mechanisms needed for the process require extra personnel and maintenance, and they have an associated extra cost. Containers may have dangerous loads, and physical measures are needed to detect them, such as X-rays and radiation portal monitors that may be costly.

4.2. Cargo Port Loading and Unloading

The cargo port loading and unloading pattern includes several use cases, including *UC6: Move Container to Ship*; UC7: Load Container to Ship; *UC8: Unload Container from Ship;* and *UC9: Move Container to Storage*. In this subsection, we discuss *UC8: Unload Container from Ship* and UC9: Move Container to Storage in detail. This description is an update of [16].

4.2.1. Intent

This pattern provides the typical functions of a container terminal facility: loading and unloading of containers to/from a ship.

4.2.2. Context

Port operations and logistics have a significant impact on the economy and international trade. As international trade increases, ports face pressure to improve their infrastructure in order to maintain their operations and respond to market demands. In addition, given the high level of competition, ports must use their resources efficiently and effectively, which has resulted in increased automation. Port operations refer to the activities and processes necessary to manage and control a port. Container terminals are commonly open systems of material flow. In a container terminal, there are three main areas of operation: seaside operations, yardside operations, and landside operations. A port logistic system must keep track of many container positions. A log to keep track of all the container movements must also be kept.

4.2.3. Problem

Traditional ship unloading in ports and terminals varies. There are fully integrated ship-to-stockyard systems, incorporating STS cranes, AGVs, and automated stacking cranes for fully dedicated berths. Alternatively, the system for multi-cargo berths operates cranes, trailers, trucks, and hoppers in the specific stockyards. The main issue is how to do it in an efficient, safe, and environmentally friendly manner.

The solution to this problem is guided by the following forces:

- Flexibility: a variety of users are involved in ports. They are required to operate the facility and have assigned roles. Some external users can also have access for administrative or management purposes. We need to accommodate this variety of roles;
- Safety: the equipment has physical limits that cannot be exceeded because they would get damaged or endanger the port workers;
- Records: we need to record all activities that may need to be audited later to record any type of violation.

4.2.4. Solution

Ports include loading/unloading facilities such as cranes and storage units. During unloading, operators move containers from ships to storage units, and during loading, operators move containers from storage units to ships.

4.2.5. Structure

Figure 8 shows the class diagram of a port loading system, Port-Load. The system is composed of a set of cranes, several crane operators, and some storage units (warehouses, bins). In a given moment, this system loads or unloads a set of ships, each one carrying a set of containers.

4.2.6. Dynamics

We describe the dynamic aspects of the load/unload container process in a cargo port terminal using an activity diagram for the following use cases: unload container from ship and move container to storage.

Activity: Unload Container from a Ship and Move to Storage

Summary: this activity diagram presents all the steps necessary to unload a container from a ship and move it to an appropriate location in storage. See Figure 9.

Actors: crane operator, trailer driver.

Precondition: the location of the ship and which crane to operate have been previously entered in the system.

Description:

1. Crane operator activates the crane.
2. Crane operator picks up container from the ship.
3. Crane operator deposits the container in the trailer.
4. Trailer driver receives the container.
5. Trailer driver takes the container to the stacking crane transfer point.

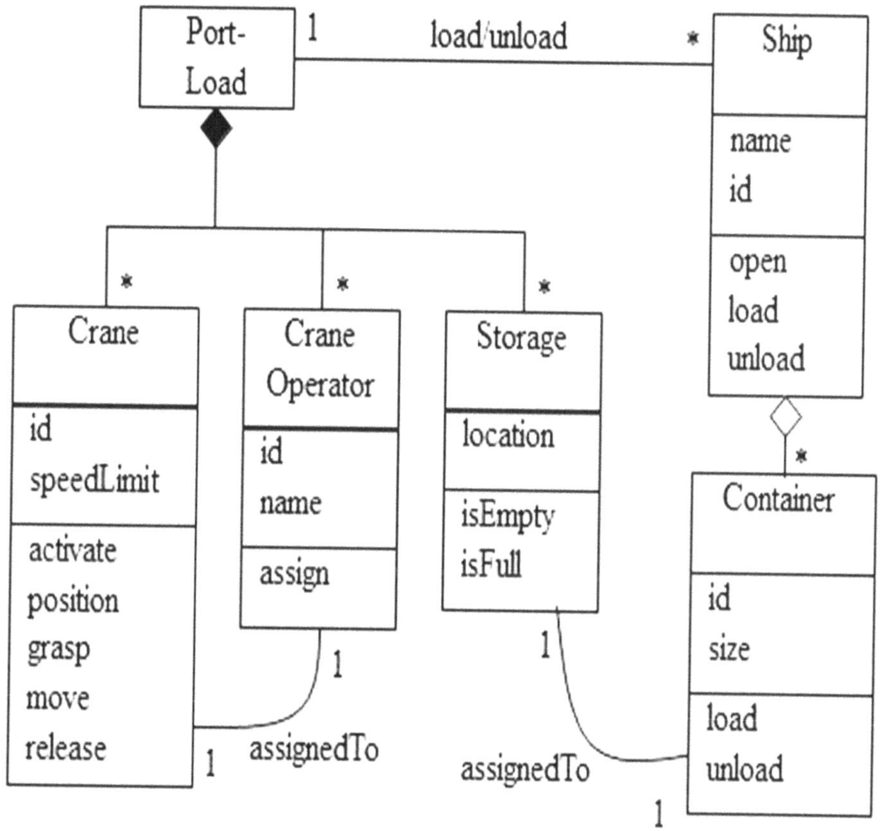

Figure 8. *Class diagram of port loading system.*

6. ASC picks up the container and finds its location in storage.
7. ASC deposits container in an appropriate place in storage.
8. Transaction is logged.
9. Postcondition: a container has been unloaded from a ship and deposited in its corresponding place in storage. The log is updated.

4.2.7. Implementation

We can first consider the physical units. We need to give unique identifiers to all the cranes, operators, ships, and storage units. These components are relatively unchanging.

After this, we need to define rights for operators and other roles. Their specific members and rights may change frequently.

While RBAC is not the only option, it is the most used security model in industry because of its general advantages [52].

Ships are the most variable components; they come and go. Their identifiers may be outside our control, but we need to know their IDs to authenticate them when they arrive.

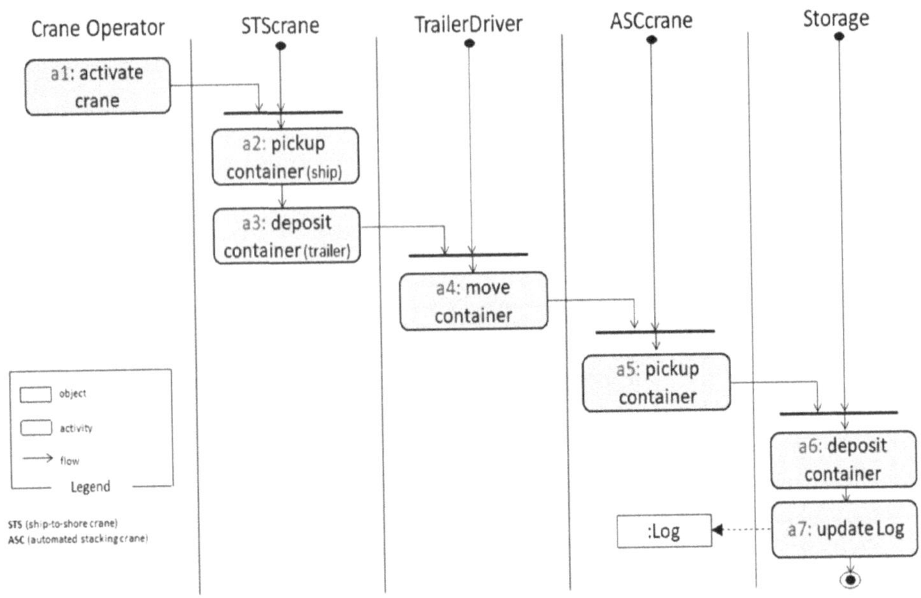

Figure 9. *Activity diagram for UCs unload container and move to Storage.*

4.2.8. Known Uses

Most ports worldwide use this process for the loading and unloading of containers. In automated container terminals, "trailers" have been replaced by automated guided vehicles (AGVs), and most container terminal cranes are unmanned and operated by crane operators from a control room.

4.2.9. Consequences

The advantages of this pattern include:

- Flexibility: RBAC allows for accommodating any kinds of roles appropriate for all kinds of users;
- Safety: The equipment's physical limits can be controlled with safety assertions. Other assertions can prevent situations that would be dangerous for humans;
- Records: We can use the security logger/auditor pattern to record security-relevant activities.
- The pattern also has some liabilities:
- There is the possibility that containers have dangerous loads, and we need physical measures to detect them.

4.3. Container Terminal Physical Structure

4.3.1. Intent

Describe the structure of a container terminal in a cargo port, including its zones and their subdivisions.

4.3.2. Context

Due to the operational characteristics of maritime transportation, port locations are constrained to a limited array of sites defined mostly by geography. Container terminal infrastructure, on the other hand, is determined by the cost and size of the port. Container terminals work day and night, and they need specific areas to load and unload ships. They also need areas for storage of the containers and an area to accommodate transshipment activities.

4.3.3. Problem

To define an infrastructure to keep track of the containers while they are being loaded or unloaded to/from ships and/or transshipped to other places for distribution. We need to identify the different areas and their functions. We also need to have a way to control access.

The solution to this problem is guided by the following forces:

- Flexibility: the description of the physical structure should be flexible and scalable. The physical units may change their functions, within restrictions, and we may need more physical units of some type;
- Scalability: the number of buildings and their divisions can change up or down depending on the amount of work required;
- Upgradability: should allow for changes in the structure to add or remove port functions;
- Physical Security: need to provide a good level of security for cargo port functions.

4.3.4. Solution

Once "maritime access", which refers to the physical capacity of the site to accommodate ship operations, and "maritime interface", which indicates the amount of space that is available to support these operations, are satisfied by a proposed port site, the surrounding land is divided for the construction of the port. The port site must have infrastructure such as docks, stacking or storage areas, warehouses, and equipment such as cranes, some by the water to unload the ships and some by the storage areas and warehouses to move and store the containers. Depending on the size of the port and the services provided, we determine the number of employees and the number of buildings needed for offices.

4.3.5. Structure

Figure 10 shows the static structure of a container terminal and the relationships among its components. Class *Port* represents all the terminals or docks in our cargo port. *Terminal, TransportMean, Area,* and *Crane* classes represent the physical components. Terminal represents the docks of the cargo port, and TransportMean represents all the transportation means within the port: trucks moving containers in and out of the terminal, railroad trains carrying the containers in and out of the terminal, or possibly another ship. Class Area represents several surfaces in the terminal; StorageArea represents the area to park the containers; BufferArea, a temporary storage area; Crane represents all types of cranes: ship-to-shore cranes that service the ship; YardCrane and StackingCrane employed to stack the containers.

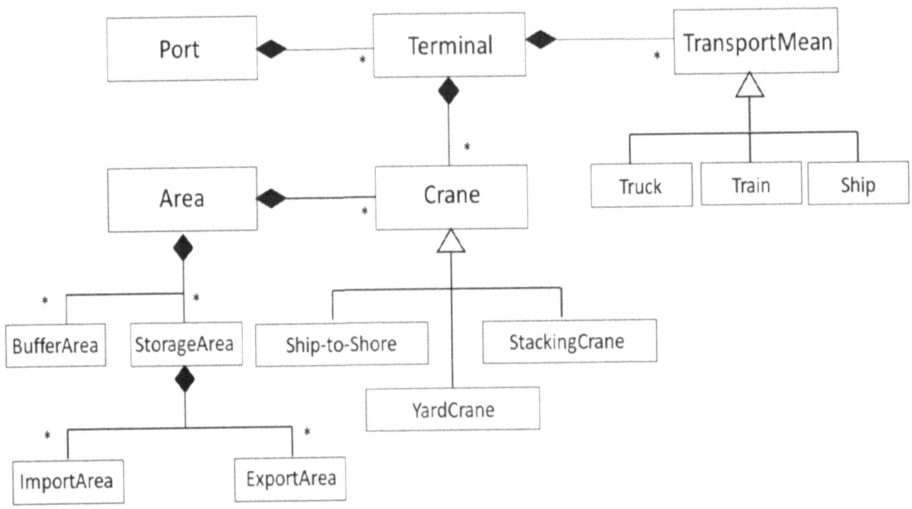

Figure 10. *Class dagram of a container terminal's physical structure.*

4.3.6. Implementation

This pattern can be implemented for all existing and new container terminals. There are a variety of possible implementations to take advantage of technological advances.

Remote crane operations are becoming more popular in yard operations and in ship-to-shore cranes. The cranes are being fitted with advanced automation and remote control.

4.3.7. Known Uses

Most ports worldwide have the infrastructure and container handling equipment presented in this pattern. The ports of Los Angeles [50] and Long Beach in California, USA [51], as well as the ports of Miami and Port Everglades in Fort Lauderdale [53], present the described infrastructure.

Port Everglades is the #1 seaport in Florida by revenue as well as one of the top container ports in the state. The port is also South Florida's main seaport for petroleum products, including gasoline, jet fuel, and alternative fuels. It has 44 available berths with a draft depth of 44 ft. Port Everglades' existing seven gantry cranes in the Southport area, where most of the containerized cargo operations take place, are 151 feet high and can reach containers stacked six high on deck and sixteen containers across [53]. As part of its expansion program (fall 2020), they have currently purchased an additional six Super Post-Panamax container gantry cranes, the largest of their kind worldwide. On November 17, Broward County's Port Everglades received three. In comparison with the existing ones, the three 175-foot-high gantry cranes, valued at $13.8 million each, have the ability to handle containers stacked nine high from a ship's deck and reach 22 containers across. This will enable the port to handle larger ships and gain economies of scale [53].

4.3.8. Consequences

The advantages of this pattern include:

- Flexibility: this pattern is flexible enough that we can add other types of cranes and other types of areas to store the containers, i.e., areas to handle refrigerated containers, areas to store liquids, etc;
- Scalability: this pattern provides for any amount of container handling equipment, vehicles used to move the containers around the terminal, and storage areas, depending on the amount of work required;
- Upgradability: this pattern allows for any changes in functionality, such as adding or removing port functions;
- Physical Security: access of personnel to specific areas can be controlled using authorization models [**17,52**].
- The pattern has some liabilities:
- This pattern does not detect dangerous loads, and physical measures are needed to detect them, such as X-rays and radiation portal monitors that may be costly.

4.4. Ship Arrivals and Departures

Ship arrival and departure patterns include several use cases, including *UC15: Assign Harbor Master*; *UC16: Assign Marine Pilot*; *UC17: Assign Tugboat Pilot*; *UC18: Ship Arrival;* and *UC19: Ship Departure*. In this subsection, we discuss *UC18: Ship Arrival in detail*. The activities pertaining to the ship and the ship crew prior to arrival are not covered in this pattern; only the actions required when arriving at port are.

4.4.1. Intent

Arriving and departing ships must be taken to/from their predefined port locations (berths) in a safe manner.

4.4.2. Context

Cargo ports have several locations to hold ships for loading/unloading (berths). These locations must be assigned in advance but can be changed depending on the conditions of the port or berth availability.

4.4.3. Problem

Arrival at a port and departure from a port are two separate but very important aspects of a ship's voyage. Both of these procedures are critical and involve a number of different complexities in them. Arriving at some ports may require going through narrow or busy seaways where navigation may be difficult.

The solution to this problem is guided by the following forces:

Availability: berth facilities must be assigned in advance and in such a way as to minimize ship stays in the port;

Scalability: the number of provisioned, designated locations to dock in a port should vary depending on the traffic of vessels to the port. New berths can be provisioned for loading/unloading as needed;

Berth Scheduling: berth scheduling should be based on the priority of resource allocation and a comparison of the vessel and berth data. How to assign its berth time and position is of great importance to improving the level of port services and reducing the port's cost of production;

Safety and Port Integrity: arrival and departure procedures should ensure the safety of the ship's arrival and departure, both for the people on board as well as those at the seaport. Most of the collisions and groundings of ships are reported during the maneuvering of the vessels in port, and hence the maneuvering operation at port arrival or departure is considered the most crucial time a ship faces in her voyage. The availability of marine pilots and tugboats should ensure the avoidance of physical damage [54].

4.4.4. Solution

Before entering the port area, the ship is required to contact the port authority regarding its estimated time of arrival (ETA), give particulars of the ship and her cargo, check for berth availability, and request a marine pilot to direct it to the port area. The ship may also require the assistance of a tugboat for maneuvering inside the port area and completing entrance and mooring operations. Similar operations are carried out when the ship leaves the port area [53].

All berthing, unberthing, and maneuvering of vessels within the port limits must be done under the supervision of a marine pilot. It is the duty of the Harbor Master to inform the pilot of any special conditions, difficulties, or weather conditions.

4.4.5. Structure

In Figure 11, the class *Port* contains a set of berths—the multiple areas for docking in a port. For each object of class *Berth,* there are multiple stays of a ship in a berth. Each instance of *ShipStay* can be an *Arrival* or a *Departure,* representing either a ship's arrival or departure. Multiple objects of the class *Tugboat* can be associated with multiple objects of the class Ship. The associations between ships and tugboats are temporary and described by instances of the mediator pattern [8]. Each object of Berth is assigned a schedule, *Berth-Schedule.*

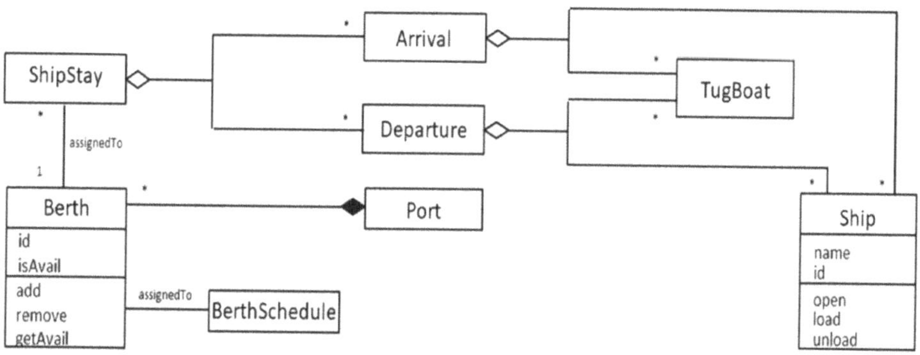

Figure 11. *Class diagram of ship arrival and departure operations.*

4.4.6. Dynamics

UC "ship arrival" is executed as a series of steps described below. Figure 12 shows the sequence diagram for the arrival of the ship at the port.

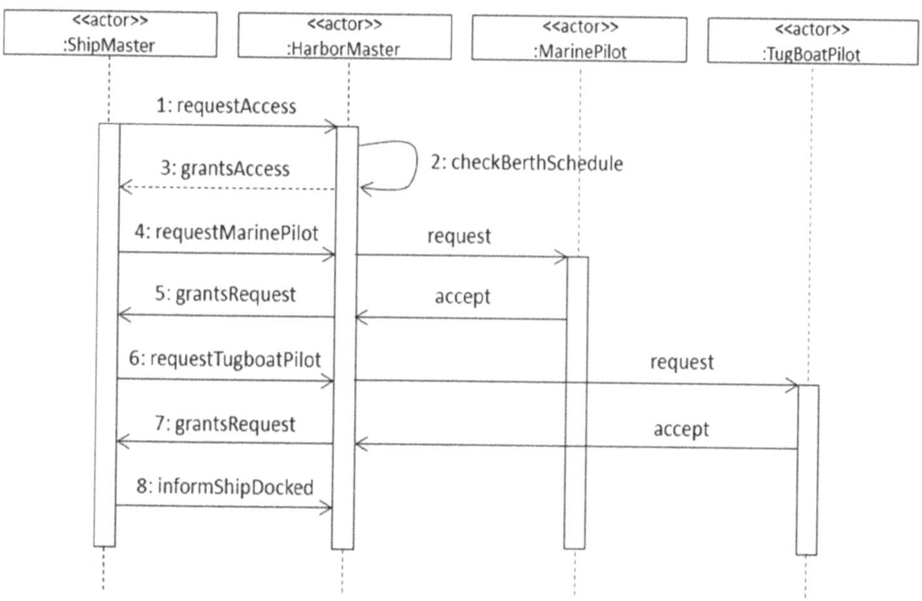

Figure 12. *Sequence diagram for UC ship arrival.*

UC: Ship Arrival

Summary: a ship arrives at a port.

Actors: ship master, harbor master, marine pilot, and tugboat pilot.

Preconditions: ships' times of arrival are known. Assignments of berths to arriving ships, of pilots to ships, and/or of tugboats must be predefined.

Description:

1. Ship master contacts harbor master and requests access to the port.
2. Harbor master checks the berthing schedule.
3. Harbor master grants access for ships to berth.
4. Ship master requests a marine pilot to bring the ship to port.
5. Harbor master grants request to send marine pilot.
6. Ship master requests a tugboat for maneuvering vessel in port.
7. Harbor master grants request to send tugboat pilot and tugboat to assist ship.
8. Ship master informs ship docked in berth.
9. Postcondition: A ship successfully docks at the port.

4.4.7. Implementation

Starting in 2001, ships and ports started to install automatic identification system (AIS) equipment. AIS transmitters on board the ships automatically report the arrival and departure times to the port authorities. This technology is primarily used to avoid collisions and increase port security. At the present time, AIS devices have not been installed in all ships and ports, but most major ports and the largest ships are included. AIS provides a means for ships to electronically send data, including vessel identification, position, speed, and course. AIS uses global positioning systems (GPS) in conjunction with shipboard sensors and digital VHF radio communication equipment to automatically exchange navigation information electronically and to monitor vessel location and movement primarily for traffic management, collision avoidance, and other safety applications [55,56,57].

4.4.8. Known Uses

The ports of Los Angeles and Long Beach in the USA use this process for ship arrivals and departures. Most ports worldwide, such as the port of Shanghai and the port of Singapore, also use this process for ship arrivals and departures.

Additionally, Port Everglades and the Port of Miami use AIS for vessel tracking [53]. AIS tracking websites (e.g., Marine Traffic) operate over 2000 AIS stations in over 165 countries across the globe.

4.4.9. Consequences

The advantages of this pattern include:

- Availability: port facilities work 24 h a day. By previously scheduling with the harbor master, vessels can arrive at a port at any time. Ports are always open and available to vessels when needed;

- Scalability: ports and harbor masters provision berths for vessels to dock depending on traffic and the number of vessels entering the port;
- Berth scheduling: proper berth assignment and scheduling ensure that the vessel characteristics are compatible with the characteristics of the berth to which the vessel is scheduled. It improves the level of port services and reduces the port's cost of production;
- Safety and port Integrity: arrival and departure procedures ensure that the essential safety steps are followed both for the people on board as well as those at the seaport. Most of the collisions and consequent grounding of ships are avoided by utilizing experienced marine pilots and tugboat pilots to aid in the mooring of a vessel.
- The pattern has some liabilities:
- Having experienced marine pilots and tugboats reserved and waiting to help the ships dock increases the cost of running the port.

5. A REFERENCE ARCHITECTURE FOR A CARGO PORT

Now that we have seen the different conceptual units that together make up a cargo port system, we can show its corresponding reference architecture. See Figure 13. This RA combines the patterns shown earlier and includes the process patterns, stakeholders, and use cases described above. In this RA, we included the patterns for: loading and unloading of containers to/from a ship at a cargo port loading facility [16]; a pattern for the delivery of cargo port containers (drayage) [15]; a pattern for ship arrivals and departures; and a pattern for a container terminal physical structure [58]. Patterns may exhibit overlapping classes. As indicated earlier, this is not a complete RA for cargo ports.

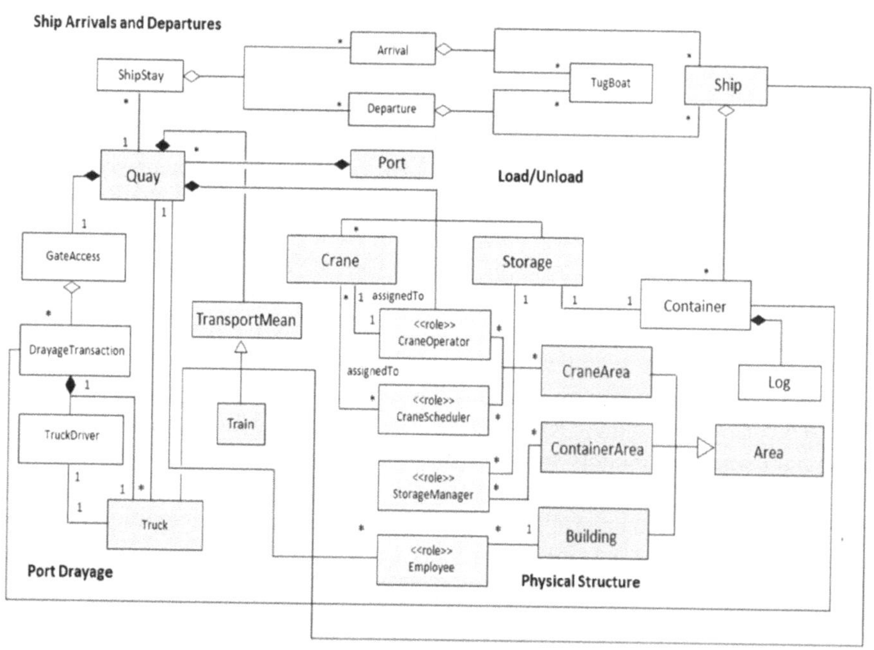

Figure 13. *Partial reference architecture for a cargo port.*

A brief explanation of the architecture follows: a typical cargo port, *Port*, contains many container terminals, *Quay*. Each container terminal should include port security and access control functions. Drivers arriving at the entrance, *GateAccess*, intend to either pick up or drop off a container, destined for import or export, respectively. The movement of containers between a port terminal and an inland distribution point or rail terminal is referred to as port drayage. Each single port drayage transaction is an instance of *DrayageTransaction*. For each drayage transaction, there is one *TruckDriver* in charge of a *Truck* and a *Container*. Ports include loading and unloading facilities such as *Crane*s and *Storage* units. During unloading, *CraneOperator*s move containers from *Ship*s to storage units; during loading, crane operators move containers to ships. *CraneScheduler*s are in charge of assigning crane operators to cranes. *StorageManager* is in charge of assigning storage locations and warehousing facilities. *Employee*s are members of the Port Authority organization. Arrival at a port and departure from a port are two separate but very important aspects of a ship's voyage. Both of these procedures are critical and involve a number of different complexities. *ShipStay* represents each stay of a *Ship* in a berth, either an *Arrival* or a *Departure*. The ship may also require the assistance of a *Tugboat* for maneuvering inside the port area and completing entrance and mooring operations. Similar operations are carried out when the ship leaves the port area. Multiple tugboats can be associated with multiple ships. The associations between ships and tugboats are temporary. *TransportMean* represents all the transportation means: trucks moving containers in and out of the terminal, railroad trains carrying the containers in and out of the terminal, or possibly other ships. We do not claim completeness in this RA, but it can be used as a framework to guide the design of the cargo port system.

6. VALIDATION OF THE RA

RA and patterns are abstract models, not software, and such models cannot be evaluated in terms of performance, reliability, or security using traditional experimentation or testing methods [20,26]; it does not make sense to experimentally validate an instance of a pattern or RA because evaluating its properties verifies them only for that instance, not for all the instances of the pattern or RA. On the other hand, other criteria can be used to validate RA and patterns. We can validate abstract models' completeness by comparing an RA model to models of commercial cargo port systems, explicitly comparing the fundamental components of our cargo port RA with the elements of these systems. If the RA includes all the fundamental features and functions offered by existing cargo port architectures, it means the abstractions faithfully represent their key aspects. Our patterns include "Known Uses" subsections, which indicate examples of commercial cargo port models where we have verified, pattern by pattern, that all the components of our RA are present in these ports. We studied descriptions of the ports of Shanghai, China; Singapore; Rotterdam, Netherlands; Los Angeles, Long Beach, Miami, and Everglades, USA; all these ports represent modern ports, including automation and other advances. Following a standard way to find patterns [7,8], the first four ports were used to build the patterns, and the other three were selected to confirm that real ports used these patterns. None of these descriptions used patterns or UML, so we needed to abstract their components and functions.

In general, specific patterns are normally evaluated by submitting them to a pattern conference. In these conferences, a pattern paper is developed with the help of a shepherd and then discussed in a workshop of about 8–12 subject matter experts. The pattern is exposed for criticism, revised, and then published. In terms of precision, we created our RA through UML modeling and a well-defined pattern template for describing the pattern solutions. This method is widely used in the software engineering discipline and is regarded as more precise as compared to other representations such as textual or block diagram-based representations.

In addition, we have presented our work in several architecture-oriented meetings, such as Pattern Languages of Programming (PLoP) conferences, open Web application security project (OWASP) meetings, and in the pattern language community, where we received significant feedback on the usefulness and benefits of this work. Ultimately, the final validation of this work will come from practitioners using it to build concrete architectures.

Another type of validation refers to the usefulness of the RA [23,26,59], shown in Section 7.

7. VALUE AND USE OF THE REFERENCE ARCHITECTURE IN CARGO PORT DESIGN AND OPERATION

The design of a cargo port is a major endeavor and is performed in three basic stages: A conceptual design is done first for deciding aspects such as location, capacity, personnel, cost, regulations, etc. [60]. This is followed by a detailed design indicating the number and placement of buildings, number of cranes, level of automation, functional distribution of zones, interactions with external organizations, compliance with regulations, security requirements, etc. [61,62]. Finally, the software design of the complete facility is done, for which our reference architecture would be used. This stage involves experts in ports, CPSs, and software (architects and developers). The increasing automation of ports requires an increasing amount of software to coordinate and monitor operations as well as describe the functions of the components [63].

As indicated earlier, Martínez-Fernández et al. [59] addressed the benefits and drawbacks of using RAs in industry practice from the perspective of different stakeholders involved in their design and usage. They estimated the value and risks of building and using an RA for a particular domain. Their study revealed that not all organizations realize the theoretical benefits of RAs; most only realize a subset of these benefits. They revealed that one main drawback was the additional learning curve experienced by application developers who use the RA, causing higher time to market and high complexity when they do not aim to facilitate the development of applications. They concluded that different stakeholders present different concerns. For software architects, standardization and reliability were the key benefits of using RAs, whereas for application developers, it was the use of the latest technologies. Buccaioni et al. [23] give a list of the advantages of having a reference architecture for enterprises.

Some of the practical uses of a model like ours include:

1. RAs aggregate knowledge and design expertise in a specific domain [23]. This knowledge can be reused to build new ports and improve existing ones.

2. Holistic and unified models such as RAs are useful for gaining understanding of a complex system and its limitations. Cargo ports are very complex systems, and having an abstract architecture free of implementation details is very important to understanding the system.
3. An RA serves as a way to decompose a complex system into simpler and more coherent parts that can be assigned to different types of specialists.
4. Automation is now a basic objective to increase efficiency and reduce costs, and an RA can provide a perspective on how the automation of parts of the system contributes to the efficiency of port operation.
5. UML models used to represent the RA provide more precision as compared to other descriptions such as block diagrams.
6. We can contribute to the standardization of concrete architectures by using the RA as a template. Standardization can accelerate automation and reduce costs [63].
7. An RA serves as a way to unify terminology and make it easier to train the developers, system architects, and administrators so they become familiar with the platforms and processes.
8. Abstract representational models of the cargo port components in the RA help address heterogeneity; this is accomplished by focusing on the general categories of the system and grouping them on the basis of similarities in structure and functionality.
9. We can identify common functionalities and configurations to encourage reuse in new projects.
10. RAs support interoperability among different applications and their components by establishing common mechanisms for information exchange.
11. RAs can serve as a guideline to indicate where to add defenses for the expected threats and for adding security monitoring in the physical port.
12. Security, compliance, privacy, safety, reliability, and governance can all be improved by the use of these models. For example, the analysis of threats in complex systems can be done systematically with the help of RAs [17].
13. RAs can guarantee the compliance of real architectures with RAs and regulatory requirements [3,23]. Once a new architecture is built, we can test if it follows the struc-ture and policies of its corresponding RA.

8. RELATED WORK

We enumerate now work that helps provide a perspective on our work.

8.1. Papers That Use Patterns and/or UML Models to Describe CPS Architectures

Maidl et al. [64] describe a UML RA for CPSs oriented to analyze security. It describes typical aspects of CPSs such as sensors, interfaces, and networks, from which a deployment diagram for a control system is derived.

Gottschalk et al. [4], using UML models and MDE, defined an RA for smart grids. Another paper in that group relates CPSs to systems of systems [25].

Hehenberger et al. [65] is a survey paper that provides an overview of different types of CPSs and the transition process from mechatronics to CPS. It indicates that methodologies for CPS design should be part of a multi-disciplinary development process within which designers should focus not only on the separate physical and computational components but also on their integration and interaction.

Tan et al. [66] use a service-oriented architecture (SOA) style to design CPSs, including patterns, but their approach is oriented toward the dynamic composition of services [67]. They also propose a SOA architecture described using a block diagram.

La et al. [68] present a 3-tier Software Oriented Architecture (SOA) and define in detail the components of this architecture. Their tiers include the environmental tier for the target physical environment, the control tier for making decisions about networked physical devices, and the service tier for managing reusable services. They use UML sequence diagrams to describe the system components.

RAs using patterns have been very effective in describing other complex non-CPS and IoT systems. A pattern-based RA for a cloud system is presented in [69,70], and a network function virtualization (NFV) RA is presented in [71]. In Syed et al. [47], the authors defined an RA for a cloud ecosystem that described at an abstract level the main features of the cloud system and showed the relationships between containers, the cloud, and IoT ecosystems.

Seiger et al. [14] consider models for CPS processes, described using UML class diagrams; they could be used in the structure of a CPS RA.

None of these architectures considers cargo ports, but [64] has aspects that would apply to the implementation aspects of ports. Seiger's models [14] could be useful to extend the RA.

8.2. Architectures of Transportation Systems

A work discusses the construction of the infrastructure as a collaborative network of stakeholders offering business services based on a single contract [72]. Other authors discuss a service-oriented approach to modeling Port Community Systems (PCS) as an electronic platform linking the multiple systems operated by private and public organizations (enterprises and government institutions) in a complex service system network required to coordinate flows of merchandise, property rights, payments, and information in the global supply chain. Their approach is to provide tools to describe the service-based relationships of the actors, and to understand how each contributes to improving the efficiency of the network [73,74,75]. The last two studies use a collaborative multi-agent framework. Thurston et al. [67] present a port layout to reduce the time spent by ships in a port. This last work can be considered a complement to the port structure pattern shown in this paper, but the others have very different objectives.

Cargo ports are part of maritime transportation systems, and these papers could be useful to extend our models, although none of them provides architectural models.

8.3. Architectures Proposed as Part of Descriptions of CPSs or Development Methodologies

The literature mentions the possibility of using domain models for CPSs, but no details are given. Refs. [76,77,78,79,80,81] all give basic block diagram architectures with short descriptions of the units. Spichkova in [82] introduces a development methodology for

CPSs focusing on the abstraction levels of the system model, based on the idea of refinement-based development of complex, interactive systems. Krotofil and Gollmann describe the layered structure of CPSs [83].

8.4. Architectures for Other Types of CPSs

Microsoft defined an RA for smart grids [84], and a more up-to-date architecture for smart grids is given in [4], mentioned above. Verissimo et al. [73] describe a security architecture for smart grids emphasizing reliability. Autosar and Industry 4.0 are important architectures for vehicles and manufacturing, respectively. Several RAs for IoT are given in [13], while Guth et al. used an RA to compare IoT platforms [85].

All these architectures are too general (CPSs or IoT in general) or apply to other CPSs, not cargo ports.

8.5. Papers about Quality Aspects of CPSs

A very important quality aspect is security. Our complementary paper [17] shows the design of a security reference architecture (SRA). We discussed a variety of threats, including:

Drop container, which could destroy the container, kill humans, and damage the ship.

Misplace container, which would make it very difficult to find it later.

Accelerate the crane beyond its limit, which could destroy it.

Denial of service [17,86], where the system is flooded with messages that do not allow it to perform useful work.

Other threats, indicated by one of the referees, complement our threats and include:

False data injection [87,88], can affect sensor readings and provide a crane operator with incorrect crane positions that could cause a crane to place a container in the wrong place or cause a crane crash.

Preventing actuation attacks [89] prevents the exchange of information between a controller and its actuators.

More threats are discussed in [90].

Martínez-Fernández et al. [59] addressed the benefits and drawbacks of using RAs in industry practice from the perspective of different stakeholders involved in their design and usage. We used their work in Section 7.

8.6. Papers about Development Methodologies for CPSs

Reference architectures and patterns are artifacts used in model-driven development methodologies. Some authors have proposed specific methodologies for CPSs: Hehenberger et al. [65] use SysML (a variant of UML), J.C. Jensen et al. [91] propose a model-based approach, Spichkova et al. [82] propose a multilevel semiformal model, Jacky uses formal models [92], and Karsai et al. propose a layered architecture [93].

In summary, our patterns and RA, as well as the methodology we used to build them, are original and useful for architects and developers building software for cargo ports. The RA captures semantic aspects unique to this type of CPS and has been the basis of a security RA [17].

9. CONCLUSIONS AND FUTURE WORK

We have applied architectural modeling using patterns and reference architectures to build the reference architecture of a specific cyber-physical system, a transportation system, namely, a cargo port. We have verified that no other reference architecture exists for cargo ports. Existing RAs model the lower levels of a CPS but do not capture the semantics of specific port operations, a basic feature of our models. The method presented in this paper of using patterns to build the reference architecture provides not only a holistic and unified view of the system to users, developers, and researchers, but it is also fundamental to handling the cross-domain complexity of this type of system. A reference architecture is an ongoing process. The model presented here is just a step toward achieving a precise architectural representation; other use cases and components need to be defined as patterns to complete the architecture. Building a complete RA is a very large enterprise, not feasible for a two-person team; therefore, we just intended to define the methodology to build such architectures and build part of it to illustrate this methodology. The complete architecture has many similar components that just differ from others in small details, so building a complete architecture, while important for an enterprise, is not particularly interesting from a research point of view. In parallel work, we have built a security reference architecture for cargo ports [17]. Our architecture considers only the application level, but it can be extended to become a multilevel architecture such as Industry 4.0 or SGAM. RAs can be used as part of a model-driven methodology to build systems; we have proposed such a methodology for distributed systems [94], where our patterns and RAs could be used to build CPSs. We have used UML for our models, a semi-formal approach, but ontologies can be derived from patterns and RAs to have formal descriptions of this RA, which may be in our future work.

REFERENCES

1. Lee, E.A. Cyber Physical Systems: Design Challenges. In Proceedings of the 11th IEEE International Symposium on Object and Component-Oriented Real-Time Distributed Computing, Orlando, FL, USA, 5–7 May 2008; pp. 363–369.

2. Denker, G.; Dutt, N.; Mehrotra, S.; Stehr, M.-O.; Talcott, C.; Venkatasubramanian, N. Resilient dependable cyber-physical systems: A middleware perspective. *J. Internet Serv. Appl.* **2012**, *3*, 41–49.

3. Yimam, D.; Fernandez, E.B. Building Compliance and Security Reference Architectures for cloud systems. In Proceedings of the IEEE International Conference on Cloud Engineering (IC2E) 2016, Berlin, Germany, 4–8 April 2016.

4. Gottschalk, M.; Delfs, C.; Model, T.S.G.A. *The Use Case and Smart Grid Architecture Model Approach*; Springer: Berlin, Germany, 2017; pp. 41–61.

5. Avgeriou, P. Describing, Instantiating and Evaluating a Reference Architecture: A Case Study. *Enterp. Archit. J.* **2003**, *342*, 1–24.

6. Taylor, R.N.; Medvidovic, N.; Dashofy, E.M. *Software Architecture: Foundations, Theory, and Practice*; Wiley: London, UK, 2009; ISBN 0470167742, 9780470167748.

7. Buschmann, F.; Meunier, R.; Rohnert, H.; Sommerlad, P.; Stal, M.; Architecture, P.-O.S. *A System of Patterns*; John Wiley Sons Ltd.: New York, NY, USA, 1996; Volume 1.

8. Gamma, E.; Helm, R.; Johnson, R.; Vlissides, J. *Design Patterns: Elements of Reusable Object-Oriented Software*; Addison-Wesley: Boston, CM, USA, 1994.

9. Romero, V.M.; Fernandez, E.B. Towards a Security Reference Architecture for Cyber Physical Systems. In Proceedings of the Fifthteen Latin American and Caribbean Conference for Engineering and Technology, Boca Raton, FL, USA, 19–21 July 2017.

10. Warmer, J.; Kleppe, A. *The Object Constraint Language: Getting Your Models Ready for MDA*, 2nd ed.; Addison-Wesley Longman: Boston, MA, USA, 2013; ISBN 0321179366.

11. Garavel, H.; Graf, S. Formal methods for safe and secure computer systems. In *Technical Report. BSI Study 875*; Federal Office for Information Security: Bonn, Germany, 2013.

12. Pereira-Vale, A.; Fernandez, E.B. An Ontology for Security Patterns. In Proceedings of the 38th International Conference of the Chilean Computer Science Society (SCCC 2019), Concepción, Chile, 4–8 November 2019.

13. Weyrich, M.; Ebert, C. Reference architectures for the internet of things. *IEEE Softw.* **2015**, *33*, 112–116.

14. Seiger, R. Modelling complex and flexible processes for smart cyber-physical environments. *J. Comput. Sci.* **2015**, *10*, 137–148.

15. Romero, V.M.; Fernandez, E.B. A Pattern for Secure Cargo Port Drayage. In Proceedings of the 7th Asian Conference on Pattern Languages of Programs, Asian PLoP'18, Tokyo, Japan, 1–2 March 2018; p. 9.

16. Fernandez, E.B.; Monge, R.; Carvajal, R. A pattern for a secure and safe port loading facility. In Proceedings of the 10th Latin American Conference on Pattern Languages of Programs—SugarLoafPLoP, Rio de Janeiro, Brazil, 20–23 November 2014.

17. Fernandez, E.B.; Romero, V. A security reference architecture for cargo ports. *Internet Things Cyber-Phys. Syst.* **2022**, *2*, 120–137.

18. Buschmann, F.; Henney, K.; Schmidt, D.C. *Architecture, Pattern-Oriented Software Architecture: On Patterns and Pattern Languages*; John Wiley Sons Ltd.: Hoboken, NJ, USA, 2007; Volume 5.

19. Wieringa, R. Design Science as nested problem solving. In Proceedings of the 4th International Conference on Design Science Research in Information Systems and Technology, New York, NY, USA, 7–8 May 2009.

20. Fernandez, E.B. *Security Patterns in Practice: Building Secure Architectures Using Software Patterns*; Wiley Series on Software Design Patterns: Chichester, UK, 2013.

21. Fernandez, E.B.; Pelaez, J.; Larrondo-Petrie, M. Attack Patterns: A New Forensic and Design Tool. In *Advances in Digital Forensics III*; Springer: New York, NY, USA, 2007; pp. 345–357.

22. Fernandez, E.B.; Yoshioka, N.; Washizaki, H. Modeling Misuse Patterns. In Proceedings of the 2009 International Conference on Availability, Reliability, and Security (ARES 2009), Fukuoka, Japan, 16–19 March 2009; pp. 566–571.

23. Bucaioni, A. Reference architectures modelling and compliance checking. *Softw. Syst. Model.* **2022**, 1–27.

24. Stricker, V.; Lauenroth, K.; Corte, P.; Gittler, F.; Panfilis, S.D.; Pohl, K. Creating a reference architecture for service-based systems—A pattern-based approach. In *Towards the Future Internet—Emerging Trends from European Research*; IOS Press: Amsterdam, The Netherlands, 2010; pp. 149–160.

25. Uslar, M.; Rohjans, S.; Neureiter, C.; Andrén, F.P.; Velasquez, J.; Steinbrink, C.; Efthymiou, V.; Migliavacca, G.; Horsmanheimo, S.; Brunner, H.; et al. Applying the Smart Grid Archi-

tecture Model for Designing and Validating System-of-Systems in the Power and Energy Domain: A European Perspective. *Energies* **2019**, *12*, 258.

26. Fernandez, E.B.; Monge, R.; Hashizume, K. Building a security reference architecture for cloud systems. *Requir. Eng.* **2016**, *21*, 225–249.
27. Romero, V.M.; Fernandez, E.B. Building a Reference Architecture for Cargo Ports using Patterns. In Proceedings of the Sixteenth Latin American and Caribbean Conference for Engineering and Technology, Lima, Peru, 18–20 July 2018.
28. US Dept. of Transportation, Research and Innovative Technology Administration. In *Freight Transportation: Global Highlights*; Bureau of Transportation Statistics: Washington, DC, USA, 2010.
29. Available online: **http://www.rita.dot.gov/bts/sites/rita.dot.gov.bts/files/publications/freight_transportation/pdf/entire.pdf** (accessed on 14 February 2023).
30. Bureau, P.R. 2011 World Population Data Sheet. (Washington, DC: 2012). Available online: **http://www.prb.org/pdf11/2011population-data-sheet_eng.pdf** (accessed on 1 May 2013).
31. The World Bank, Container Port Traffic. Available online: **https://data.worldbank.org/indicator/IS.SHP.GOOD.TU** (accessed on 14 February 2023).
32. US Dept. of Transportation, Research and Innovative Technology Administration, Bureau of Transportation Statistics. America's Container Ports: Linking Markets at Home and Abroad; Washington, DC, USA. 2011. Available online: **https://www.bts.gov/publications/americas_container_ports/2011/pdf/entire.pdf** (accessed on 1 May 2013).
33. Magazine, F. Internet Reference. Available online: **http://fortune.com/2018/01/30/port-automation-robots-container-ships/** (accessed on 14 February 2023).
34. Yang, C.H.; Choi, Y.S.; Ha, T.Y. Simulation-based performance evaluation of transport vehicles at automated container terminals. *OR Spectr.* **2004**, *26*, 149–170.
35. YorkWallischeck, E. *ICS Security in Maritime Transportation A White Paper Examining the Security and Resiliency of Critical Transportation Infrastructure*; DOT-VNTSC-MARAD-13-01; Bureau of Transportation Statistics: Washington, DC, USA, 2013.
36. Bielli, M.; Boulmakoul, A.; Rida, M. Object Oriented Model for Container Terminal Distributed Simulation. *Eur. J. Oper. Res.* **2006**, *175*, 1731–1751.
37. Boulmakoul, A.; Rida, M. Discrete event simulation component specification for container terminals operational management. In Proceedings of the 2nd IEEE ISSPIT (International Symposium on Signal Processing and Information Technology), Marrakesh, Morocco, 18–21 December 2002; pp. 266–271.
38. Organization, P. Internet Reference. Available online: **https://www.pema.org/wp-content/uploads/downloads/2016/06/PEMA-IP12-Container-Terminal-Automation.pdf** (accessed on 14 February 2023).
39. Steenken, D.; Voß, S.; Stahlbock, R. Container terminal operation and operations research-a classification and literature review. *OR Spectr.* **2004**, *26*, 3.
40. Cheng, Y.L.; Sen, H.C.; Natarajan, K.; Teo, C.P.; Tan, K.C. Dispatching Automated Guided Vehicles in a Container Terminal. In *Supply Chain Optimization. Applied Optimizatio*; Geunes, J., Pardalos, P.M., Eds.; Springer: Boston, MA, USA, 2005; Volume 98.
41. Zaffalon, M.; Rizzoli, A.E.; Gambardella, L.M.; Mastrolilli, M. Resource allocation and scheduling of operations in an intermodal terminal. In Proceedings of the 10th European Simulation Symposium and Exhibition, Simulation in Industry, ESS98, Nottingham, UK, 26–28 October 1998; pp. 520–528.

42. Günther, H.; Kim, K.H. *Container Terminals and Automated Transport Systems: Logistics Control Issues and Quantitative Decision Support*; Springer Science & Business Media: Berlin/Heidelberg, Germany, 2005.
43. Lethbridge, T.C.; Laganière, R. *Object-Oriented Software Engineering: Practical Software Development Using UML and Java*; McGraw-Hill: New York, NY, USA, 2001; Volume 102.
44. Angelov, S.; Grefen, P.; Greefhorst, D. A framework for analysis and design of software reference architectures. *Inf. Softw. Technol.* **2012**, *54*, 417–431.
45. Pankowska, M. Stakeholder Oriented Enterprise Architecture Modelling. In Proceedings of the 12th International Conference on e-Business, Colmar, France, 20–22 July 2015.
46. Notteboom, T.; Winkelmans, W. Stakeholder Relations Management in ports: Dealing with the interplay of forces among stakeholders in a changing competitive environment. In Proceedings of the IAME 2002 Maritime Economics: Setting the Foundations for Port and Shipping Policies, Panama City, Panama, 12–15 November 2002.
47. Syed, M.H.; Fernandez, E.B. A reference architecture for the container ecosystem. In Proceedings of the 13th International Conference on Availability, Reliability and Security (ARES 2018), Hamburg, Germany, 27–30 August 2018; ACM: New York, NY, USA, 2018.
48. Saini, K.; Kaur, S. Forensic Examination of Computer-Manipulated Documents using Image Processing Techniques. *Egypt. J. Forensic Sci.* **2016**, *6*, 317–322.
49. National Cooperative Freight Research Program, N.C.R.P. *Report 11. Truck Drayage Productivity Guide*; Grant No. DTOS59-06-00039; The National Academies Press: Washington, DC, USA, 2011.
50. The Port of Los Angeles Port Drayage. Available online: **https://www.portoflosangeles.org/** (accessed on 14 February 2023).
51. The Port of Long Beach Port Drayage. Available online: **https://www.polb.com/** (accessed on 14 February 2023).
52. Stallings, W.; Brown, L. *Computer Security: Principles and Practice*, 4th ed.; Pearson: London, UK, 2018.
53. Port Everglades Website, Broward County, FL, USA. Available online: **https://www.porteverglades.net/articles/post/port-everglades-inspects-new-super-post-panamax-cranes/** (accessed on 14 February 2023).
54. Fernandez, E.B.; Sorgente, T.; VanHilst, M. Constrained Resource Assignment Description Pattern. In Proceedings of the Nordic Conference on Pattern Languages of Programs, Viking PLoP 2005, Otaniemi, Finland, 23–25 September 2005.
55. Longo, F.; Padovano, A.; Baveja, A.; Melamed, B. Challenges and opportunities in implementing green initiatives for port terminals. In Proceedings of the 3rd International Workshop on Simulation for Energy, Sustainable Development and Environment, SESDE 2015, Bergeggi, Italy, 21–23 September 2015.
56. Marine Insight. Available online: **https://www.marineinsight.com/guidelines/general-procedure-of-preparing-ships-for-entering-ports/** (accessed on 14 February 2023).
57. Kaluza, P.; Kölzsch, A.; Gastner, M.T.; Blasius, B. The complex network of global cargo ship movements. *J. R. Soc. Interface* **2010**, *7*, 1093–1103.
58. Fernandez, E.B.; Ballesteros, J.; Desouza-Doucet, A.C.; Larrondo-Petrie, M.M. Security Patterns for Physical Access Control Systems. In *Data and Applications Security XXI. DBSec 2007. Lecture Notes in Computer Science*; Barker, S., Ahn, G.J., Eds.; Springer: Berlin/Heidelberg, Germany, 2007; Volume 4602.

59. Martínez-Fernández, S.; Ayala, C.P.; Franch, X.; Marques, H.M. Benefits and drawbacks of software reference architectures: A case study. *Inf. Softw. Technol.* **2017**, *88*, 37–52.
60. Group, T.B. Container Port Design and Planning. Available online: **https://tba.group/en/services/bulk-container-terminal-design** (accessed on 14 February 2023).
61. Wikipedia, Container Port Design Process. Available online: **https://en.wikipedia.org/wiki/Container_port_design_process** (accessed on 14 February 2023).
62. Cisco, Ports and Terminals Design Guide. Available online: **https://www.cisco.com/c/en/us/td/docs/solutions/Verticals/CCI/Ports/DG/cci-ports-dg/cci-ports-dg.html** (accessed on 14 February 2023).
63. Martin-Soberon, A.M.; Monfort, A.; Sapina, R.; Monterde, N.; Calduch, D. Automation in port container terminals. *Procedia-Soc. Behav. Sci.* **2014**, *160*, 195–204.
64. Maidl, M.; Wirtz, R.; Zhao, T.; Heisel, M.; Wagner, M. Pattern-based modeling of cyber-physical systems for analyzing security. In Proceedings of the 24th European Conference on Pattern Languages of Programs, Irsee, Germany, 3–7 July 2019; Volume 23, pp. 1–10.
65. Hehenberger, P.; Achiche, S. Design, modelling, simulation and integration of cyber physical systems: Methods and applications. *Comput. Ind.* **2016**, *82*, 273–289.
66. Tan, Y.; Goddard, S.; Perez, L.C. A prototype architecture for cyber-physical systems. *ACM Sigbed Rev.* **2008**, *5*, 1–2.
67. Thurston, T.; Hu, H. Distributed agent architecture for port automation. In Proceedings of the 26th Annual International Computer Software and Applications, Oxford, England, 26–29 August 2002.
68. La, H.J.; Kim, S.D. A service-based approach to designing cyber-physical systems. In Proceedings of the 9th IEEE/ACIS International Conference on Computer and Information Science, Washington, DC, USA, 18–20 August 2010; pp. 895–900.
69. Hashizume, K.; Fernandez, E.B.; Larrondo-Petrie, M. *Cloud infrastructure pattern. First International Symposium on Software Architecture and Patterns*; LACCEI: Panama City, Panama, 2012; pp. 23–27.
70. Hashizume, K.; Fernandez, E.B.; Larrondo-Petrie, M.M. Cloud service model patterns. In Proceedings of the 19th International Conference on Pattern Languages of Programs (PLoP2012), Tucson, AZ, USA, 19–21 October 2012.
71. Alnaim, A.K.; Alwakeel, A.M.; Fernandez, E.B. Towards a Security Reference Architecture for Network Function Virtualization. *Sensors* **2022**, *22*, 3750.
72. Nota, G.; Bisogno, M.; Saccomanno, A. A service-oriented approach to modeling and performance analysis of Port Community Systems. *Int. J. Eng. Bus. Manag* **2018**, *10*.
73. Verissimo, P. CRUTIAL: Towards a reference critical information infrastructure architecture. *Eur. CIIP Newsl.* **2007**, *3*, 6–8.
74. Osório, A.; Afsarmanesh, H.; Camarinha-Matos, L. Towards a Reference Architecture for a Collaborative Intelligent Transport System Infrastructure. In *Collaborative Networks for a Sustainable World, IFIP Advances in Information and Communication Technology*; Springer: Berlin, Germany, 2010; Volume 336, pp. 469–477.
75. Zambrano, G.R.; Vera, L.O. Reference architecture for an intelligent transportation system. *Int. J. Innov. Appl. Studies* **2016**, *15*, 2028–9324.
76. Shi, J.; Wan, J.; Yan, H.; Suo, H. A survey of cyber-physical systems. In Proceedings of the International Conference on Wireless Comm. and Signal Processing, Nanjing, China, 9–11 November 2011.

77. King, M.; Zhu, B.; Tang, S. Optimal Path Planning. *Mob. Robot.* **2001**, *8*, 520–531.

78. Cardenas, A.A.; Amin, S.; Lin, Z.-S.; Huang, Y.-L.; Huang, C.H.; Sastry, S. Attacks against process control systems: Risk assessment, detection, and response. In Proceedings of the ASSIACS'11, Hong Kong, China, 22–24 March 2011; pp. 355–366.

79. Miller, A. Trends in process control systems security. *IEEE Secur. Priv.* **2005**, *3*, 57–60.

80. Rahman, H.A.; Beznosov, K. SPAPI: A security and protection architecture for physical infrastructures and its deployment strategy using wireless sensor networks. In Proceedings of the 10th IEEE International Conference on Emerging Technologies and Factory Automation (ETFA 2005), Catania, Italy, 19–22 September 2005; pp. 885–892.

81. Liu, Y.; Peng, Y.; Wang, B.; Yao, S.; Liu, Z. Review on Cyber-physical Systems. *IEEE/CAA J. Autom. Sin.* **2017**, *4*, 17–40.

82. Spichkova, M.; Schmidt, H.; Peake, I. From abstract modelling to remote cyber-physical integration/interoperability testing. *arXiv* **2014**, arXiv:1403.1005.

83. Krotofil, M.; Gollmann, D. Industrial control systems security: What is happening? In Proceedings of the 2013 11th IEEE International Conference on Industrial Informatics (INDIN), Bochum, Germany, 29–31 July 2013; pp. 664–669.

84. Microsoft Power and Utilities, Smart Energy Reference Architecture, Oct. 2009. Available online: **https://www.Microsoft.com/Utilities** (accessed on 14 February 2023).

85. Guth, J.; Breitenbücher, U.; Falkenthal, M.; Leymann, F.; Reinfurt, L. Comparison of IoT platform architectures: A field study based on a reference architecture. In Proceedings of the 2016 Cloudification of the Internet of Things (CIoT), Paris, France, 23–25 November 2016.

86. Prakash, A.; Satish, M.; Bhargav, T.S.S.; Bhalaji, N. *Detection and Mitigation of Denial of Service Attacks Using Stratified Architecture*; Elsevier: Amsterdam, The Netherlands, 2016.

87. Ganjkhani, M.; Fallah, S.N.; Badakhshan, S.; Shamshirband, S.; Chau, K.-W. A Novel Detection Algorithm to Identify False Data Injection Attacks on Power System State Estimation. *Energies* **2019**, *12*, 2209.

88. Liang, G.; Zhao, J.; Luo, F.; Weller, S.R.; Dong, Z.Y. A Review of False Data Injection Attacks Against Modern Power Systems. *IEEE Trans Smart Grid* **2016**, *8*, 1630–1638.

89. Hosseinzadeh, M.; Sinopoli, B. Active Attack Detection and Control in Constrained Cyber-Physical Systems Under Prevented Actuation Attack. In Proceedings of the American Control Conference (ACC), New Orleans, LA, USA, 25–28 May 2021.

90. Loukas, G.; Vuong, D.G.T. A Review of Cyber Threats and Defence Approaches in Emergency Management. *Future Internet* **2013**, *5*, 205–236.

91. Jensen, J.C. A model-based design methodology for cyber-physical systems. In Proceedings of the 7th International Wireless Communications and Mobile Computing Conference, Istanbul, Turkey, 4–8 July 2011; pp. 1666–1671.

92. Jacky, J. Specifying a Safety-Critical Control System in Z. *IEEE Trans. Softw. Eng.* **1995**, *21*, 388–402.

93. Karsai, G.; Balasubramanian, D.; Dubey, A.; Otte, W.R. Distributed and managed: Research challenges and opportunities of the next generation cyber-physical systems. In Proceedings of the 2014 IEEE 17th International Symposium on Object/Component/Service-Oriented Real-Time Distributed Computing, Reno, NV, USA, 10–12 June 2014; pp. 1–8.

94. Uzunov, A.; Fernandez, E.B.; Falkner, K. ASE: A Comprehensive Pattern-Driven Security Methodology for Distributed Systems. *J. Comput. Stand. Interfaces* **2015**, *41*, 112–137.

Chapter 10

Next-Generation Smart Response Web (NG-SRW): An Operational Spatial Decision Support System for Maritime Oil Spill Emergency Response in the Gulf of Finland (Baltic Sea)

Mihhail Fetissov[1], Robert Aps[1], Floris Goerlandt[2], Holger Jänes[1], Jonne Kotta[1], Pentti Kujala[3] and Robert Szava-Kovats[1]

[1] *Estonian Marine Institute, University of Tartu, 12618 Tallinn, Estonia*

[2] *Department of Industrial Engineering, Dalhousie University, Halifax, NS B3H 4R2, Canada*

[3] *Marine Technology, School of Engineering, Aalto University, FI-00076 Aalto, Finland*

ABSTRACT

The Baltic Sea is a unique and sensitive brackish-water ecosystem vulnerable to damage from shipping activities. Despite high levels of maritime safety in the area, there is a continued risk of oil spills and associated harmful environmental impacts. Achieving common situational awareness between oil spill response decision makers and other actors, such as merchant vessel and Vessel Traffic Service center operators, is an important step to minimizing detrimental effects. This paper presents the Next-Generation Smart Response Web (NG-SRW), a web-based application to aid decision making concerning oil spill response. This tool aims to provide, dynamically and interactively, relevant information on oil spills. By integrating the analysis and visualization of dynamic spill features with the sensitivity of environmental elements and value of human uses, the benefits of potential response actions can be compared, helping to develop an appropriate response strategy. The oil spill process simulation enables the response authorities to judge better the complexity and dynamic behavior of the systems and processes behind the potential environmental impact assessment and thereby better control the oil combat action.

Keywords: *decision support system; emergency response; marine pollution*

1. INTRODUCTION

Oil transport in the Baltic Sea has grown rapidly over the last decade, the Gulf of Finland being one of the busiest oil shipping routes in the world [1]. The coastal waters of the Gulf of Finland also host environmentally valuable ecosystems, which face risk from heavy vessel traffic [2].

Accident statistics indicate a high level of maritime safety in the Baltic Sea, and, while accidents typically result in minor consequences in terms of oil pollution [1,3], risks of major oil spills remain both in open water [4,5] and during winter [6,7], for example, the spill caused by the Runner-4 accident in 2006 [8], which led to large oil patches drifting into the shallow waters around Tallinn with detrimental ecological and economic consequences.

Because the effectiveness of mechanical recovery is limited both in open water [9] and in icy conditions [10], appropriate plans and tools are needed to minimize damage to ecosystems in case of a sizeable spill. This task is arduous, because the impact of spilled oil varies significantly by shoreline type and many cleanup methods are shoreline specific [11]. Therefore, strategies to contain or mitigate oil spills on or near sensitive shorelines need to establish if a response is necessary, and, if so, the nature and extent of the response. These strategies also affect significantly the final cost of mitigation [12].

Spatial Decision Support Systems (SDSS) play a role in almost all emergency situations including marine oil spills [13,14]. SDSS integrate spatial and non-spatial data, Geographic Information Systems (GIS), and decision-making algorithms to analyze emergency situations and evaluate remedial actions [15].

Decision support systems (DSS) represent another aid to tactical oil spill response planning. Most DSS employ oil trajectory and weathering models to calculate how processes such as evaporation, dissolution, emulsification, dispersion, and biodegradation affect movement of oil slicks and changes in oil slick properties [16,17,18].

It is stated [19] that the mathematical and numerical formulations of oil transport, diffusion, and transformation processes together with many simplifying hypotheses and parameterizations form the basis of a new, open source Lagrangian surface oil spill model, the so-called MEDSLIK-II, based on its precursor MEDSLIK [20]. The MEDSLIK II is used to simulate oil slick transport and transformation processes for realistic oceanic cases, where data from satellite or drifting buoys are available for verification. The model is also coupled with operational oceanographic currents, atmospheric analyses winds, and remote sensing data for initialization [21].

As a new generation of DSS for oil spill response operations and similar systems, the WITOIL DSS is based on environmental monitoring, state-of-the-art oil drift modelling, and innovative technology platforms. It is an innovative, web-based tool based on MEDSLIK-II oil spill code coupled with the operational oceanographic and atmospheric services and has been implemented to support emergency organization in the case of oil spill accidents in the Mediterranean Sea [22].

Moreover, current DSS do not fully integrate with information on the intrinsic value and environmental sensitivity to oil spills, such as the probabilistic approach presented by [23], or with operational models to estimate the spatial distribution of oil spills [24], and an initial conceptual outline of a more integrated spill response DSS has been generated [25].

This study introduces the Next-Generation Smart Response Web (NG-SRW) tool in order to alleviate the limitations of existing DSS for oil spill response planning—with particular focus in the Gulf of Finland. NG-SRW is a web-based application intended to provide a quasi-real-time dynamic assessment of the potential oil spill impact on sensitive shorelines and biological and human use resources. We examined state of the NG-SRW development, its scientific basis, conceptual framework, implementation, and future development.

2. STUDY AREA AND INTENDED END USERS OF THE DSS

The Gulf of Finland hosts environmentally valuable ecosystems within the Baltic Sea while also playing a vital role in the economic and social prosperity of its adjoining countries [2]. Concern continues to mount regarding the detrimental effects of accidental oil spills on both the Gulf of Finland's ecosystems and its economic and social prosperity. Consequently, national authorities around the Baltic Sea have committed to implementing a pollution response system to respond to pollution incidents that threaten the marine environment. This agreement includes engaging in contingency planning, surveillance, sharing information, and providing appropriate mutual assistance [26].

In the Gulf of Finland region, this response system is operational and continues to be advanced with, for instance, the recent development of a maritime simulator for oil spill response training [27] and the addition of the Mandatory Ship Reporting System in the Gulf of Finland Traffic Area (GOFREP)—established by the International Maritime Organization (IMO) [28] in 2003 and in operation since 2004—to improve navigation safety to prevent accidental ship-based spills. GOFREP provides mandatory ship reporting, including identification and monitoring of vessels, strategic planning of vessel movements, and provision of navigational information and assistance in both international and national (VTS—vessel traffic services) waters within the Gulf of Finland IMO [29]. The mandatory ship reporting system in the Gulf of Finland is shown in Figure 1.

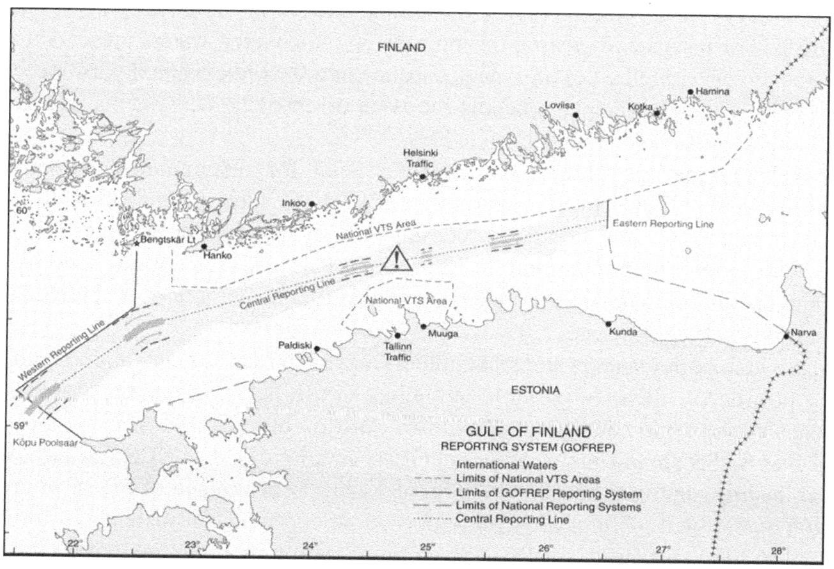

Figure 1. *The mandatory ship reporting system in the Gulf of Finland (Baltic Sea). Source: Estonian Maritime Administration.*

GOFREP is managed jointly by the Finnish Transport Agency, the Estonian Maritime Administration, and the Federal Agency for Maritime and River Transport of Russian Federation and under the auspices of GOFREP Traffic Centers of Estonia (Tallinn Traffic), Finland (Helsinki Traffic), and the Russian Federation VTMIS Centre in Petrodvorets (Saint Petersburg Traffic).

The targeted end users for SDSS oil spill response are primarily the national maritime administrations responsible for oil spill response operations and the regional response secretariat. In addition, the SDSS involve VTS operators as well as stakeholders representing the societal, environmental, economic, and social interests, whose interests can be affected by oil spills. Given the numerous user types and their wide geographic locations, a web-based platform is the optimal strategy for improving oil spill response and related issues.

3. SCIENTIFIC BASIS

Emergency response is an integral application of SDSS application, given the urgent life-threatening nature of emergencies and their complexity. This complexity demands that modelling methods be available and usable to decision makers, especially in challenging circumstances [17].

Emergency response systems demand rapid decision making, clarity of information display, and the viability for complex spatial decision making to be successful [30]. Oil spill impact assessment integrates risk and vulnerability, which have economic, ecological, cultural, political, geographical, and environmental consequences. The wide range of impacts suggests a modular framework for SDSS for oil spill response, i.e., that a tool should enable incorporation of one or multiple information layers with respect to the spatial impact of oil spills.

Other development needs may be more fundamental to the underlying structure of the models. For instance, increased oil transport in ice-covered waters presents the need to incorporate the complexities of dynamic sea ice into the pre-existing hydro-mechanical models [31]. Another example concerns the issue of changing ecosystems due to direct human activities [32] and climate change [33].

Stakeholder participation is considered crucial for sustainable development in natural resource and disaster risk management [34]. As such, participatory modelling (PM) is an interactive and iterative process in which stakeholder involvement is supported with modelling and communication tools. PM involves not only the development and use of computer-based models, but also the participation of stakeholders in the decision-making process.

SDSS help policymakers and stakeholders access to interpret and understand information in order to identify possible strategies in the decision-making process [35]. However, because intended users sometimes fail to embrace SDSS, it has been suggested that SDSS should incorporate social aspects involved in SDSS design, development, testing, and use. Socio-psychological aspects of the use of SDSS in creating common operational pictures are also important, concerning, for instance, information overload and the suitability of different communication methods to convey the information in the SDSS [36].

4. OIL SPILL EMERGENCY RESPONSE: CONCEPTUAL FRAMEWORK

It is stated [37] that "oil spill response is an extremely complex and challenging cross-disciplinary activity. In the decision-making process, it combines a wide range of issues and activities under emergency conditions that include the nature of the material spilled, changes in physical and chemical properties (weathering) and biodegradation, local environmental conditions, sensitivity of impacted natural resources, and effectiveness of response/clean-up technologies".

As a complex, dynamic, and distributed operation, oil spill response involves multiple actors [38].

Referring to the work by [39] and Figure 2, the scope of emergency oil spill response SDSS usually covers the short-term predictions of spill behavior and movement for tactical response planning and marine and coastal impacts, usually restricted to acute ecosystem impacts. Hence, based on a review by [18], it was found that most SDSS focus on the elements 1B, 2A, and 3A, whereas some tools also consider elements 1A and 2B as a part of the tools.

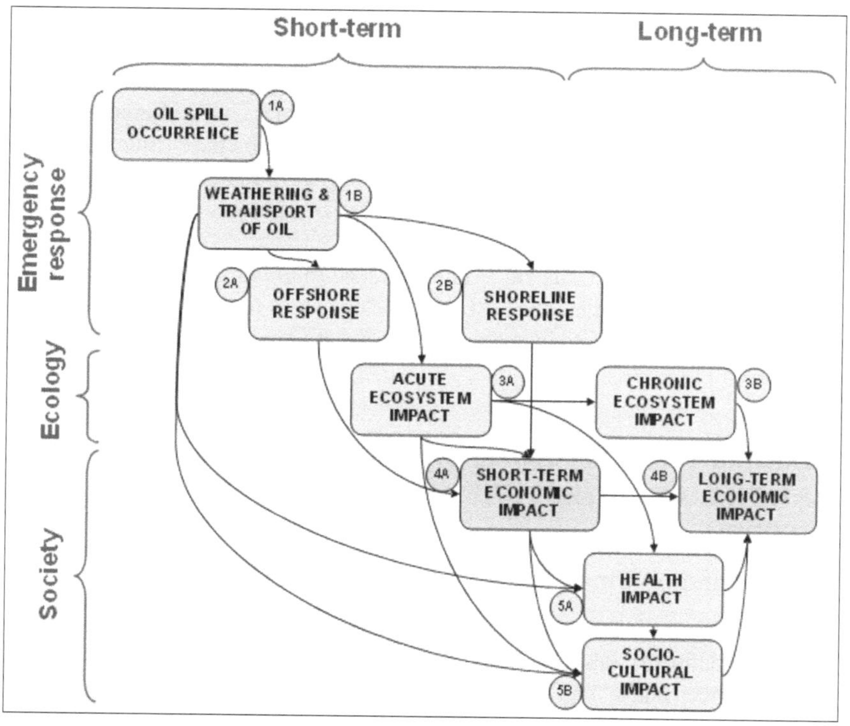

Figure 2. *Framework for classifying oil spill response and impacts (adapted from [**39**]).*

Referring to IPIECA 2005 report [40], "once oil has been spilled, urgent decisions need to be made about the options available for clean-up, so that environmental and socio-economic impacts are kept to the minimum. Getting the correct balance is always a difficult process and conflicts inevitably arise which need to be resolved in the best practicable manner. The advantages and disadvantages of different responses need to be weighed up and compared both with each other and with the advantages and disadvantages of natural clean-up, a process sometimes known as Net Environmental Benefit Analysis (NEBA)". NEBA is one of the considerations used to select spill response equipment that will effectively remove oil, are feasible to use safely in particular conditions, and will minimize the impact of the spill on the environment.

To support NEBA, knowledge about the marine and coastal ecosystem and human use values is required for a decision maker. Environmental Sensitivity Index (ESI) maps have been an integral component of oil-spill contingency planning and response in the United States since 1979, serving as a quick reference for oil spill responders [11]. The ESI maps rank shorelines into 10 classes in relation to sensitivity, natural persistence of oil, and ease of cleanup. They also provide information on coastal biological and human-use resources sensitive to oil spills.

The ESI framework was developed to reduce the environmental consequences of a spill and to help prioritize the placement and allocation of resources during cleanup efforts [41]. One of the primary objectives of oil spill planning and response, after protecting human life, is to reduce the environmental consequences of the spill and the cleanup efforts. This objective is best achieved if the location of sensitive resources are identified in advance.

Some countries outside the US have adopted the ESI approach to classify their own shorelines for similar oil spill contingency planning. Within a Gulf of Finland context, the resulting maps are referred to as Regional Environmental Sensitivity Index (RESI) maps [25]. Shores cannot be handled as static systems, implying that shore classification systems, including ESI and RESI classifications, must take into account the shore dynamics [42].

In the Baltic Sea context, the Bayesian inference to oil spill-related situation assessment is used in order to facilitate the NEBA-based decisions in selecting the best available oil spill response alternative and in evaluating the threat or probable overall environmental impact of the spill [43,44].

5. COMMON SITUATIONAL AWARENESS FOR OIL SPILL RESPONSE OPERATIONS: ARCHITECTURE AND IMPLEMENTATION OF THE NG-SRW

5.1. Next-Generation Smart Response Web (NG-SRW)

The NG-SRW application is based on operational concepts and technical implementations for user-defined Common Situational Awareness (CSA) [45]. The tool addresses oil spill occurrence (phase 1A, see Figure 2), weathering and transport of oil (1B), offshore response (2A), shoreline response (2B), acute ecosystem impact (3A), short-term economic impact (4A), and socio-cultural impact (5B). Specifically, the tool aims to utilize CSA in an online operational environment, by utilizing all available information into the decision-making process.

The NG-SRW was developed and implemented as an information collecting and sharing facility of the CSA system [45]. The basic configuration of the NG-SRW SDSS application is presented in Figure 3. NG-SRW directly interacts with Seatrack Web [46] in

Sweden, and the user can define the area of response and the type and quantity of leaked oil. Seatrack Web uses that info with real-time weather data to predict oil spill trajectory in the marine environment and inform the user where an oil spill could spread in the coming hours through the NG-SRW application. Furthermore, NG-SRW contains additional data layers such as coastal vulnerability, nature assets in the marine environments, its vulnerability, and human uses, which can be observed with overlay analysis to predict oil spill spread and avoid oil spills in sensitive marine areas. Such prioritization is required because, in the case of a more significant oil spill, we will be likely unable to act and clean all the affected marine areas. As such, we need to select areas that are sensitive to oil pollution and where large amounts of oil accumulate. The user can also switch on and off some data layers in the web application, and the application calculates summary statistics with associated oil spill polygon.

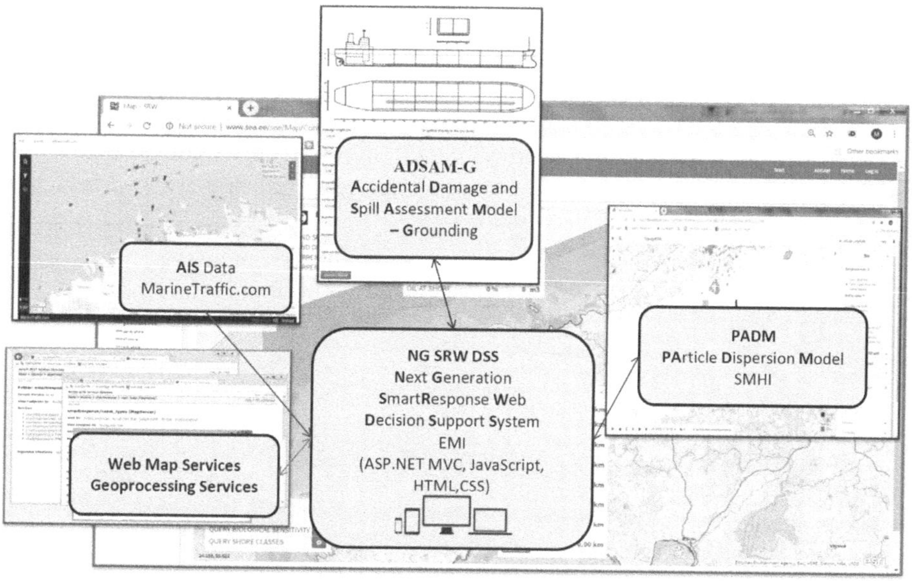

Figure 3. *Basic configuration of the NG-SRW application.*

The tool employs the NET MVC with MS SQL database engine, JavaScript, ESRI ArcGIS API for JavaScript, ESRI ArcGIS Server and ArcInfo, HTML5, and CSS technologies, allowing its use on any device with Windows, iOS, or Android operating systems. The Single Page Application (SPA) allows users to interact dynamically with all controls, data, and elements on a single page, without the need to reload the page after each action.

The NG-SRW tool consists of Server and Client/User Interface. The GIS data are prepared, analyzed, and stored in the geodatabase with ArcGIS Desktop and Python scripts. ArcGIS Server is used to share GIS data as Web Map Services (WMS) and the Geoprocessing tool Services. Auxiliary tables store oil spill model results, information on WMS layers, model parameters, and user interface tables based on MS SQL database. WMSs are portrayed with ESRI ArcGIS API for Javascript. User interface implemented on the client side supports execution of the relevant geoprocessing services.

Initially, NG-SRW connects to the ADSAM-G module [23], which provides initial information on the magnitude of accident damage and the extent of an oil spill resulting from a shipping accident. Subsequently, NG-SRW accesses the Seatrack Web application through direct connection to Particle Dispersion Model (PADM) [46], which enables independent analysis of spill drift and weathering by providing its own user interface for calculation of oil spill scenarios. Thus, NG-SRW SDSS takes advantage of the development of calculation kernel of Seatrack Web, which, in the current version, includes recent improvements of the ice code [31] but presents it within a different user interface.

MarineTraffic.com provides real-time Automatic Information System (AIS) information, such as the current position, speed, and direction of vessels. Necessary basic information on the size of vessels is obtainable for use in the ADSAM-G model. In addition, there are lights', stations', and ports, databases. All necessary information is readily obtained from the MarineTraffic map, which is embedded into NG-SRW. The NG-SRW SDSS platform is also linked to Environmental Sensitivity Index (ESI) map layers, implemented through the imported Web Map Services (WMS).

NG-SRW serves as a platform that collects and enables interaction with different streams of oil spill response-related online information [47], by which authorities can access, filter, visualize, and share information collected during an emergency response, as shown in Figure 4.

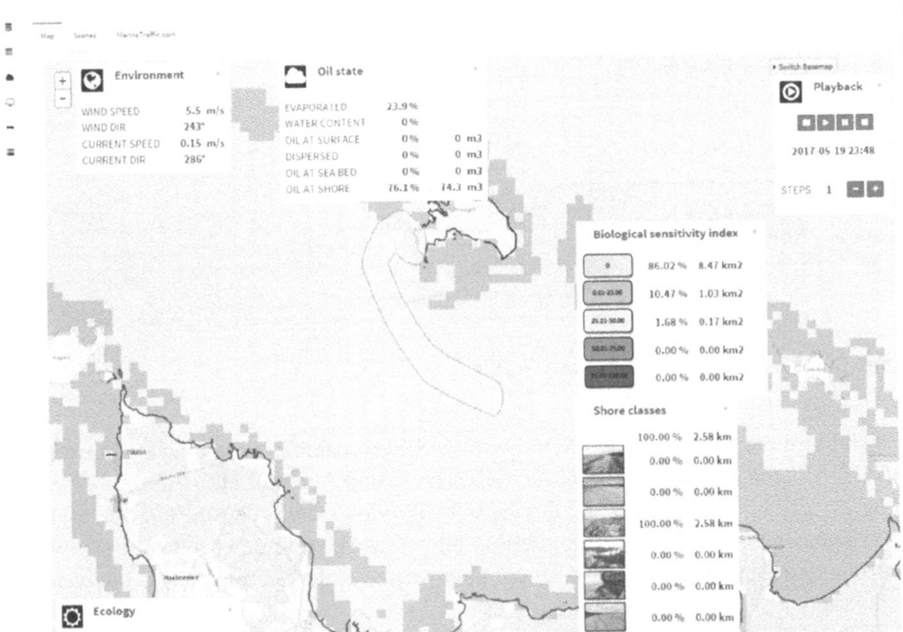

Figure 4. *NG-SRW application enables the integration of spill monitoring and evaluation functions directly into oil spill preparedness and response management processes.*

Users control the content to be included in and excluded from an oil spill response scenario, thereby allowing users to select a set of criteria to address a particular oil spill accident response unit according to their needs. This ability is essential to cater to the different decision makers and stakeholders [47].

5.2. Seatrack Web (STW)

The Seatrack Web is a fully operational, web-based, oil drift forecasting system for the Baltic and part of the North Sea [48], which allows users to simulate an oil drift on the server with the results displayed on their computer within a minute. The server has access to the most recent weather and ocean forecasts, thereby providing the user the optimal decision tool to assess oil spills. The oil weathering and transport data are simulated by the STW PADM [46], which includes the latest ice code improvements [33]. The Seatrack Web oil drift calculation system is the official HELCOM drift model/forecasting and hindcasting system used by national authorities and research organizations to simulate oil spills [49].

Seatrack Web is used to predict the location of oil spills after some hours, thereby enabling optimal allocation of oil recovery equipment and shoreline protection. As such, STW addresses 1A oil spill occurrence and weathering and transport of oil (phases 1A and 1B, Figure 2) of oil spill response. Future development of Seatrack Web would see improvement of the interaction between oil and complex, dynamic ice conditions, in particular, ridges and the movement of oil under the ice sheet, as well as refinement to the display of model and parameter uncertainties [31].

5.3. Accidental Damage and Spill Assessment Model—Grounding (ADSAM-G)

ADSAM-G is an online platform to assess rapid oil outflow from grounded tankers [23], which addresses oil spill occurrence (Phase 1A, Figure 2). ADSAM-G incorporates tanker size and configuration (as accessible from AIS) to estimate the amount and duration of oil leakage by inputs of the size of the rupture and load. This information is integrated with impact conditions, including vessel speed and bottom profile.

Currently, for ADSAM-G to oil tankers and leaks from vessel grounding, future development will see inclusion of other types of vessels and collision. In addition, oil outflow calculations are rather simple; refinement is needed to include the effects of wave action and currents [23].

5.4. Environmental Sensitivity Index (ESI) Map Layers

Mapping of the environment sensitivity to accidental oil pollution is vital to oil pollution preparedness, response, and cooperation. Referring to NOAA 2002 report [11], Environmental Sensitivity Index (ESI) maps provide information on shoreline classification (ranked according to sensitivity, natural persistence of oil, and ease of cleanup), biological resources (an assessment of the vulnerability of organisms to oil), and human-use resources (an assessment of sensitivity to oil and value from human use).

NG-SRW uses ESI maps as a visual background for the results of the simulations based on ADSAM-G and STW. As such, this application is of universal use to the geographical area concerned. GIS data are stored on a GIS server in the MS SQL Server geo-database and shared as Web Map Services (WMS). The ESI maps currently available for NG-SRW are for Baltic Sea areas under Estonian jurisdiction.

6. DISCUSSION

The proposal of a new DSS demands consideration of its expected effectiveness in practical response operations [50]. This effectiveness is difficult to assess because maritime oil spills are rare and because the large range of conceivable scenarios limit quality of assessment information. A comprehensive evaluation of the performance of the SDSS is, therefore, not yet available. However, current assessment is positive.

First, the development of the NG-SRW is rooted in the theoretical basis of SDSS, recognizing the importance of CSA in a dynamic context of decision makers, actors, and stakeholders. Furthermore, the components of NG-SRW have been used extensively by different users in a variety of contexts. For instance, the PADM model as implemented in STW is the operational tool for oil spill drift in the Baltic Sea as recommended by HELCOM [49], and AIS data are used extensively for navigation support and real-time situational awareness of shipping activity [51]. Finally, a preliminary test of the integrated NS-SRW tool has been executed in a stakeholder workshop with maritime professionals [52]. Results indicate that the tool may be useful for specific organizations and that it is relatively easy to understand and use.

While the current implementation of the NG-SRW is considered a significant step forward in NEBA-based oil spill response, addressing NG-SRW's limitations is important both for preventing over-reliance on the tool and for guiding future research and development. NG-SRW is based largely on distributed databases and the imported Web Map Services (WMS). Therefore, NG-SRW is usable in numerous geographical areas and can complement most national or regional accidental oil spill response systems. However, the lack of harmonized ESI/RESI map layers limits wider and cross-border application of the NG-SRW. Incorporation of harmonized ESI/RESI maps for the whole Baltic Sea region is necessary for future development.

A dynamic CSA that identifies shorelines sensitive to oil spills is critical in determining the kind and extent of response that may be appropriate. These choices ultimately dictate cleanup costs. Therefore, Baltic Sea ESI/RESI maps are needed that display detailed ecological and socio-economic values of shorelines and coastal waters. In addition, NG-SRW would improve if maps were incorporated that assess response performance under specific meteorological and sea ice conditions.

Finally, the functionality and effectiveness of the NG-SRW need to be evaluated in simulation-based testing, response training exercises, and in real operations. Simulation-based testing would serve as the first approach, e.g., using cross-border simulator environments [27]. Further evaluation could then be focused on genuine exercises, which would examine the social context of the emergency response, the interactions of end users with NG-SRW, and how NG-SRW supports wider communication and information exchange [36]. Importantly, the current application does not include the use of Satellite Aperture Radar and other remote sensing data. However, in the future, a broader range of information sources both at model training and product validation stage may further improve the functioning of the NG-SRW. Such research can inform design updates that consider human–machine interaction issues, team resource management, and the development of learning-oriented training programs.

7. CONCLUSIONS

This study introduced NG-SRW for building common situational awareness, providing a quasi-real-time dynamic assessment of the potential oil spill impact on sensitive shorelines, ecological values, and the human use areas. However, the SG-SRW system has yet to be practically tested in case of a future oil spill. NG-SRW enables authorities then to better simulate the complexity and dynamic behavior of the systems and processes underlying ecological risk assessment and thereby undertake oil spill mitigation more effectively. As a platform for creating a shared situational awareness, the tool also has possible use for VTS operators and it can be used in strategic pollution preparedness and response risk management.

Implementing the oil spill process simulation as an element of the NEBA processes also ensures continuous information feedback in addressing response operations, i.e., the expected drift and fate of the spilled oil, the feasibility of mitigation at sea, and the potential threat to sensitive resources.

While NG-SRW is currently operational only in Estonian marine and coastal areas and needs further evaluation and refinement, it can be used to minimize harmful effects of marine oil spills outside this area.

REFERENCES

1. HELCOM. *Assessment on Maritime Activities in the Baltic Sea 2018. Baltic Sea Environment Proceedings No. 152*; Helsinki Commission: Helsinki, Finland, 2018; 253p.
2. Sonninen, S.; Nuutinen, M.; Rosqvist, T. *Development Process of the Gulf of Finland Mandatory Ship Reporting System—Reflections on the Methods*; VTT Publications 614: Espoo, Finland, 2006.
3. Goerlandt, F.; Goite, H.; Valdez, B.O.A.; Höglund, A.; Ahonen-Rainio, P.; Lensu, M. An analysis of wintertime navigational systems in the Northern Baltic Sea. *Saf. Sci.* **2017**, *92*, 66–84.
4. Goerlandt, F.; Montewka, J. A framework for risk analysis of maritime transportation systems: A case study for oil spill from tankers in a ship–ship collision. *Saf. Sci.* **2015**, *76*, 42–66.
5. Helle, I.; Ahtiainen, H.; Luomaa, E.; Hänninen, M.; Kuikka, S. A probabilistic approach for a cost-benefit analysis of oil spill management under uncertainty: A Bayesian network model for the Gulf of Finland. *J. Environ. Manag.* **2015**, *158*, 122–132.
6. Valdez, B.O.A.; Goerlandt, F.; Kuzmin, V.; Kujalaa, P.; Montewka, J. Risk management model of winter navigation operations. *Mar. Pollut. Bull.* **2016**, *108*, 242–262.
7. Valdez, B.O.A.; Goerlandt, F.; Montewka, J.; Kujala, P. A risk analysis of winter navigation in Finnish sea areas. *Accid. Anal. Prev.* **2015**, *79*, 100–116.
8. Wang, K.; Leppäranta, M.; Gästgifvars, M.; Vainio, J.; Wang, C. The drift and spreading of the Runner 4 oil spill and the ice conditions in the Gulf of Finland, winter 2006. *Est. J. Earth Sci.* **2008**, *57*, 181–191.
9. Lehikoinen, A.; Luoma, E.; Mäntyniemi, S.; Kuikka, S. Optimizing the Recovery Efficiency of Finnish Oil Combating Vessels in the Gulf of Finland Using Bayesian Networks. *Environ. Sci. Technol.* **2013**, *47*, 1792–1799.

10. Lu, L.; Goerlandt, F.; Valdez, B.O.A.; Kujalaa, P.; Höglund, A.; Arneborg, L. A Bayesian Network risk model for assessing oil spill recovery effectiveness in the ice-covered Northern Baltic Sea. *Mar. Pollut. Bull.* **2019**, *139*, 440–458.

11. *Environmental Sensitivity Index Guidelines, Version 3*; NOAA Technical Memorandum Nos OR&R 11; Hazardous Materials Response Division, National Ocean Service: Seattle, WA, USA, 2002.

12. Montewka, J.; Weckström, M.; Kujala, P. A probabilistic model estimating oil spill clean-up costs—A case study for the Gulf of Finland. *Mar. Pollut. Bull.* **2013**, *76*, 61–71.

13. Pourvakhshouri, S.Z.; Mansor, S. Decision support system in oil spill cases (literature review). *Disaster Prev. Manag.* **2003**, *12*, 217–221.

14. Pourvakhshouri, S.Z.; Shattri, B.M.; Zelina, Z.I.; Noordin, A. Decision support system in oil spill management. In Proceedings of the ISPRS Technical Commission II Symposium, Vienna, Austria, 12–14 July 2006; pp. 93–96.

15. Keenan, P.B.; Jankowski, P. Spatial Decision Support Systems: Three decades on. *Decis. Support Syst.* **2019**, *116*, 64–76.

16. Li, P.; Cai, Q.; Lin, W.; Chen, B.; Zhang, B. Offshore oil spill response practices and emerging challenges. *Mar. Pollut. Bull.* **2016**, *110*, 6–27.

17. Ciappa, A.C. Reverse trajectory study of oil spill risk in Cyclades Islands of the Aegean Sea. *Reg. Stud. Mar. Sci.* **2021**, *41*, 101580.

18. Amir-Heidari, P.; Raie, M. Response planning for accidental oil spills in Persian Gulf: A decision support system (DSS) based on consequence modeling. *Mar. Pollut. Bull.* **2019**, *140*, 116–128.

19. De Dominicis, M.; Pinardi, N.; Zodiatis, G.; Lardner, R. MEDSLIK-II, a Lagrangian marine surface oil spill model for short-term forecasting—Part 1: Theory. *Geosci. Model Dev.* **2013**, *6*, 1851–1869.

20. Zodiatis, G.; Lardner, R.; Hayes, D.; Georgiou, G.; Pinardi, N.; De Dominicis, M.; Panayidou, X. The Mediterranean oil spill and trajectory prediction model in assisting the EU response agencies. In Proceedings of the Congreso Nacional de Salvamento en la Mar, Cadiz, Spain, 2–4 October 2008; pp. 535–547.

21. De Dominicis, M.; Pinardi, N.; Zodiatis, G.; Archetti, R. MEDSLIK-II, a Lagrangian marine surface oil spill model for short-term forecasting—Part 2: Numerical simulations and validations. *Geosci. Model Dev.* **2013**, *6*, 1871–1888.

22. Liubartseva, S.; Coppini, G.; Pinardi, N.; De Dominicis, M.; Lecci, R.; Turrisi, G.; Cretì, S.; Martinelli, S.; Agostini, P.; Marra, P.; et al. Decision support system for emergency management of oil spill accidents in the Mediterranean Sea. *Nat. Hazards Earth Syst. Sci.* **2016**, *16*, 2009–2020.

23. Kokkonen, T.; Ihaksi, T.; Jolmaa, A.; Kuikka, S. Dynamic mapping of nature values to support prioritization of coastal oil combating. *Environ. Model. Softw.* **2010**, *25*, 248–257.

24. Haapasaari, P.; Dahlbo, K.; Aps, R.; Brunila, O.-P.; Fransas, A.; Goerlandt, F.; Hänninen, M.; Jönsson, A.; Laurila-Pant, M.; Lehikoinen, A.; et al. *Minimizing Risks of Maritime Oil Transport by Holistic Safety Strategies* (*MIMIC*); Kotka Maritime Research Centre, Merikotka: Kotka, Finland, 2014; 64p.

25. Aps, R.; Tõnisson, H.; Suursaar, Ü.; Orviku, K. Regional Environmental Sensitivity Index (RESI) Classification of Estonian Shoreline (Baltic Sea). *J. Coast. Res.* **2016**, *75*, 972–976.

26. HELCOM. *Convention on the Protection of the Marine Environment of the Baltic Sea Area, 1992* (*Helsinki Convention*); as amended; HELCOM: Helsinki, Finland, 2013.

27. Halonen, J.; Lanki, A. Efficiency of maritime simulator training in oil spill response competence development. *TransNav Int. J. Mar. Navig. Saf. Sea Transp.* **2019**, *13*, 199–204.
28. IMO. *Mandatory Ship Reporting Systems*; International Maritime Organization: London, UK, 2003.
29. IMO. *Mandatory Ship Reporting Systems. SN. 1/Circ. 258*; International Maritime Organization: London, UK, 2006.
30. Andrienko, G.; Andrienko, N.; Jankowski, P.; Keim, D.; Kraak, M.J.; MacEachren, A.; Wrobel, S. Geovisual analytics for spatial decision support: Setting the research agenda. *Int. J. Geogr. Inf. Sci.* **2007**, *21*, 839–857.
31. Arneborg, L.; Höglund, A.; Axell, L.; Lensu, M.; Liungman, O.; Mattsson, J. Oil drift modelling in pack ice: Sensitivity to oil-in-ice parameters. *Ocean Eng.* **2017**, *144*, 340–350.
32. Raudsepp, U.; Laanemets, J.; Maljutenko, I.; Hongisto, M.; Jalkanen, J.-P. Impact of ship-borne nitrogen deposition on the Gulf of Finland ecosystem: An evaluation. *Oceanologia* **2013**, *55*, 837–857.
33. Isaev, A.V.; Eremina, T.R.; Ryabchenko, V.A.; Savchuk, O.P. Model estimates of the impact of bioirrigation activity of Marenzelleria spp. on the Gulf of Finland ecosystem in a changing climate. *J. Mar. Sys.* **2017**, *171*, 81–88.
34. Hedelin, B.; Evers, M.; Alkan-Olsson, J.; Jonsson, A. Participatory modelling for sustainable development: Key issues derived from five cases of natural resource and disaster risk management. *Environ. Sci. Policy* **2017**, *76*, 185–196.
35. Rodel, R.; Bregt, A.K.; Ligtenberg, A.; Pérez-Soba, M.; Verweij, P. The social side of spatial decision support systems: Investigating knowledge integration and learning. *Environ. Sci. Policy* **2017**, *76*, 177–184.
36. Luokkala, P.; Nikander, J.; Korpi, J.; Virrantaus, K.; Torkki, P. Developing a concept of a context-aware operational picture. *Saf. Sci.* **2017**, *93*, 277–295.
37. Ivshina, I.B.; Kuyukina, M.S.; Krivoruchko, A.V.; Elkin, A.A.; Makarov, S.O.; Cunningham, C.J.; Peshkur, T.A.; Ronald, M.; Atlasd, R.M.; Philp, J.C. Oil spill problems and sustainable response strategies through new technologies. *Environ. Sci. Process. Impacts* **2015**, *17*, 1201–1219.
38. Luokkala, P.; Virrantaus, K. Developing information systems to support situational awareness and interaction in time-pressuring crisis situations. *Saf. Sci.* **2014**, *63*, 191–203.
39. Chang, S.E.; Stone, J.; Demes, K.; Piscitelli, M. Consequences of oil spills: A review and framework for informing planning. *Ecol. Soc.* **2014**, *19*, 26.
40. IPIECA. *Choosing Spill Response Options to Minimize Damage: Net Environmental Benefit Analysis*; (IPIECA Report Series, 10); International Petroleum Industry Environmental Conservation Association: London, UK, 2015.
41. Jensen, J.R.; Halls, J.N.; Michel, J. A Systems Approach to Environmental Sensitivity Index (ESI) Mapping for Oil Spill Contingency Planning and Response. *PE RS* **1998**, *64*, 1003–1014.
42. Aps, R.; Tõnisson, H.; Anfuso, G.; Perales, J.A.; Orviku, K.; Suursaar, Ü. Incorporating dynamics factor to the Environmental Sensitivity Index (ESI) shoreline classification—Estonian and Spanish example. *J. Coast. Res.* **2014**, *70*, 235–240.
43. Aps, R.; Herkül, K.; Kotta, J.; Kotta, I.; Kopti, M.; Leiger, R.; Mander, Ü.; Suursaar, Ü. Bayesian inference for oil spill related Net Environmental Benefit Analysis. In *Coastal Processes*; Brebbia, C., Benassai, G., Rodriguez, G., Eds.; WIT Press: Southampton, UK; Boston, MA, USA, 2009; pp. 235–246.

44. Aps, R.; Fetissov, M.; Herkül, K.; Kotta, J.; Leiger, R.; Mander, Ü.; Suursaar, Ü. Bayesian inference for predicting potential oil spill related ecological risk. In *Safety and Security Engineering*; Guarascio, M., Brebbia, C.A., Garzia, F., Eds.; WIT Press: Southampton, UK, 2009; pp. 149–159.

45. Aps, R.; Fetissov, M.; Goerlandt, F.; Kopti, M.; Kujala, P. STAMP—Mar based safety management of maritime navigation in the Gulf of Finland (Baltic Sea). In Proceedings of the European Navigation Conference, Helsinki, Finland, 30 May–2 June 2016; pp. 1–8.

46. Ambjörn, C.; Liungman, O.; Mattsson, J.; Håkansson, B. Seatrack Web: The HELCOM Tool for Oil Spill Prediction and Identification of Illegal Polluters. In *Oil Pollution in the Baltic Sea. The Handbook of Environmental Chemistry*; Kostianoy, A., Lavrova, O., Eds.; Springer: Berlin/Heidelberg, Germany, 2011; Volume 27, pp. 155–184.

47. HELCOM. *Openrisk Guideline for Regional Risk Management to Improve European Preparedness and Response at Sea*; HELCOM: Helsinki, Finland, 2018.

48. Ambjorn, C. SeatrackWeb, forecasts of oil spills, a new version. In Proceedings of the 2006 IEEE US/EU Baltic International Symposium, Klaipeda, Lithuania, 23–26 May 2006.

49. HELCOM. *HELCOM STW Oil Drift Modelling*; HELCOM: Helsinki, Finland, 2020.

50. Nordström, J.; Goerlandt, F.; Sarsama, J.; Leppänen, P.; Nissilä, M.; Ruponen, P.; Lubcke, T.; Sonninen, S. Vessel TRIAGE: A method for assessing and communicating the safety status in maritime distress situations. *Saf. Sci.* **2016**, *85*, 117–129.

51. Fournier, M.; Hilliard, R.C.; Rezaee, S.; Pelot, R. Past, present, and future of the satellite-based automatic identification system: Areas of applications (2004–2016). *WMU J. Marit. Aff.* **2018**, *17*, 311–345.

52. Goerlandt, F.; Laine, V.; Bal Beşikçi, E.; Baldauf, M.; Al-Quhali, M.A.; Koldenhof, Y. End-user and stakeholder views on selected risk assessment tools for marine oil spill preparedness and response, including future research and development needs. *TransNav* **2019**, *13*, 213–220.

Chapter 11

An Integrated Approach to Assessing the Wave Potential for the Energy Supply of Ports: A Case Study

Victor Ramos[1,2], Gianmaria Giannini[1,2], Tomás Calheiros-Cabral[1,2], Paulo Rosa-Santos[1,2] and Francisco Taveira-Pinto [1,2]

[1] *Department of Civil Engineering (FEUP), Faculty of Engineering, University of Porto, Rua Dr. Roberto Frias, s/n, 4200-465 Porto, Portugal*

[2] *Interdisciplinary Centre of Marine and Environmental Research (CIIMAR), University of Porto, Avenida General Norton de Matos, s/n, 4450-208 Matosinhos, Portugal*

ABSTRACT

In recent years, seaports have faced increasing pressure to transition towards a low-carbon and more sustainable energy model. In this context, the exploitation of the local wave energy resource may appear as a promising alternative. Therefore, the objective of this work is to present a methodology to select the best WEC-site combination to supply the energy demands of ports. To illustrate this methodology, the Port of Leixões (Portugal) is used as a case study. For the selection of wave energy sites in port areas, the methodology proposes a detailed spatial characterisation of both the wave resource and marine uses. For the area of study, having considered the main marine uses (sediment disposal, biodiversity, aquaculture, recreational and navigation), two exploitable wave energy sites (Areas I and II) with average annual energy resources of 24 and 17 kWm−1−1, respectively, were found. Next, the methodology proposes a techno-economic optimisation of WECs, based on the local wave conditions of Areas I and II, to minimise their associated Levelised Cost of Energy (LCoE). The results obtained confirm the effectiveness of the methodology, with the novel oCECO device, appearing as the most feasible option (with an LCoE of EUR 387.6/MWh) to exploit the wave potential in the surrounding areas of the port.

Keywords: *wave energy; marine spatial planning; WEC performance assessment; levelised cost of energy; techno-economic optimisation; Port of Leixões*

1. INTRODUCTION

Over the last decade, growing concerns regarding environmental pollution, climate change, and long-term sustainability of the fossil fuel-based energy model have prompted different climate and energy policies (e.g., *Paris Agreement* [1], *2030 Agenda for Sustainable Development* [2] and *EU's Green Deal* [3]), whose main goals are to foment a fair transition towards a sustainable and carbon-neutral economy.

As a consequence of these policies, seaports are subject to increasing pressure to reduce their environmental footprint and transition towards a low-carbon energy model while increasing their competitiveness [4]. Seaports are essential for global trade and economic growth since more than 80% of the world's goods are transported by cargo vessels [5]. In addition to cargo handling, seaports may also house a large number of activities among which, passenger transportation, fishing, aquaculture and shipbuilding, stand out. As a result, seaports present high-energy demands, which as of to date are mainly fulfilled by means of fossil-fuel energy sources [6]. All these factors have turned seaports into significant sources of pollution, mainly in terms of noise, air/water contaminants [7] and greenhouse gas (GHG) emissions [8,9].

Therefore, port authorities are developing ambitious plans to reduce carbon emissions and environmental pollution (e.g., *World Ports Sustainability Program* [10]). In this sense, relevant ports such as Helsinki (Finland) [11], Busan (Korea) [12] and Houston (USA) [13] have already presented their action plans to achieve carbon neutrality in the 2035–2050 horizon. For this purpose, efficient energy use and electrification of port equipment and activities, including onshore power supply of vessels, become essential [6,14]. However, port electrification together with the expected growth of maritime transport will result in a significant increase in the energy requirements of ports [6,15], which must be fulfilled by renewable and/or low-carbon energy sources to achieve the above-mentioned sustainability and emission targets.

On these grounds, considering the convergence of marine resources, infrastructures and facilities available at seaports, Marine Renewable Energy (MRE) appears as a promising energy solution [16]. Among the different types of MRE sources, wave energy, which presents a large and globally diverse resource of roughly 2 TW [17], has attracted the greatest deal of interest from the port managing authorities [18]. Despite its large energy potential, wave energy has yet to reach commercial viability [19]. This fact is closely related to the low conversion efficiencies of most of Wave Energy Converter (WEC) technologies [20]. In addition, WECs commonly operate under very harsh marine conditions, jeopardising their structural integrity and increasing the difficulty in terms of logistics, deployment, operation and maintenance tasks [21]. As a result, wave energy presents values of Levelised Costs of Energy (LCoE) that can be up to ten times higher than other traditional renewable sources such as onshore wind and solar PV [22].

Nonetheless, for the specific case of seaports, the integration of certain types of WECs (e.g., Oscillating Water Columns (OWCs), Overtopping Devices (OTDs) and some Wave Activated Devices (WADs) [23]) within seaport structures (e.g., breakwaters, piers and jetties) could help those WEC concepts to increase their cost-competitiveness and, consequently, progress more rapidly towards commercialisation [18]. This approach presents several advantages. First of all, capital and construction expenditure could be shared between the breakwater and WEC [24]. Secondly, the breakwater could serve as a foundation and shelter for the WEC, improving as a result its reliability and operating life span [18]. Thirdly, the complexity and costs in terms of accessibility and maintenance are significantly lower than in the case of offshore wave farms [24]. Similarly, the difficulty and costs of the WEC-grid connection are highly reduced [25]. However, WEC-breakwater integration is not without downsides. On the one hand, not all breakwaters are suitable for WEC integration, which depends mainly on their location, orientation and typology [26]. On the other hand, the nearshore wave transformation processes could reduce significantly the available wave energy resource, in comparison with offshore locations [27]. Furthermore, the functionality of breakwaters, sheltering port areas from the action of the waves, cannot be compromised by the presence of WECs. Full-scale examples of WEC-breakwater integration are the OWCs of ports of Mutriku (Spain) [28] and Civitavecchia (Italy) [29], the overtopping device OBREC in the port of Naples (Italy) [30] and the Eco Wave Power hinged arm point absorber in Gibraltar [31]. In addition, extensive research has dealt with WEC-breakwater integration in different ports worldwide, addressing in detail relevant issues such as hydrodynamic and hydraulic performance enhancement of OWCs [32,33] and OTDs [34,35], respectively; PTO optimisation [36,37,38], wave loading distribution and structural integrity [39,40,41], economic and life-cycle environmental impact assessment [42,43] and hybridisation of OWC and OTD concepts [26,44].

Alternatively, WEC concepts such as attenuators, point absorbers and oscillating wave surge devices, which are not compatible for breakwater integration, could also increase their cost-competitiveness if they were to be deployed in the surrounding marine areas of seaports, to fulfil part of their energy demands. This approach also presents important advantages in comparison with traditional offshore wave energy farms. Among them, the vicinity between the generation and consumption points stands out [6]. Therefore, lower energy losses and costs of installation, operation and maintenance are expected, due to the reduced length of electric export cables and the simpler requirements in terms of mooring systems and farm accessibility [45]. Additionally, the presence of nearshore wave farms in the vicinity of seaports could help to mitigate damaging effects on coastal structures [46] and hazards such as erosion [47] and flooding [48]. Another possible positive outcome, which lies beyond the scope of this work and still requires extensive research, relates to the potential impacts of wave farm operation on harbour oscillations (induced by infragravity waves) that can create significant problems to ships moored inside the harbour [49,50,51] and cargo operations [52,53]. On the other hand, the main disadvantage of this approach are the potential disruptions with the large amount of maritime uses (navigation, fishing, aquaculture and recreational activities) that are concentrated in the vicinity of seaports, which could limit the sea-space availability for the installation of WECs [54]. Furthermore, nearshore processes such as wave breaking and seabed dynamics may jeopardise WEC's operating lifespan [55]. Finally, it is worth mentioning that this approach could be also suitable for naturally sheltered ports, whose defence structures are not subject to intense wave action.

In this context, it becomes apparent that selecting the most appropriate WEC technology (e.g., OWC vs. OTD) and location of installation (breakwater-integrated vs. nearshore) to supply the energy demands of a particular port, depends on multiple factors, which include: (i) main port characteristics, considering its location, typology and orientation of coastal defence structures, energy consumption patterns and maritime activities in its jurisdiction and surrounding areas; (ii) available wave energy resource in the port and its vicinity areas [56]; (iii) WEC typology and performance in terms of energy production, capacity factor and capture width ratio [57]; and finally, (iv) WEC associated energy costs [42]. Previous research dealing with WEC integration in ports has mainly focused on selecting and optimising WEC technologies according to specific breakwater characteristics [24,58] and assessing the resulting annual energy production [59,60]. However, an integrated decision-making approach, which takes into account all the aforementioned factors, must be considered when planning the optimum WEC-site combination within seaports areas.

Against the foregoing backdrop, the objective of this paper is to propose an integrated approach to assess the feasibility of exploiting the wave potential to fulfil the energy demands of ports. To this end, a methodology that takes into account the key characteristics of ports (i.e., location, breakwater typology and available wave resource) and WECs (site–specific performance and techno-economic optimisation) is developed to find the optimum WEC–site combination within seaport areas. To illustrate this methodology, the Port of Leixões is used as a case study (Section 3). Located on the Northern coast of Portugal facing the North Atlantic (Figure 1), the Port of Leixões is subject to one of the most energetic wave regimes in continental Europe [56] and, therefore, appears as a compelling case study to exploit the wave energy resource to fulfil its energy needs.

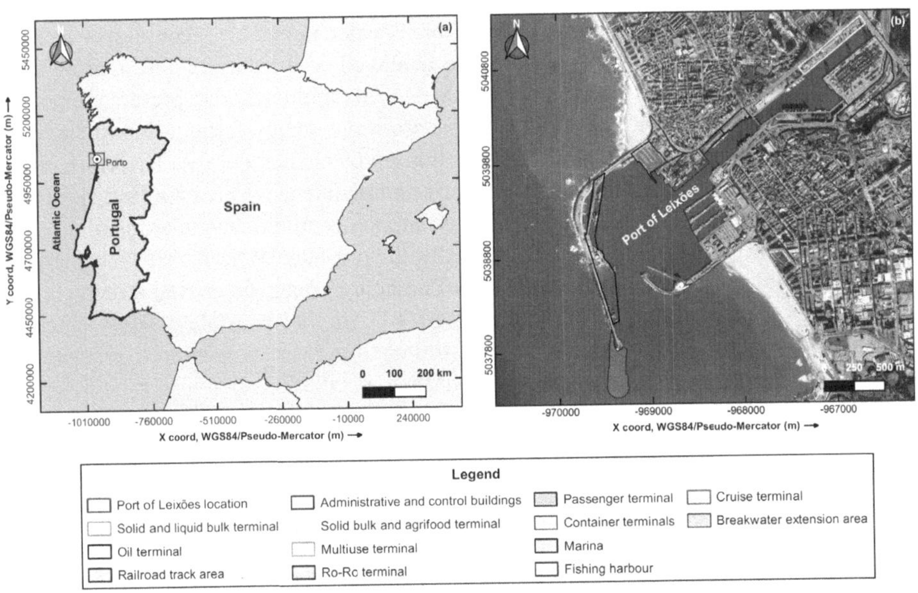

Figure 1. *Location (**a**) and main terminals (**b**) of Port of Leixões.*

The remainder of this paper is structured as follows. Section 2 presents the methodology developed to find the best WEC-site combination for seaport areas. Section 3 and Section 4 describe in detail the main characteristics of the Port of Leixões and the case-study WECs considered for this work. Section 5 and Section 6 show the results obtained for the present case study and discuss the main aspects of the proposed methodology, respectively. Finally, conclusions are drawn in Section 7.

2. AN INTEGRATED APPROACH FOR OPTIMUM WEC-SITE SELECTION IN PORT AREAS

In this scction, an integrated methodology is proposed to determine the optimum WEC-site combination for port energy supply. To this end, the proposed methodology is composed of four different stages, namely: (i) detailed analysis of main port characteristics (Section 2.1); (ii) selection of wave energy exploitation sites and WEC technologies (Section 2.2); (iii) site-specific WEC performance (Section 2.3); (iv) and techno-economic WEC optimisation (Section 2.4). Figure 2 presents a holistic view of the methodology, highlighting the main characteristics of the proposed stages and the linkages between them.

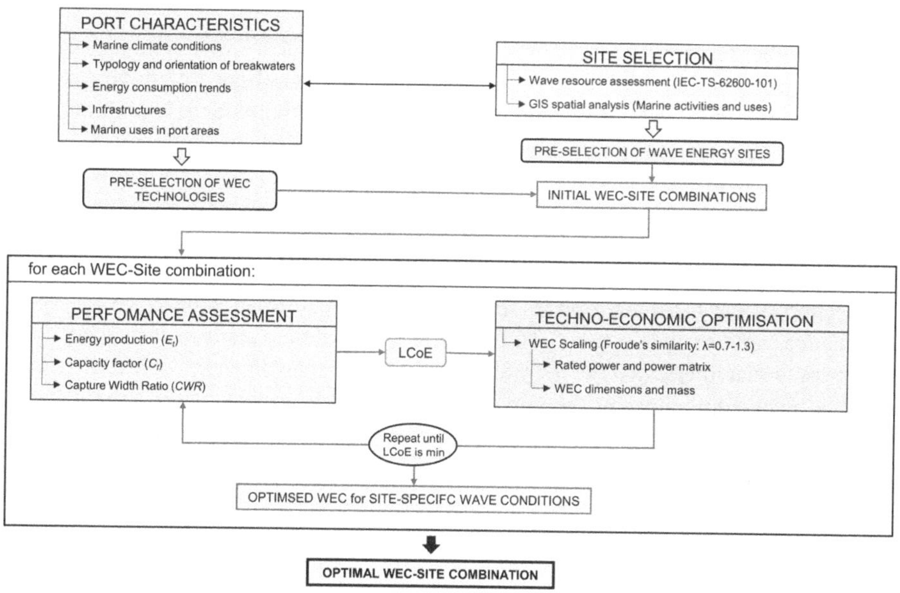

Figure 2. *General overview of the integrated methodology to select the optimal WEC-site combination for the energy supply of seaports.*

2.1. Analysis of Port Characteristics

As expected, the main port characteristics are going to influence which type of WEC integration (breakwater-integrated vs. nearshore farms) is potentially feasible. Among them, local wave conditions, breakwater orientation and typology, energy consumption trends, availability of electrical infrastructures, and supply chains should be considered.

On these grounds, the level of exposure to wave action and the orientation and typology of breakwaters are of crucial importance [23].

On the one hand, for ports subject to intense wave action (e.g., outer ports) and whose main breakwaters are aligned with the predominant wave direction, the use of breakwater-integrated WECs appears as the most viable and preferred option for many authors [18,23]. In this case, the suitable WEC typology is mainly limited to some WAD, OWC, OTD and hybrid OWC-OTD concepts [18]. Concerning breakwater typology, ports adopt two main approaches (vertical vs. rubble mound), which diverge on their wave energy dissipation principles. However, no criteria or recommendations for the integration of specific WEC technologies and breakwater typologies are yet defined in the literature. As a result, examples of integration of OWCs and OTDs can be found for both rubble mound (Mutriku's OWC [28] and OBREC [30]) and vertical (REWEC3 [29] and SSG [61]) breakwaters. Therefore, as long as the hydraulic performance and stability of the breakwater are not compromised, the selection of the WEC typology to be integrated into a breakwater will be governed by its site-specific performance, installation and construction limitations, and associated costs.

On the other hand, for ports naturally sheltered from intense wave climates or whose breakwater sections, susceptible to WEC integration, are not aligned with the main wave direction, the use of offshore wave energy farms (from approximately 10 to 40 m of water depth [18]) stands out as the only viable option to harvest the local wave energy resource. In this case, the limitations in terms of WEC concept selection are mainly governed by water depth availability. Nonetheless, the variety of WEC concepts available for this approach is significantly larger (e.g., point absorbers, attenuators, oscillating surge devices and floating OWCs and OTDs) in comparison with the WEC-breakwater integration approach. Again, the selection of the most appropriate technology will require an exhaustive site-specific performance assessment (Section 2.3). Furthermore, a detailed analysis is needed, considering the potential interactions of the wave farm with the maritime uses and activities present in the surrounding areas of the port. On these grounds, special attention must be paid to marine navigation routes (ensuring port accessibility), fishing, aquaculture, recreational activities and, biodiversity and environmentally protected areas.

In addition, marine climate conditions in the port area, such as tidal ranges and the occurrence of extreme sea events are also noteworthy. In this context, previous research has found that the tidal range could significantly impact the performance of breakwater-integrated WECs. For the case of OTDs, it becomes apparent that large tidal ranges may have a resounding impact on the volume of water that reaches the OTD's reservoirs, reducing as a result its overall hydraulic efficiency [62]. For OWCs, even though the resonant period shifts with the water level variations, the efficiency is more stable and less affected than the OTDs [60]. Conversely, water level variations are expected to have considerably fewer effects on the performance of nearshore WEC deployments (i.e., between 10 and 40 m of water depth). On the other hand, a high frequency of extreme sea events could jeopardise the operating life of both nearshore and breakwater-integrated WECs. For example, the OWC of Port of Mutriku suffered important structural damages in the walls and roof of four chambers during intense storm conditions [63]. Consequently, a detailed characterisation of the magnitude and frequency of occurrence of extreme sea events must be considered in the decision making for the selection of the most appropriate WEC-site combination within the port area.

2.2. Site Selection for WEC Installation

In order to select the optimum WEC installation location, a detailed spatial and temporal characterisation of the wave energy resource in the port area is required [57]. For this purpose, the present methodology proposes the use of high-resolution spectral wave modelling combined with the recommendations of the technical standard IEC-62600-101 [64], which according to previous research offers an accurate estimation of the wave energy resource [65,66]. The IEC-62600-101 standard divides wave resource assessment in three distinct classes, namely: reconnaissance (Class 1), feasibility (Class 2) and design (Class 3). For the specific case of seaports, Class 3 assessments are recommended, since they are focused on relatively small coastal areas (<25 km of shoreline), offering a detailed characterisation of the spatial distribution of the wave energy resource, with a low degree of uncertainty. For further details on the IEC-62600-101 standard and the implementation procedure of Class 3 wave resource assessments, the readers are referred to Ref. [64]. However, the estimation of the spatial distribution of the wave energy resource does not suffice to determine the optimum deployment location of WECs. In addition, spatial analysis, using Geographical Information System (GIS) techniques, is required for the identification of port areas housing maritime activities and uses that are not compatible with WEC operation. It is worth noting that GIS spatial analysis has been successfully applied for coastal site selection of wave [54] and other marine renewable deployments [67,68].

2.3. Site-Specific WEC Performance Assessment

With the aim of assessing WEC performance at specific port areas, the parametric approach proposed by Carballo et al. [57] is adopted. This methodology computes different key performance indicators of WECs, including energy production, capacity factor and capture width ratio.

Therefore, the energy output produced by a WEC for a specific location, Et, can be calculated according to:

$$E_t = \sum_{i=1}^{n} P_i O_i, \tag{1}$$

where n is the total number of sea states considered, Pi is the power generated by the WEC for the i-th sea state (Figure 5), and Oi is the occurrence (number of hours) for a certain reference period (year or month) of the i-th sea state.

The capacity factor, Cf, which represents the percentage of time over a reference period, T (year or month), that the WEC operates as its rated power, PR, can be computed as:

$$C_f = \frac{E_t}{TP_R}, \tag{2}$$

Finally, the capture width ratio, CWR, measures the hydrodynamic efficiency of the WEC, representing the fraction of available wave energy captured by the WEC [69], and can be computed as:

$$CWR = \frac{P_{out}}{JB}, \tag{3}$$

where Pout is the power output of the device, J is the available wave power per meter of wave front and B is the characteristic capture dimension of the device.

2.4. Techno-Economic WEC Optimisation

The final step of the proposed methodology (Figure 2), consists of optimising the main WEC characteristics (dimensions, mass and power absorption) according to the local wave conditions, with the aim of reducing the associated Levelised Cost of Energy (LCoE) for all the considered WEC-site combinations. This techno-economic optimisation is necessary since most WEC concepts are developed for specific applications and wave climates and, consequently, present operating windows that may not be suited for the wave conditions of seaport areas. Therefore, the present methodology proposes to scale the considered WEC technologies using the well-known Froude similarity criterion, for which wave heights scale linearly with the geometric scale λ, wave periods with λ12, mass with λ3 and absorbed power with λ72[70]. Therefore, the devices are upscaled or downscaled for values of λ lower or higher than 1, respectively. Furthermore, to ensure that the original and scaled WECs present an equivalent hydrodynamic behaviour, the variation of the geometric scale (λ) is restricted to a maximum value of 30% [71]. In consequence, a constrained optimisation problem, which aims to find the value of the geometric scale (λ) that minimises the LCoE for each WEC-site combination is solved.

However, owing to the lack of techno-economic information shared by WEC developers, an accurate estimation of the LCoE of wave energy deployments is not straightforward [45]. As a result, several authors have proposed different models to estimate the costs during the life cycle of wave farms [72,73,74]. For the present work, an in-depth literature review has been carried out to adopt the cost estimations, which are most suited for wave energy deployments within port areas. In consequence, the LCoE model proposed in this work stands as a first approximation and should be revisited in the future, when new information coming from the wave energy industry is made available.

Therefore, the proposed model computes the LCoE using a life-cycle cost approach [72]:

$$LCoE = \frac{\sum_{t=0}^{T_l} \frac{CAPEX_t + OPEX_t + ABEX_t}{(1+r)^t}}{\sum_{t=0}^{T} \frac{E_t}{(1+r)^t}}, \tag{4}$$

where Tl is the lifespan of the project in years, the CAPEXt, OPEXt and ABEXt stand for the capital, operational and abandonment expenditures for the t-th year of the project, respectively; Et is the energy output for the t-th year of the project (Equation (1)), and r is the discount rate, which is obtained as:

$$r = \frac{r_l + r_i}{1 - r_i}, \tag{5}$$

where rl and ri are, respectively, the loan and inflation rates. Based on previous studies, a 10% loan rate [72] and a 2% inflation rate [71] were considered.

Regarding costs, the CAPEX accounts for all the expenditure required prior to the operating stage of the project, and can be estimated as:

$$CAPEX = C_{WECs} + C_{base} + C_{allow} + C_{cont}, \tag{6}$$

where CWECs represents the cost of design, engineering and manufacturing of WECs and can be computed according to Equation (7); Cbase are the base costs of the project, which are estimated as indicated in Equation (9); and finally, Callow and Ccontg stand for the allowance and contingency costs, with each of them representing up to 10% of the base cost (Cbase) according to the work of Oliveira-Pinto et al. [71].

The cost of the WECs (CWECs) can be estimated as:

$$C_{WECs} = C_{PTO} + C_{desig} + C_{struct}, \tag{7}$$

where CPTO account for the expenses related to the Power Take-Off (PTO) systems, which according to Oliveira-Pinto et al. [71] represent 25% of the base costs (Cbase); Cdesig stands for the engineering and design costs of the WECs, for which Pecher and Kofoed [73] proposed a unitary cost of EUR 68,945 per MW; and finally, Cstruct is the structural cost of WECs, whose unitary value can be computed as:

$$C_{struct} = fC_m m_{WEC}, \tag{8}$$

where *f* is the complexity factor in WEC manufacturing, which was introduced by Oliveira-Pinto et al. [71] and ranges between 1 and 1.20 depending on the structural design of the WEC (Table 2). Cm is the unitary cost of the structural material. For the case of reinforced concrete, the price per ton ranges between EUR 100 and 150 [75,76], while for the case of steel, several authors have indicated that the price could range between EUR 1000 and 1400 per ton [71,75]. Finally, mWEC represents the structural mass of the WEC.

The base costs (Cbase) can be computed as:

$$C_{base} = C_{signal} + C_{grid} + C_{m\&f} + C_{inst} + C_{pman}, \tag{9}$$

where Csignal represents the wave farm signalling costs, which according to Pecher and Kofoed [73], present a unitary cost of EUR 15,000 per MW; Cgrid are the grid connection costs, which can be computed as indicated in Equation (10); Cm&f stands for the mooring and foundation costs, whose unitary costs are summarised in Table 1; Cinst are the installation costs, which are obtained by means of Equation (11); and finally, Cpman represents the project management costs, which can be estimated as 5% of the combined costs of WECs (CWECs) and installation (Cinst).

Table 1. Summary of unitary mooring and foundation costs.

	Parameter	Cost	Reference
Moorings	Catenary Steel Wire	EUR 22.5/m	[71,77]
	Vertical Steel Wire	EUR 67.0/m	[71,77]
	Tether	EUR 440.0/m	[71,77]
Foundations	Suction pile	EUR 51,227/unit	[71,77]
	Concrete Mounting Plate	EUR 75,000/unit	[71,77]
	Embedded anchor	EUR 165,000/unit	[71,77]

The grid connection costs (Cgrid) can be computed as:

$$C_{grid} = C_{array\text{-}cab} + C_{export\text{-}cab} + C_{land\text{-}cab} + C_{subst}, \quad (10)$$

where Carray–cab and Cexport–cab are the costs of the inter-array and export submarine electrical cables, which according to Pecher and Kofoed [73] present a unitary cost of EUR 62,860 and 184,222 per MW, respectively; Cland–cab represents the costs of terrestrial electrical cables, which based on the work of Astariz and Iglesias [45] are in the order of EUR 70 per meter; and finally, Csubst is the cost of the electrical substation, which according to Oliveira-Pinto et al. [71] is in the order of EUR 1.1 M.

The installation costs (Cinst) can be estimated as:

$$C_{inst} = C_{WEC\text{-}inst} + C_{subst\text{-}inst} + C_{array\text{-}cab\text{-}inst} + C_{export\text{-}cab\text{-}inst} + C_{m\&f\text{-}inst}, \quad (11)$$

where, CWEC–inst and Csubst–inst are respectively the WEC and substation installation costs, whose unitary costs, according to Pecher and Kofoed [73], can be estimated as EUR 101,339 and 31,389 per installed MW, respectively; Carray–cab–instand Cexport–cab–inst stand for the installation costs of inter-array and export cables, which for offshore deployments can be assumed as EUR 281,000 and 443,000 per km [77], respectively; and, Cm&f–inst represents the mooring and foundation installation costs, which according to Oliveira-Pinto et al. [71] can be estimated in EUR 78,125 per day of work.

On the other hand, the OPEX accounts for all the operating and maintenance costs during the lifespan of the project. However, the estimation of the OPEX for wave energy projects appears as a challenging task due to the lack of long-term data from wave farms in operation [71]. In consequence, the OPEX is usually obtained as a percentage of the CAPEX. For example, Astariz and Iglesias [45] estimated that the OPEX ranges between 1.5% and 5.0% of the CAPEX. In this work, a conservative approach was adopted and, therefore, the OPEX was considered 5% of the CAPEX. Similarly, the ABEX is usually estimated as a percentage of the CAPEX, which according to Astariz and Iglesias [45], ranges between 0.5% and 1%. For the present work, the ABEX was set at 1% of the CAPEX.

The effect of economies of scale was also considered in the model, since the unitary costs of fabrication and installation of WECs would decrease with the number of units. For this purpose, the methodology proposed by O'Connor et al. [78] to estimate the cumulative cost of fabrication and installation of multiple WECs was used:

$$C_t = \sum_{i=1}^{n} N^{\left(\frac{\ln \chi}{\ln 2}\right)} IC_n, \quad (12)$$

where N is the total number of WECs, ICn is the unitary cost, Ct is the cumulative cost and χ is the bulk discount factor, which usually varies from 0.85 and 0.95. Based on the work of Oliveira-Pinto et al. [71], χ was set to 0.90.

Finally, it is worth mentioning that the effects of external factors such as the fluctuations in the price of raw materials and inflation rates are beyond the scope of this work and were not considered for the cost model proposed.

3. PORT OF LEIXÕES

The Port of Leixões is located in the district of Porto (Northern Portugal) with direct access to the Atlantic Ocean (Figure 1a). This privileged location makes the Port of Leixões the second largest port in Portugal in traffic of goods, supplying commodities to a hinterland of more than 14 million inhabitants [79]. As a result of its economic and strategic importance, the port authority is developing ambitious plans to ensure its long-term competitiveness and sustainability. Consequently, with the purpose of providing better berthing and mooring conditions as well as facilitating access to the port, a 300 m extension (with a 20° rotation to the West) of the outer breakwater is planned (Figure 1b) [60]. As a result, the proposed extension could offer the possibility of integrating different WEC technologies into the breakwater and, therefore, harnessing the local wave energy resource [60]. On the other hand, the managing authority of the port has recently announced the intention of achieving carbon neutrality and energy self-sufficiency by 2035 [80]. In this context, wave energy is expected to play a significant role in achieving the aforementioned goals as shown by the concession agreement granted to *Eco Wave Power* for the installation and exploitation of a 1 MW wave power plant integrated within the jurisdiction area of the port [81].

3.1. Energy Consumption Patterns

Based on the latest available information (years 2017 and 2018), the average annual energy consumption of the Port of Leixões is approximately 24.6 GWh per year, comprising three main energy sources, natural gas (0.52%), diesel (37%) and electricity (62.48%) [82]. In addition, the intra-annual energy consumption is presented in Figure 3. Overall, it can be observed that the energy consumption varies from 1.8 (June) to 2.3 (December) GWh per month, with the second part of the year presenting higher energy demands (above 2 GWh per month). Regarding each individual energy source, it can be observed that diesel consumption spikes in the last part of the year (from September to December), while the electricity demands are higher at the beginning of the year (from January to March). Lastly, it is worth mentioning that the energy consumption patterns reported in this section only include the facilities (administrative buildings, port and terminal lighting) and activities (tow boating and container and bulk cargo handling) that are managed by the port authority.

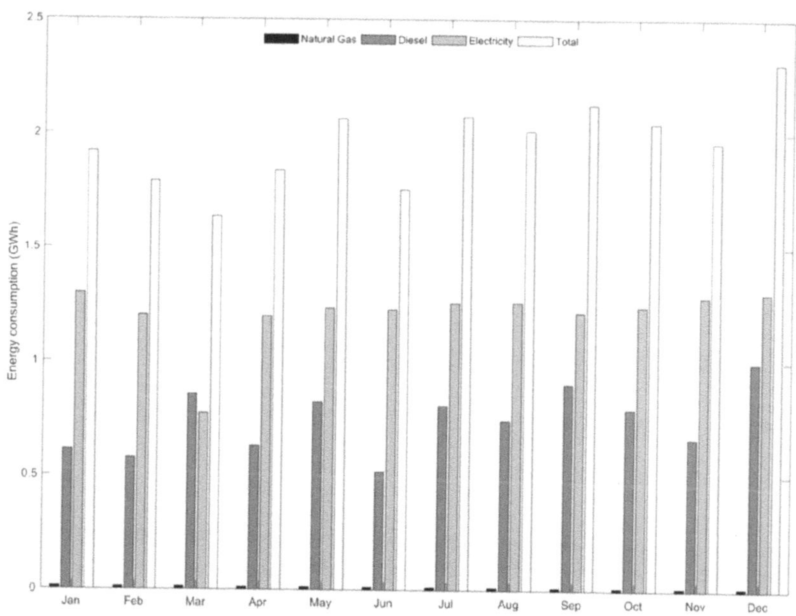

Figure 3. *Energy consumption patterns of Port of Leixões.*

3.2. Main Maritime Uses and Activities

In Portugal, marine spatial planning follows an adaptative approach, favouring a seamless coexistence of multiple maritime uses in the same sea-space while avoiding any potential disruptions to the marine environment [83]. As a result, a planning instrument, the so-called *Situation Plan*, was created to identify the temporal and spatial distribution of present and potential maritime uses [84]. For the case of the Port of Leixões and its surroundings, the areas contemplated by the *Situation Plan* are highlighted in Figure 4. Overall, it can be observed that the vicinity of the port presents priority areas for different marine activities, among which artificial feeding of beaches, disposal of dredged material, aquaculture, marine navigation and surf stand out (Figure 4a). Regarding the presence of biodiversity, conservation and cultural heritage regions, four coastal protected areas, which span a total surface of approximately 200 ha, are found within the jurisdiction area of the port (Figure 4b), while the southern approach to the port is marked as a Site of Community Importance (Figure 4a). Finally, it is worth mentioning that potential areas for future uses such as multi-use platforms and reef, sport and leisure activities are also considered.

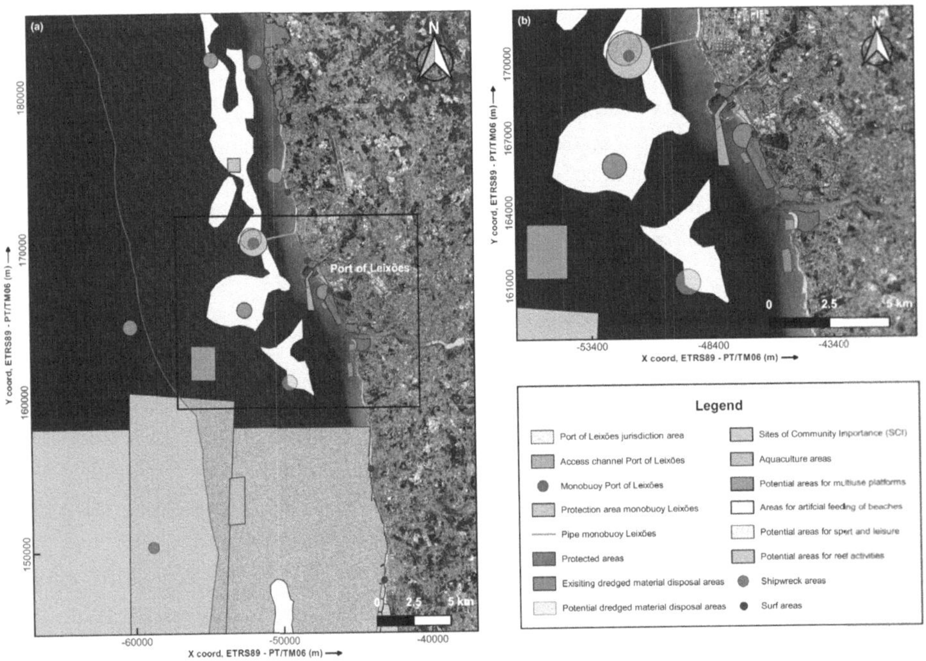

Figure 4. *Marine uses in the surrounding areas of Port of Leixões [84].*

4. CASE-STUDY WECS

As mentioned in Section 2.1, the exploitation of the wave potential to supply the energy needs of ports can be achieved using either breakwater-integrated or offshore WECs.

Owing to its privileged location and layout (Figure 1), both possibilities are suitable for the Port of Leixões. However, for the present work, the use of breakwater-integrated WECs was discarded, owing to the lack of information to accurately estimate the different costs (manufacturing, installation and decommissioning) and LCoE of these types of devices. Therefore, only offshore WEC concepts that have reached a relevant level of technology readiness (TRL ≥ 4) were considered [71], namely, F-OWC, F-2HB, Bref-SHB, B-HBA, B-OF and oCECO. The main characteristics of those devices are summarised in Table 2, while their corresponding power matrices are shown in Figure 5.

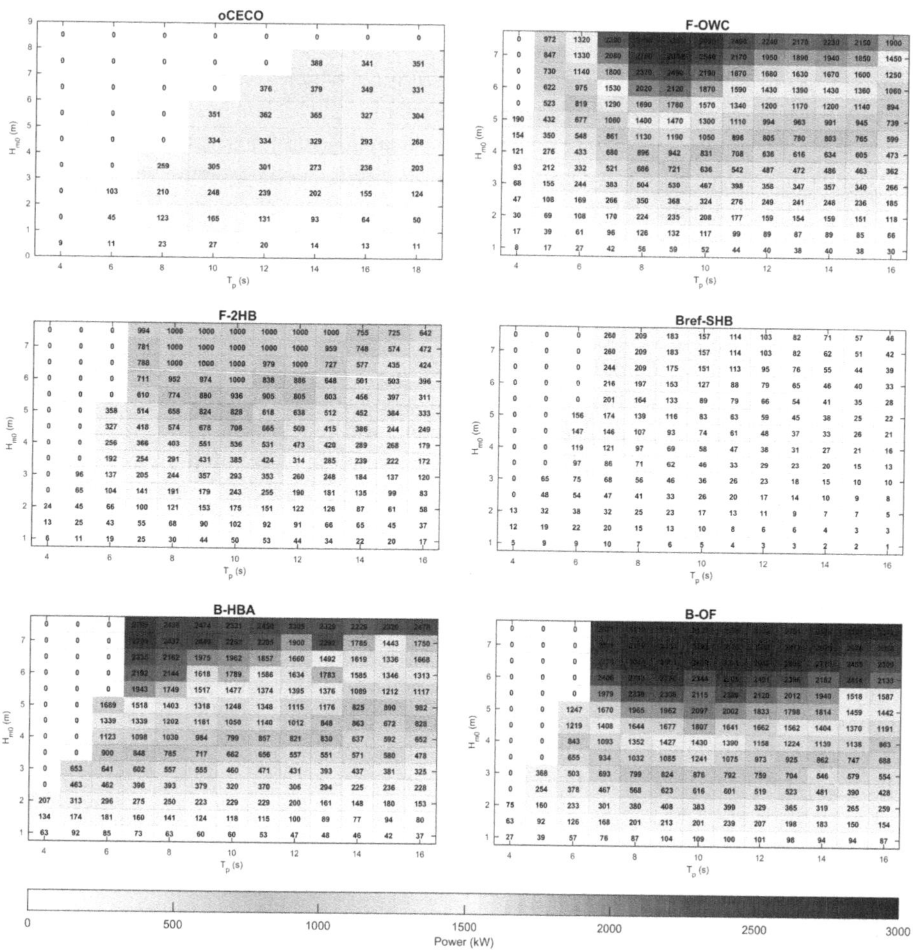

Figure 5. *Wave power matrices of the case-study WECs (Numbers represent the mean power production for each sea-state bin. Colourmap represents the power produced for the different sea states).*

Table 2. Main characteristics of case-study WECs.

	F-OWC	F-2HB	Bref-SHB	B-OF	B-HBA	oCECO
Rated Power (kW)	2880	1000	260	3330	2710	500
Capture width, B (m)	24	20	7	26	100	12
Mass (ton)	1565	718	35	3800	1600	358
Foundation	Suction pile	Suction pile	Suction pile	Steel frame	Jack-up structure	Suction pile
Mooring	Catenary	Catenary	Tether	-	-	Tether
Mooring lines	3	3	1	-	-	3
Manufacturing complexity factor, f	1.05	1.05	1.10	1.20	1.20	1.10

F-OWC, as its name indicates, is a floating OWC [85], which is inspired by the OE-Buoy device [86]. It consists of a partially submerged structure with an opening aligned with the incoming wave direction connected to a single air chamber. As the water column oscillates, the air is forced to flow through an air turbine, which drives an electric generator.

F-2HB is a floating two-body heaving converter [85], which is based on the WaveBob device [87]. Its operating principle consists of a heaving point absorber with a hydraulic PTO that is driven by the relative motion of a torus sliding along a vertical float.

Bref-SHB is a bottom-referenced submerged heave-buoy [85], which takes inspiration from the CETO device [88]. It is formed by a submerged axi-symmetric buoy, which is forced to heave as a result of the pressure difference caused by wave motions. The heaving motions are captured by a hydraulic PTO, which drives an electric generator located inside the buoy.

B-OF is a bottom oscillating flap device [85], which is derived from the Oyster device [89]. It consists of a pitching flap fixed to the seabed, whose oscillations are converted into hydraulic energy, which is then used to drive an electric generator.

B-HBA is a bottom-heaving buoy array [85], which is based on the Wavestar device [90]. B-HBA is formed by several semi-spherical floaters connected to a jack-up structure by means of hinged arms. The motions of the floaters are captured by a hydraulic PTO, which drives an electric generator housed in the jack-up structure.

Finally, oCECO is a novel WEC composed of two floating modules joined by a frame, which present their motions restrained to an inclined direction to capture the vertical and horizontal force components of the waves [91]. The longitudinal motions of the floating modules are transformed by a rack pinion system to drive a rotatory electric generator, which is housed in the floating supporting structure. It is important to point out that oCECO was developed and optimised for the wave conditions of the area of study (Northern coast of Portugal) [92,93]. Consequently, for the present work, oCECO was also used as a benchmark to verify the validity of the techno-economic optimisation model proposed in Section 2.4.

5. RESULTS

5.1. Wave Energy Resource Characterisation and Site Selection

As indicated in Section 2.2, for the characterisation of the wave energy resource in port areas, the use of high-resolution numerical modelling is recommended. Consequently, in the present study, a wave model spanning the jurisdiction and surrounding areas of Port of Leixões was implemented following the recommendations of the IEC-62600-101 (Class

3 assessment). For this purpose, the open–source spectral wave model SWAN (Simulating WAves Nearshore) [94] was used. Further details on the physics and the implementation procedure of the wave model can be found in Appendix A.

Upon implementation, the wave model was run for a period of 31 years (from 1990 to 2020) to capture the seasonal variations of the wave resource. Then, the results obtained were processed to determine the spatial distribution of the wave energy resource in the port area and its surroundings. Figure 6 shows the results obtained for the annual scenario. Overall, for the more offshore regions (i.e., around 40 m of water depth) the wave resource is homogeneously distributed, with average values ranging between 23 and 25 kWm−1−1. In the nearshore regions, due to their associated wave dissipation processes, the available resource progressively reduces. As a result, the available resource in the surrounding areas of the port is in the order of 15 to 18 kWm−1−1.

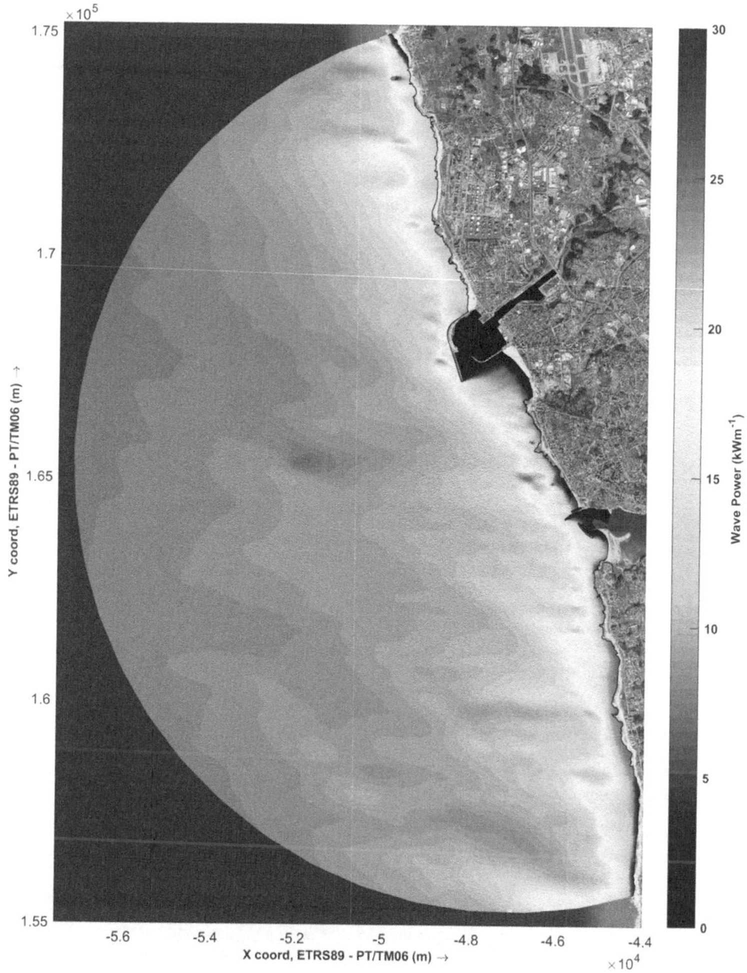

Figure 6. *Spatial distribution of the wave annual energy resource.*

Next, following the methodology presented in Section 2.2, the site selection for wave energy exploitation was performed by combining the spatial distribution of both the wave energy resource (Figure 6) and marine uses in the port area (Figure 4). As indicated in Section 3.2, Portugal's marine spatial planning follows an adaptative approach (i.e., several marine uses can coexist in the same sea space as long as the priority use of the area is not compromised); however, for the present work, a conservative approach was used and, therefore, all marine priority areas were discarded for wave energy exploitation to avoid potential disruptions (Figure 7). As a result, Areas I and II, highlighted in Figure 7, were considered potential candidates for the exploitation of the wave resource. Area I, located 5.5 km offshore from the port at a water depth of approximately 40 m, exemplifies the use of offshore floating wave farms for the energy supply of ports. Conversely, Area II, located in the vicinity of the port, illustrates the exploitation of the wave resource by means of nearshore wave farms.

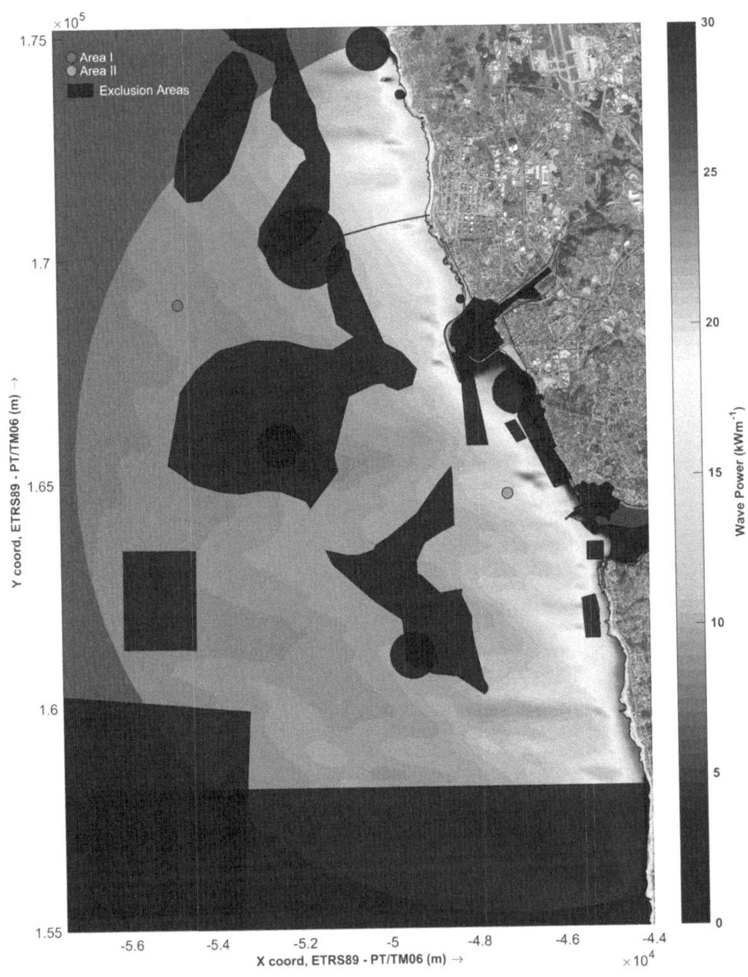

Figure 7. *Example of suitable areas for the exploitation of the wave energy resource in the vicinity of Port of Leixões.*

Once the potential wave energy sites were identified, the hourly records of wave data (spanning from 1 January 1990 to 31 December 2020) computed by the wave model were used to assess the distribution of the energy resource among the most recurrent sea states. For this purpose, the omni-directional wave energy matrices (Figure 8) were constructed in terms of significant wave height (Hm0), peak period (TP) and annual number of hours of occurrence (Oi). For Area I, the most recurrent sea states are concentrated in the range of 1–3 m of Hm0 and 8–14 s of TP. Nonetheless, the bulk of wave energy is displaced towards the region ranging between 1–5 m of Hm0 and 10–16 s of TP, with some sea states exceeding mean annual values of 20 MWhm−1−1. For Area II, owing to the nearshore wave dissipation processes, the most recurrent sea states are concentrated in the region of 1–3 m of Hm0 and 6–12 s of TP, which results in a significant reduction of the available resource, with only a limited number of sea states exceeding 10 MWhm−1−1.

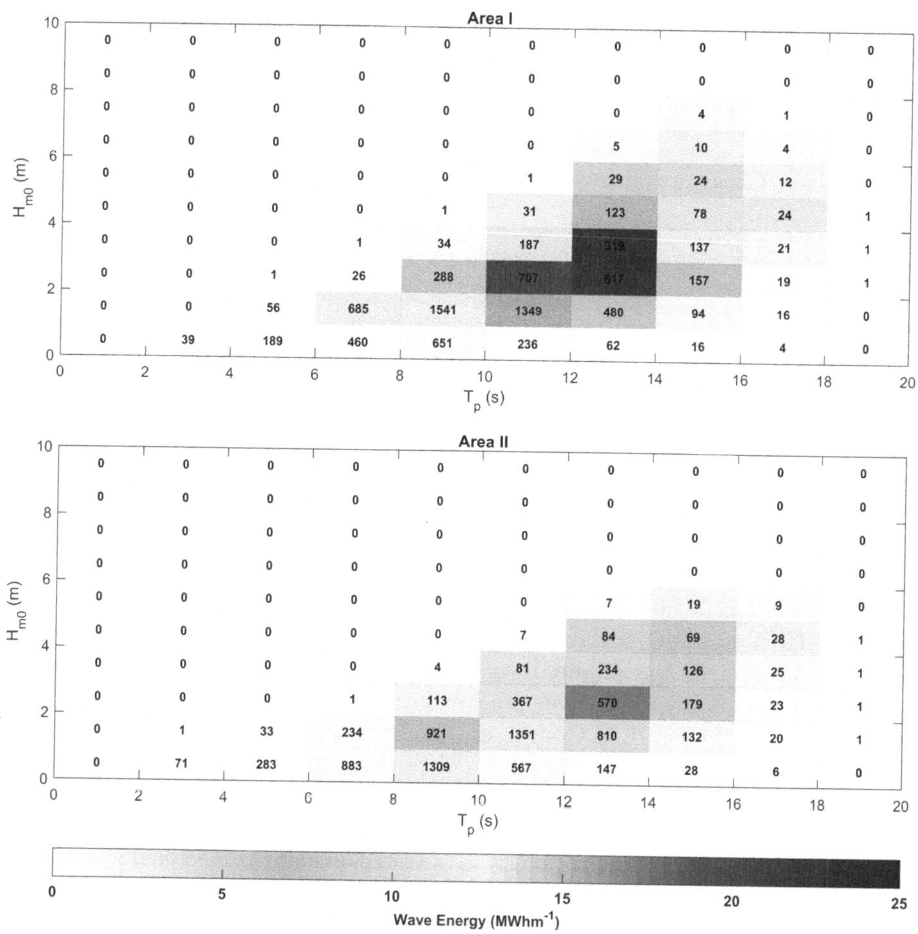

Figure 8. *Wave energy resource matrices for the potential wave energy sites (Colourmap represents the annual energy per meter of wave front and the numbers represent the occurrence in annual hours for each sea state (Oi).*

Therefore, taking into account the main characteristics (water depth and energy resource) of the potential wave energy sites and the case-study WECs (operating range) presented in Section 4 (Table 2), an initial WEC-site combination is proposed. As a consequence, for Area I, only floating devices are suitable and, therefore, oCECO, F-OWC, F-2HB and Bref-SHB were considered, while the characteristics of B-HBA and B-OF make them appropriate for Area II.

5.2. Techno-Economic Optimisation of Case-Study WECs

The techno-economic optimisation of the case-study WECs was performed following the methodology presented in Section 2.4. For this purpose, an algorithm was developed to compute the geometric scale (λ) of Froude's similarity criterion, which minimises the associated LCoE for each WEC-site combination. As a result, the main WEC characteristics (dimensions, mass and power production), as well as the number of devices required to meet a certain energy target were optimised for each deployment location. For the present case study, owing to the limited space of Area II for WEC installation (Figure 7), the energy target was set at 50% of the annual energy demand of the port (i.e., approximately 12 GWh per year). In addition, Table 3 summarises the main input values used for the LCoE model proposed in Section 2.4. It is worth noting that for the WEC deployments, the length of the inter-array electric cable was determined assuming a separation among WECs of 5 times their characteristic capture dimension [71].

Table 3. Input parameters for the LCoE model.

Parameter	Value
Lifespan (years)	20
Inflation rate, r_i (%)	2
Loan rate, r_l (%)	10
$OPEX$ (% $CAPEX$)	5
Discount factor (χ)	0.9

The results of the techno-economic model for the considered WEC-site combinations are presented in Table 4. Overall, the λ value, which minimises the associated LCoE, varies significantly for each WEC-site combination. For instance, F-OWC, F-2HB, B-HBA and B-OF present λ values exceeding 1, highlighting that a downscaling of the devices was required. As a result, the power capacity of the device is reduced and shifted towards the most recurrent and lower values of Hm0 and TP present in Areas I and II (Figure 8). Conversely, Bref-SHB had to be significantly upscaled (λ=0.76), increasing and shifting its rated power capacity towards higher values of Hm0 and TP. Finally, for the case of oCECO, the optimal λ value of 0.99 indicates that the optimised device has almost the same characteristics as the reference one. As indicated in Section 4, this behaviour was expected since oCECO was developed and optimised around the wave conditions present in the area of study. In consequence, the results obtained seem to ratify the validity of the techno-economic optimisation proposed in this work.

Table 4. Techno-economic optimisation of case-study WECs.

Location	WEC	λ	B (m)	Mass (ton)	Rated Power (kW)	LCoE (EUR/MWh)
Area I	oCECO	0.99	12.05	362.2	507	387.6
	F-OWC	1.26	19.05	782.5	1283	594.2
	F-2HB	1.09	18.43	561.5	751	470.6
	Bref-SHB	0.75	9.26	81.7	699	555.2
Area II	B-HBA	1.16	86.21	1025.2	1612	621.8
	B-OF	1.26	20.71	1920.1	1503	714.9

Regarding the LCoE values obtained for each WEC-site combination (Table 4), those range between EUR 387.6 (oCECO) and 714.9 (B-OF)/MWh, being in good accordance with the values found in the literature for wave energy deployments [45]. In addition, with the aim of assessing the performance of the techno-economic optimisation, a comparison between the LCoE for the optimised and reference WECs is presented in Table 5. Overall, it can be observed that the techno-economic optimisation of the WECs results in significant reductions of their associated LCoE values. This is especially noticeable for the case of F-OWC, Bref-SHB, B-HBA and B-OF, for which LCoE reductions ranging from EUR 25 to 130/MWh were found. As expected, for the case of oCECO, the variations of LCoE between the reference and optimised device are negligible. Again, these results confirm that the techno-economic optimisation model proposed performs well, reducing the LCoE for different WEC-site combinations.

Table 5. Input parameters for the LCoE model.

Location	WEC	$LCoE_{ref}$ (EUR/MWh)	$LCoE_{opt}$ (EUR/MWh)
Area I	oCECO	389.3	387.6
	F-OWC	704.5	594.2
	F-2HB	495.6	470.6
	Bref-SHB	603.6	555.2
Area II	B-HBA	705.4	621.8
	B-OF	843.3	714.9

On the other hand, Table 6 presents the results of the annual energy production (Et), capacity factor (Cf) and capture width ratio (CWR) for the optimised devices. As expected, oCECO presents the best performance indicators, especially in terms of annual energy production (1.33 GWh) and CWR (49%), which are in good agreement with the results obtained in previous works [93,95]. For the remaining WECs, F-2HB and F-OWC perform relatively well in Area I with the annual energy production and CWR exceeding 0.8 GWh and 22%, respectively. However, the low values of Cf highlight that the rated capacity of the electric generator is still too high for the available resource of Area I. Finally, for the case of B-HBA and B-OF, despite presenting annual energy productions in the order of 1 GWh, the values of Cf are lower than 10%, which indicates again that the capacity of the generator is still oversized for the wave regime of Area II.

Table 6. Performance assessment of the optimised WECs.

Location	WEC	E_t (GWh/year)	C_f (%)	CWR (%)
Area I	oCECO	1.33	30.14	49.41
	F-OWC	0.92	8.15	23.34
	F-2HB	0.86	13.10	22.66
	Bref-SHB	0.24	3.93	12.59
Area II	B-HBA	0.93	6.56	6.87
	B-OF	1.19	9.03	44.20

All in all, the results obtained show that the most viable WEC-site combination to fulfil 50% of the energy needs of Port of Leixões is the deployment of oCECO in Area I. This fact is underpinned by the results of oCECO both in terms of LCoE (EUR 387.6/ MWh) and device performance (i.e., oCECO presents the largest annual energy production, capacity factor and capture width ratio). Nonetheless, it is worth pointing out that the results obtained are site-specific (i.e., oCECO was optimised for the wave conditions of the area of study) and, therefore, under different wave conditions, the relative performance of the case-study WECs could differ completely.

6. DISCUSSION

This section presents a discussion on the main aspects of the methodology proposed to find an optimal WEC-site combination to supply the energy needs of seaports. First of all, the results obtained in Section 5.1 highlight that the selection of the areas to exploit the wave energy resource in the vicinity and/or port areas should be studied from a holistic perspective, considering not only the available wave energy resource but the main port characteristics, including energy consumption patterns, typology and orientation of coastal defence structures, as well as the socio-economic constrains imposed by the different maritime uses and biodiversity protected areas present in the surroundings of the port. For the case of Port of Leixões, this fact becomes apparent, with a significant amount of sea space discarded for WEC deployment due to the presence of maritime priority areas (marine navigation, disposal of dredged material and activities such as aquaculture and surf) and biodiversity areas (SCIs). As a result, the available sea surface will determine the size and capacity of the wave farm. For the present case study, due to the space limitations of Area II and the considered separation between WECs (5 times the characteristic capture width), the capacity of the farm was set to supply 50% of the annual energy demands of Port of Leixões (approx. 12 GWh). It is noteworthy to point out that despite not being considered for the present study, the proposed methodology is flexible enough to include a wave farm layout optimisation. In consequence, the size and capacity of the wave farm could be adapted even better to the port characteristics.

On the other hand, the results obtained in Section 5.2 confirm the need of conducting a techno-economic optimisation of WECs based on the characteristics of the local wave conditions, since most WEC technologies are usually designed for specific applications and/or wave climates. In consequence, most of the case-study WECs had to be upscaled or downscaled (according to Froude's similarity criterion, Section 2.4) to be better suited to the wave characteristics nearby the Port of Leixões and, therefore, reduce their associated values of LCoE (from 5 to 15% as indicated in Table 5). The only exception was the case of oCECO, for which negligible differences were found between the optimised and reference device. The latter confirms the validity of the techno-economic optimisation model proposed, since oCECO was developed and optimised for the wave conditions present in the area of study. However, it is worth noting that the LCoE model proposed in the present work is based on a deep literature review of cost estimations of the different stages of a wave energy project (manufacturing, installation, operation, maintenance and decommissioning). In addition, the cost model proposed does not take into account external factors such as spikes in the price of raw materials or inflation rates. Therefore, the proposed LCoE model should be considered as a first approximation and must be revisited when new information coming from the wave energy industry becomes available. Consequently, a conservative approach was taken for the estimation of the different costs. For example, the OPEX was estimated as 5% of the CAPEX, which corresponds to the maximum limit proposed in the literature [45]. As a result, the model may overestimate the LCoE, especially for the nearshore WECs (B-HBA and B-OF in Area II), where the operation and maintenance costs are expected to be lower. Nonetheless, the methodology proposed is flexible enough to adjust the cost estimation according to the characteristics of the considered WEC-site combination. Finally, another aspect that may have an influence on the results obtained for the LCoE is that the wave farm layout was not optimised, which could also contribute to the overestimations of the LCoE.

7. CONCLUSIONS

In recent years, regulatory and social pressure have forced port authorities to reduce their environmental footprint and transition towards a low-carbon and more sustainable energy model. For this purpose, the inclusion of renewable sources in the energy mix of ports becomes essential. In this context, due to the availability and proximity of the resource, wave energy appears as a promising option. Therefore, exploiting the wave energy resource to supply the energy demands of ports requires an appropriate selection of WEC technologies and installation locations. However, this selection is not straightforward as depends on multiple factors, which include, among others, port exposure to wave action, typology and orientation of port sheltering structures, potential interactions with other maritime uses and associated energy costs. On these grounds, the objective of this paper is to propose an integrated methodology to select the most appropriate WEC-site combination to supply the energy demands of ports. To illustrate this methodology, the Port of Leixões, located on the Northern coast of Portugal, was used as a case study.

For the present case study, the methodology proposed has proven to be robust and coherent, being capable of identifying the optimal WEC-site combination to fulfil the energy needs of the port. Regarding the identification of wave energy sites, it was found that in addition to an accurate estimation of the wave resource, a detailed spatial characterisation of the different maritime uses in the port area is essential to avoid potential disruptions between them. For this purpose, the use of marine spatial planning instruments is highly recommended. Another key aspect of the methodology is the techno-economic optimisation proposed to adapt the WECs to the local wave conditions of ports. This stage of the methodology has proven to be highly effective, with most of the case study WECs presenting noticeable reductions (ranging from 5 to 15%) in their associated LCoE values. However, the cost estimations proposed for the LCoE model must be considered as a first approximation and should be revisited when more accurate cost information is shared by the wave energy industry. Finally, it is important to note that the methodology proposed can be applied elsewhere, using different case-study WECs.

In summary, the results obtained confirm that the methodology proposed performs well, identifying the most suitable locations for wave farm installation and optimising the devices to minimise their associated LCoE, according to the site-specific wave conditions. Nonetheless, additional aspects relevant to WEC-site selection, such as wave farm layout optimisation and seabed geology, are not considered in this work and will be addressed in detail in future research.

Appendix A. SWAN Numerical Model

Appendix A.1. SWAN Physics

SWAN is an open-source third-generation spectral wave model, which accounts for the different wave transformation processes by solving the wave action balance equation:

$$\frac{\partial N}{\partial t} + \nabla \cdot \left(\vec{C} N \right) + \frac{\partial c_\sigma N}{\partial \sigma} + \frac{\partial c_\theta N}{\partial \theta} = \frac{S_{tot}}{\sigma}, \tag{A1}$$

where C stands for the propagation velocity in the geographical space, σ and θ represent the relative frequency and direction of waves, respectively; cσ and cθ stand for the propagation velocities in spectral σ and θ space, respectively; and finally, N is the wave action density, which is defined as the variance density spectrum E(σ,θ) divided by the relative frequency:

$$N(\sigma, \theta) = \frac{E(\sigma, \theta)}{\sigma}. \tag{A2}$$

On the other hand, Stot represents the source and sink terms, which take into account the physical processes of generation, dissipation and non-linear wave-wave interactions:

$$S_{tot} = S_{in} + S_{nl3} + S_{nl4} + S_{ds,w} + S_{ds,b} + S_{ds,br}, \tag{A3}$$

where Sin denotes the wave growth by wind; Snl3 and Snl4 refer to the non-linear transfer of wave energy through three-wave (triads) and four-wave (quadruplets) interactions, respectively; and lastly, Sds,w Sds,b and Sds,br represent the wave decay due to whitecapping [94], bottom friction [96] and depth-induced wave breaking [97], respectively.

SWAN computes the components of wave power per meter of wave front, Jx and Jy from the full wave spectrum E(σ,θ), according to the following expressions:

$$J_x = \rho g \int_0^{2\pi} \int_0^{\infty} c_g(\sigma, d) E(\sigma, \theta) \cos(\theta)\, d\sigma d\theta,$$

$$J_y = \rho g \int_0^{2\pi} \int_0^{\infty} c_g(\sigma, d) E(\sigma, \theta) \sin(\theta)\, d\sigma d\theta,$$

where c represents the group velocity, g is the acceleration due to gravity, ρ is the water density, and x, y are the grid coordinate directions. As a result, the wave power per meter of wave front, J, is calculated as:

$$J = \sqrt{J_x^2 + J_y^2}. \tag{A6}$$

Appendix A.2. SWAN Model Implementation

As mentioned in Section 2.2, the wave numerical model implementation was performed following the recommendations of the IEC-62600-101 technical standard. Therefore, a regional model, which was implemented and successfully validated in [98], was used to generate high-resolution space-varying 2D wave spectrum boundary conditions for a local (Class 3) model. The local model spans the surrounding area of the Port of Leixões (Figure A1) and was implemented onto an unstructured grid, with a varying grid size, ranging from 5 to 265 m. This allows for an accurate representation of the bathymetry and coastlines (including the breakwaters of the port) of the area of study. In order to capture the seasonal variations of the wave resource, the model was run for a period of 30 years (from 1990 to 2020) with a frequency interval of 1 h, which translates into 263,000 sea states analysed. In addition, the local model was set to account for the tidal variations and, therefore, the values recorded by the tidal gauge of the port were fed into the model with a 3 h frequency. Finally, Table A1 summarises the main implementation characteristics of the local wave model.

Table A1. Input parameters for the LCoE model.

Parameter	Value
IEC Class	Class 3 (Feasibility)
Grid type	Unstructured (Triangular shape)
Grid resolution	5–265 m
Boundary conditions	2D spectral. Spatially varying
Forcing factors	Wind and water level variations
Run period	1990–2020
Time resolution	1 h

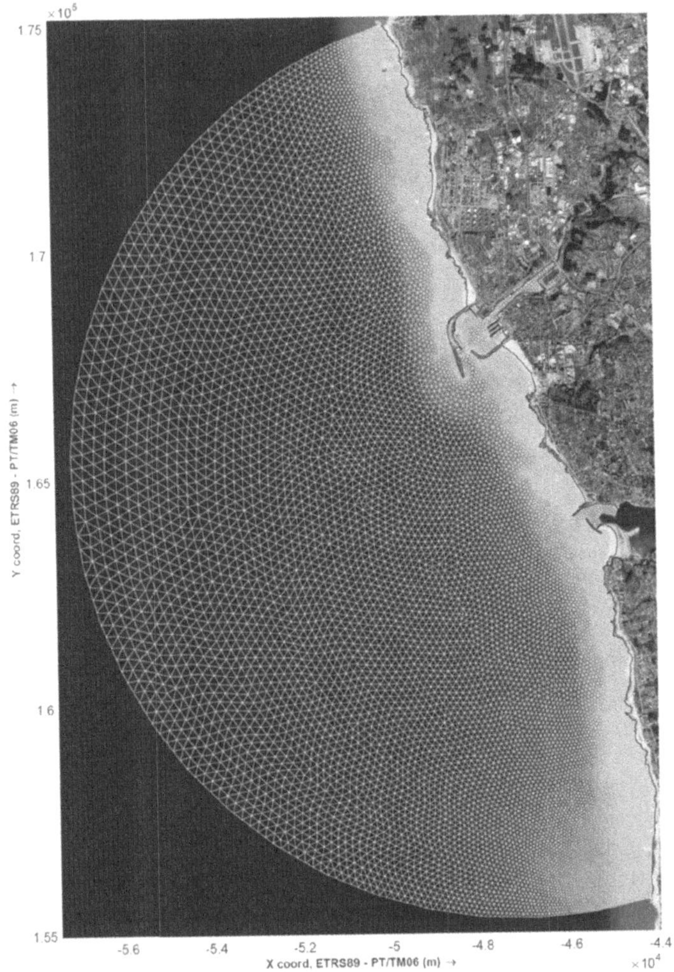

Figure A1. *SWAN computational grid.*

REFERENCES

1. United Nations. Adoption of the Paris Agreement. United Nations Framework Convention on Climate Change (UNFCCC). 2015. Available online: **https://unfccc.int/resource/docs/2015/cop21/eng/l09r01.pdf** (accessed on 19 October 2019).
2. United Nations. Transforming Our World, the 2030 Agenda for Sustainable Development. 2015. Available online: **https://sdgs.un.org/2030agenda** (accessed on 21 October 2019).
3. European Commission. The European Green Deal. 2020. Available online: **https://ec.europa.eu/info/sites/info/files/european-green-deal-communication_en.pdf** (accessed on 4 October 2020).
4. Alzahrani, A.; Petri, I.; Rezgui, Y.; Ghoroghi, A. Decarbonisation of seaports: A review and directions for future research. *Energy Strateg. Rev.* **2021**, *38*, 100727.

5. United Nations Conference on Trade and Development, UNCTAD. *Review of Maritime Transport 2021*; UNCTAD: Geneva, Switzerland, 2021; Available online: **https://unctad.org/system/files/official-document/rmt2021_en_0.pdf** (accessed on 14 November 2022).
6. Cascajo, R.; García, E.; Quiles, E.; Correcher, A.; Morant, F. Integration of Marine Wave Energy Converters into Seaports: A Case Study in the Port of Valencia. *Energies* **2019**, *12*, 787.
7. Bachvarova, E.; Spasova, T.; Marinski, J. Air Pollution and Specific Meteorological Conditions at the Adjacent Areas of Sea Ports. *IFAC-PapersOnLine* **2018**, *51*, 378–383.
8. Hiranandani, V. Sustainable development in seaports: A multi-case study. *WMU J. Marit. Aff.* **2014**, *13*, 127–172.
9. Winnes, H.; Styhre, L.; Fridell, E. Reducing GHG emissions from ships in port areas. *Res. Transp. Bus. Manag.* **2015**, *17*, 73–82.
10. International Association of Ports and Harbours. World Ports Sustainability Program. 2020. Available online: **https://sustainableworldports.org/** (accessed on 2 April 2022).
11. World Ports Sustainability Program—International Association of Ports and Harbours. Port of Helsinki—Carbon Neutral Port 2035. 2019. Available online: **https://sustainableworldports.org/project/port-of-helsinki-carbon-neutral-port-2035/** (accessed on 2 April 2022).
12. Ajsa Habibic—Offshore Energy. Busan Port Outlines Roadmap to CARBON neutrality by 2050. 2022. Available online: **https://www.offshore-energy.biz/busan-port-outlines-roadmap-to-carbon-neutrality-by-2050/** (accessed on 2 June 2022).
13. Port Houston. Port Houston Commits to Carbon Neutrality by 2050. 2022. Available online: **https://porthouston.com/wp-content/uploads/Port-Houston-commits-to-Carbon-Neutrality-by-2050.pdf** (accessed on 10 June 2022).
14. Spengler, T.; Tovar, B. Potential of cold-ironing for the reduction of externalities from in-port shipping emissions: The state-owned Spanish port system case. *J. Environ. Manag.* **2021**, *279*, 111807.
15. Wilmsmeier, G.; Spengler, T. *Energy Consumption and Container Terminal Efficiency*; Economic Commission for Latin America and the Caribbean—UN: Santiago, Chile, 2016.
16. Ramos, V.; Carballo, R.; Alvarez, M.; Sanchez, M.; Iglesias, G. A port towards energy self-sufficiency using tidal stream power. *Energy* **2014**, *71*, 432–444.
17. Gunn, K.; Stock-Williams, C. Quantifying the global wave power resource. *Renew. Energy* **2012**, *44*, 296–304.
18. Mustapa, M.; Yaakob, O.; Ahmed, Y.M.; Rheem, C.K.; Koh, K.; Adnan, F.A. Wave energy device and breakwater integration: A review. *Renew. Sustain. Energy Rev.* **2017**, *77*, 43–58.
19. Aderinto, T.; Li, H. Ocean Wave Energy Converters: Status and Challenges. *Energies* **2018**, *11*, 1250.
20. López, I.; Andreu, J.; Ceballos, S.; de Alegría, I.M.; Kortabarria, I. Review of wave energy technologies and the necessary power-equipment. *Renew. Sustain. Energy Rev.* **2013**, *27*, 413–434.
21. Choupin, O.; Pinheiro Andutta, F.; Etemad-Shahidi, A.; Tomlinson, R. A decision-making process for wave energy converter and location pairing. *Renew. Sustain. Energy Rev.* **2021**, *147*, 111225.
22. European Commission. Working Group Ocean Energy. SET-Plan Ocean Energy Implementation Plan. 2018. Available online: **https://setis.ec.europa.eu/system/files/set_plan_ocean_implementation_plan.pdf** (accessed on 12 September 2019).
23. Vicinanza, D.; Lauro, E.D.; Contestabile, P.; Gisonni, C.; Lara, J.L.; Losada, I.J. Review of innovative harbor breakwaters for wave-energy conversion. *J. Waterw. Port Coastal Ocean Eng.* **2019**, *145*, 03119001.

24. Vicinanza, D.; Contestabile, P.; Quvang Harck Nørgaard, J.; Lykke Andersen, T. Innovative rubble mound breakwaters for overtopping wave energy conversion. *Coast. Eng.* **2014**, *88*, 154–170.

25. Falcão, A.F.; Henriques, J.C. Oscillating-water-column wave energy converters and air turbines: A review. *Renew. Energy* **2016**, *85*, 1391–1424.

26. Koutrouveli, T.I.; Di Lauro, E.; das Neves, L.; Calheiros-Cabral, T.; Rosa-Santos, P.; Taveira-Pinto, F. Proof of Concept of a Breakwater-Integrated Hybrid Wave Energy Converter Using a Composite Modelling Approach. *J. Mar. Sci. Eng.* **2021**, *9*, 226.

27. Folley, M.; Whittaker, T. Analysis of the nearshore wave energy resource. *Renew. Energy* **2009**, *34*, 1709–1715.

28. Torre-Enciso, Y.; Ortubia, I.; De Aguileta, L.L.; Marqués, J. Mutriku wave power plant: From the thinking out to the reality. In Proceedings of the 8th European Wave and Tidal Energy Conference, Uppsala, Sweden, 7–10 September 2009; Volume 710, pp. 319–329.

29. Arena, F.; Romolo, A.; Malara, G.; Fiamma, V.; Laface, V. The first full operative U-OWC plants in the port of Civitavecchia. In Proceedings of the International Conference on Offshore Mechanics and Arctic Engineering, Trondheim, Norway, 25–30 June 2017; American Society of Mechanical Engineers: New York, NY, USA, 2017; Volume 57786, p. V010T09A022.

30. Contestabile, P.; Crispino, G.; Di Lauro, E.; Ferrante, V.; Gisonni, C.; Vicinanza, D. Overtopping breakwater for wave Energy Conversion: Review of state of art, recent advancements and what lies ahead. *Renew. Energy* **2020**, *147*, 705–718.

31. Eco Wave Power. Eco Wave Power Gibraltar Project. 2016. Available online: **https://www.ecowavepower.com/gibraltar/** (accessed on 10 June 2022).

32. Daniel Raj, D.; Sundar, V.; Sannasiraj, S. Enhancement of hydrodynamic performance of an Oscillating Water Column with harbour walls. *Renew. Energy* **2019**, *132*, 142–156.

33. Rezanejad, K.; Gadelho, J.; Guedes Soares, C. Hydrodynamic analysis of an oscillating water column wave energy converter in the stepped bottom condition using CFD. *Renew. Energy* **2019**, *135*, 1241–1259.

34. Bruce, T.; van der Meer, J.; Franco, L.; Pearson, J. Overtopping performance of different armour units for rubble mound breakwaters. *Coast. Eng.* **2009**, *56*, 166–179.

35. Palma, G.; Mizar Formentin, S.; Zanuttigh, B.; Contestabile, P.; Vicinanza, D. Numerical Simulations of the Hydraulic Performance of a Breakwater-Integrated Overtopping Wave Energy Converter. *J. Mar. Sci. Eng.* **2019**, *7*, 38.

36. López, I.; Pereiras, B.; Castro, F.; Iglesias, G. Optimisation of turbine-induced damping for an OWC wave energy converter using a RANS-VOF numerical model. *Appl. Energy* **2014**, *127*, 105–114.

37. Henriques, J.; Sheng, W.; Falcão, A.; Gato, L. A comparison of biradial and wells air turbines on the Mutriku breakwater OWC wave power plant. In Proceedings of the International Conference on Offshore Mechanics and Arctic Engineering, Trondheim, Norway, 25–30 June 2017; American Society of Mechanical Engineers: New York, NY, USA, 2017; Volume 57786.

38. Mariani, A.; Crispino, G.; Contestabile, P.; Cascetta, F.; Gisonni, C.; Vicinanza, D.; Unich, A. Optimization of Low Head Axial-Flow Turbines for an Overtopping BReakwater for Energy Conversion: A Case Study. *Energies* **2021**, *14*, 4618.

39. Thiruvenkatasamy, K.; Neelamani, S.; Sato, M. Nonbreaking wave forces on multiresonant oscillating water column wave power caisson breakwater. *J. Waterw. Port Coast. Ocean Eng.* **2005**, *131*, 77–84.

40. Kuo, Y.S.; Lin, C.S.; Chung, C.Y.; Wang, Y.K. Wave loading distribution of oscillating water column caisson breakwaters under non-breaking wave forces. *J. Mar. Sci. Technol.* **2015**, *23*, 10.

41. Di Lauro, E.; Lara, J.L.; Maza, M.; Losada, I.J.; Contestabile, P.; Vicinanza, D. Stability analysis of a non-conventional breakwater for wave energy conversion. *Coast. Eng.* **2019**, *145*, 36–52.

42. Contestabile, P.; Di Lauro, E.; Buccino, M.; Vicinanza, D. Economic Assessment of Overtopping BReakwater for Energy Conversion (OBREC): A Case Study in Western Australia. *Sustainability* **2017**, *9*, 51.

43. Patrizi, N.; Pulselli, R.M.; Neri, E.; Niccolucci, V.; Vicinanza, D.; Contestabile, P.; Bastianoni, S. Lifecycle Environmental Impact Assessment of an Overtopping Wave Energy Converter Embedded in Breakwater Systems. *Front. Energy Res.* **2019**, *7*, 32.

44. Rosa-Santos, P.; Taveira-Pinto, F.; Clemente, D.; Cabral, T.; Fiorentin, F.; Belga, F.; Morais, T. Experimental Study of a Hybrid Wave Energy Converter Integrated in a Harbor Breakwater. *J. Mar. Sci. Eng.* **2019**, *7*, 33.

45. Astariz, S.; Iglesias, G. The economics of wave energy: A review. *Renew. Sustain. Energy Rev.* **2015**, *45*, 397–408.

46. Abanades, J.; Greaves, D.; Iglesias, G. Coastal defence through wave farms. *Coast. Eng.* **2014**, *91*, 299–307.

47. Rodriguez-Delgado, C.; Bergillos, R.J.; Ortega-Sánchez, M.; Iglesias, G. Protection of gravel-dominated coasts through wave farms: Layout and shoreline evolution. *Sci. Total Environ.* **2018**, *636*, 1541–1552.

48. Rodriguez-Delgado, C.; Bergillos, R.J.; Iglesias, G. Dual wave farms for energy production and coastal protection under sea level rise. *J. Clean. Prod.* **2019**, *222*, 364–372.

49. Costas, R.; Figuero, A.; Peña, E.; Sande, J.; Rosa-Santos, P. Integrated approach to assess resonance between basin eigenmodes and moored ship motions with wavelet transform analysis and proposal of operational thresholds. *Ocean Eng.* **2022**, *247*, 110678.

50. Zheng, Z.; Ma, X.; Huang, X.; Ma, Y.; Dong, G. Wave forecasting within a port using WAVEWATCH III and artificial neural networks. *Ocean Eng.* **2022**, *255*, 111475.

51. Dong, G.; Zheng, Z.; Ma, X.; Huang, X. Characteristics of low-frequency oscillations in the Hambantota Port during the southwest monsoon. *Ocean Eng.* **2020**, *208*, 107408.

52. López, I.; López, M.; Iglesias, G. Artificial neural networks applied to port operability assessment. *Ocean Eng.* **2015**, *109*, 298–308.

53. Zheng, Z.; Ma, X.; Ma, Y.; Dong, G. Wave estimation within a port using a fully nonlinear Boussinesq wave model and artificial neural networks. *Ocean Eng.* **2020**, *216*, 108073.

54. Arean, N.; Carballo, R.; Iglesias, G. An integrated approach for the installation of a wave farm. *Energy* **2017**, *138*, 910–919.

55. Foteinis, S.; Tsoutsos, T. Strategies to improve sustainability and offset the initial high capital expenditure of wave energy converters (WECs). *Renew. Sustain. Energy Rev.* **2017**, *70*, 775–785.

56. Ramos, V.; López, M.; Taveira-Pinto, F.; Rosa-Santos, P. Influence of the wave climate seasonality on the performance of a wave energy converter: A case study. *Energy* **2017**, *135*, 303–316.

57. Carballo, R.; Arean, N.; Álvarez, M.; López, I.; Castro, A.; López, M.; Iglesias, G. Wave farm planning through high-resolution resource and performance characterization. *Renew. Energy* **2019**, *135*, 1097–1107.

58. Zhao, X.; Ning, D.; Zou, Q.; Qiao, D.; Cai, S. Hybrid floating breakwater-WEC system: A review. *Ocean Eng.* **2019**, *186*, 106126.

59. Cabral, T.; Clemente, D.; Rosa-Santos, P.; Taveira-Pinto, F.; Morais, T.; Belga, F.; Cestaro, H. Performance Assessment of a Hybrid Wave Energy Converter Integrated into a Harbor Breakwater. *Energies* **2020**, *13*, 236.

60. Calheiros-Cabral, T.; Clemente, D.; Rosa-Santos, P.; Taveira-Pinto, F.; Ramos, V.; Morais, T.; Cestaro, H. Evaluation of the annual electricity production of a hybrid breakwater-integrated wave energy converter. *Energy* **2020**, *213*, 118845.

61. Vicinanza, D.; Margheritini, L.; Kofoed, J.P.; Buccino, M. The SSG Wave Energy Converter: Performance, Status and Recent Developments. *Energies* **2012**, *5*, 193–226.

62. Margheritini, L.; Vicinanza, D.; Kofoed, J.P. Overtopping performance of Sea wave Slot cone Generator. In *Coasts, Marine Structures and Breakwaters: Adapting to Change: Proceedings of the 9th International Conference Organised by the Institution of Civil Engineers and Held in Edinburgh on 16 to 18 September 2009*; Thomas Telford Ltd.: London, UK, 2010; pp. 750–761.

63. Medina-López, E.; Allsop, W.; Dimakopoulos, A.; Bruce, T. Damage to the Mutriku OWC breakwater—Some lessons from further analysis. In *Coasts, Marine Structures and Breakwaters 2017*; Thomas Telford Ltd.: London, UK, 2018; pp. 957–967.

64. *IEC-TS 62600-101*; Marine Energy Wave, Tidal and Other Water Current Converters—Part 101: Wave Energy Resource Assessment and Characterisation. IEC: Geneva, Switzerland, 2014.

65. Ramos, V.; Ringwood, J.V. Exploring the utility and effectiveness of the IEC (International Electrotechnical Commission) wave energy resource assessment and characterisation standard: A case study. *Energy* **2016**, *107*, 668–682.

66. Cornett, A.; Baker, S.; Toupin, M.; Piche, S.; Nistor, I. Appraisal of IEC Standards for Wave and Tidal Energy Resource Assessment. In Proceedings of the International Conference on Ocean Energy (ICOE), Halifax, NS, Canada, 4–6 November 2014.

67. Álvarez, M.; Ramos, V.; Carballo, R.; López, I.; Fouz, D.; Iglesias, G. Application of Marine Spatial Planning tools for tidal stream farm micro-siting. *Ocean Coast. Manag.* **2022**, *220*, 106063.

68. Castro-Santos, L.; Lamas-Galdo, M.I.; Filgueira-Vizoso, A. Managing the oceans: Site selection of a floating offshore wind farm based on GIS spatial analysis. *Mar. Policy* **2020**, *113*, 103803.

69. Babarit, A. A database of capture width ratio of wave energy converters. *Renew. Energy* **2015**, *80*, 610–628.

70. Bozzi, S.; Besio, G.; Passoni, G. Wave power technologies for the Mediterranean offshore: Scaling and performance analysis. *Coast. Eng.* **2018**, *136*, 130–146.

71. Oliveira-Pinto, S.; Rosa-Santos, P.; Taveira-Pinto, F. Electricity supply to offshore oil and gas platforms from renewable ocean wave energy: Overview and case study analysis. *Energy Convers. Manag.* **2019**, *186*, 556–569.

72. Clark, C.E.; Miller, A.; DuPont, B. An analytical cost model for co-located floating wind-wave energy arrays. *Renew. Energy* **2019**, *132*, 885–897.

73. Pecher, A.; Kofoed, J.P. *Handbook of Ocean Wave Energy*; Springer Nature: Berlin/Heidelberg, Germany, 2017.

74. De Andres, A.; Medina-Lopez, E.; Crooks, D.; Roberts, O.; Jeffrey, H. On the reversed LCOE calculation: Design constraints for wave energy commercialization. *Int. J. Mar. Energy* **2017**, *18*, 88–108.

75. Guanche, R.; de Andrés, A.; Simal, P.; Vidal, C.; Losada, I. Uncertainty analysis of wave energy farms financial indicators. *Renew. Energy* **2014**, *68*, 570–580.

76. Giannini, G.; Rosa-Santos, P.; Ramos, V.; Taveira-Pinto, F. Wave energy converters design combining hydrodynamic performance and structural assessment. *Energy* **2022**, *249*, 123641.
77. Myhr, A.; Bjerkseter, C.; Ågotnes, A.; Nygaard, T.A. Levelised cost of energy for offshore floating wind turbines in a life cycle perspective. *Renew. Energy* **2014**, *66*, 714–728.
78. O'Connor, M.; Lewis, T.; Dalton, G. Operational expenditure costs for wave energy projects and impacts on financial returns. *Renew. Energy* **2013**, *50*, 1119–1131.
79. Ramos, V.; Calheiros-Cabral, T.; Rosa-Santos, P.; Taveira-Pinto, F. *Multiparameter Characterisation of the Port of Leixões. Deliverable 5.1 PORTOS Project*; Technical Report; Faculdade de Engenharia da Universidade do Porto: Porto, Portugal, 2019.
80. Port of Lexiões. APDL. *2035 Port of Leixões Will Be the First Port with Zero Emissions by 2035*. 2022. Available online: **http://www.apdl.pt/en_US/web/apdl/header** (accessed on 11 March 2022).
81. International Water Power and Dam Construction. Eco Wave Power in Agreement with APDL for Wave Power Development in Portugal. 2020. Available online: **https://www.waterpower-magazine.com/news/eco-wave-power-in-agreement-with-apdl-for-wave-power-development-in-portugal** (accessed on 21 September 2020).
82. Magalhães, A.; Barbosa, R.; Ferreira, V.; Cardoso, J. *Energy Consumption Characterisation—Port of Leixões. Deliverable 5.3 PORTOS Project*; Technical Report; Faculdade de Engenharia da Universidade do Porto: Porto, Portugal, 2019.
83. Ramos, V.; Giannini, G.; Calheiros-Cabral, T.; Rosa-Santos, P.; Taveira-Pinto, F. Legal framework of marine renewable energy: A review for the Atlantic region of Europe. *Renew. Sustain. Energy Rev.* **2021**, *137*, 110608.
84. General Directorate for Natural Resources, Safety and Marine Services. Situation Plan. Portugal Maritime Spatial Plan. 2022. Available online: **https://www.psoem.pt/geoportal_psoem/** (accessed on 10 July 2022).
85. Babarit, A.; Hals, J.; Muliawan, M.; Kurniawan, A.; Moan, T.; Krokstad, J. Numerical benchmarking study of a selection of wave energy converters. *Renew. Energy* **2012**, *41*, 44–63.
86. Ocean Energy Limited. OE Buoy. 2008. Available online: **https://oceanenergy.ie/oe-buoy/** (accessed on 21 May 2022).
87. Weber, J.; Mouwen, F.; Parish, A.; Robertson, D. Wavebob-research & development network and tools in the context of systems engineering. In Proceedings of the Eighth European Wave and Tidal Energy Conference, Uppsala, Sweden, 7–10 September 2009; Volume 8, pp. 416–420.
88. Carnegie Clean Energy. CETO Technology. Available online: **https://www.carnegiece.com/ceto-technology/** (accessed on 22 January 2020).
89. Whittaker, T.; Collier, D.; Folley, M.; Osterried, M.; Henry, A.; Crowley, M. The development of Oyster: A shallow water surging wave energy converter. In Proceedings of the 7th European Wave and Tidal Energy Conference, Porto, Portugal, 11–14 September 2007; pp. 11–14.
90. WAVESTAR Energy. WAVESTAR Concept. 2011. Available online: **http://wavestarenergy.com/concept** (accessed on 12 June 2022).
91. Giannini, G.; López, M.; Ramos, V.; Rodríguez, C.A.; Rosa-Santos, P.; Taveira-Pinto, F. Geometry assessment of a sloped type wave energy converter. *Renew. Energy* **2021**, *171*, 672–686.
92. Rodríguez, C.A.; Rosa-Santos, P.; Taveira-Pinto, F. Hydrodynamic optimization of the geometry of a sloped-motion wave energy converter. *Ocean Eng.* **2020**, *199*, 107046.
93. Giannini, G.; Rosa-Santos, P.; Ramos, V.; Taveira-Pinto, F. On the Development of an Offshore Version of the CECO Wave Energy Converter. *Energies* **2020**, *13*, 1036.

94. Delft University of Techonology. *SWAN User Manual*; Delft University of Techonology: Delft, The Netherlands, 1993.

95. Ramos, V.; Giannini, G.; Calheiros-Cabral, T.; López, M.; Rosa-Santos, P.; Taveira-Pinto, F. Assessing the Effectiveness of a Novel WEC Concept as a Co-Located Solution for Offshore Wind Farms. *J. Mar. Sci. Eng.* **2022**, *10*, 267.

96. Van der Westhuysen, A.J. The Application of the Numerical Wind-Wave Model SWAN to a Selected Field Case on the South African Coast. Master's Thesis, University of Stellenbosch, Stellenbosch, South Africa, 2002; pp. 123–145.

97. Booij, N.; Ris, R.C.; Holthuijsen, L.H. A third-generation wave model for coastal regions I. Model description and validation. *J. Geophys. Res. C Oceans* **1999**, *104*, 7649–7666.

98. Ramos, V.; López, M.; Taveira-Pinto, F.; Rosa-Santos, P. Performance assessment of the CECO wave energy converter: Water depth influence. *Renew. Energy* **2018**, *117*, 341–356.

INDEX